Animal Science: Biology and Technology

Animal Science: Biology and Technology

Edited by **Hector Carling**

SYRAWOOD
PUBLISHING HOUSE

New York

Published by Syrawood Publishing House,
750 Third Avenue, 9th Floor,
New York, NY 10017, USA
www.syrawoodpublishinghouse.com

Animal Science: Biology and Technology
Edited by Hector Carling

International Standard Book Number: 978-1-68286-163-9 (Hardback)

Printed in the United States of America.

Contents

Preface

The main aim of this book is to educate learners and enhance their research focus by presenting diverse topics covering this vast field. This is an advanced book which compiles significant studies by distinguished experts. This book addresses successive solutions to the challenges arising in the area of application, along with it; the book provides scope for future developments.

Animal science is an interdisciplinary approach to study agriculture, dairy management, and animal health with the help of biology. This book includes the nutritional value of feeds and its utilisation. The researches and case-studies compiled in this book discuss various agronomic and climatic factors influencing the product rate as well as reproduction and health issues of animals. Different modules, analytical and experimental methods are discussed in this text.

It was a great honour to edit this book, though there were challenges, as it involved a lot of communication and networking between me and the editorial team. However, the end result was this all-inclusive book covering diverse themes in the field.

Finally, it is important to acknowledge the efforts of the contributors for their excellent chapters, through which a wide variety of issues have been addressed. I would also like to thank my colleagues for their valuable feedback during the making of this book.

Editor

Effect of time and depth of insemination on fertility of Bharat Merino sheep inseminated trans-cervical with frozen-thawed semen

Davendra Kumar* and Syed Mohammed Khursheed Naqvi

Abstract

Background: Artificial insemination (AI) can serve as a powerful tool to the sheep owners for making rapid genetic progress of their flock. The AI in sheep is mostly performed using fresh semen with two reasons i) lambing rate following trans-cervical AI with frozen semen is limited by the inability of frozen-thawed sperm to transit the cervix and ii) the need of circumventing the cervical barrier through laparoscope aided intrauterine AI. Therefore, AI with frozen-thawed semen is not as widespread in sheep as it is in other domestic species. However, to get maximum benefits through the use of AI, frozen-thawed semen is a prerequisite because instead of high fertility, the short shelf life of fresh semen coupled with a limitation on the number of insemination doses achievable per unit time restricts the widespread use of individual sires. Therefore, in order to enhance lambing rate, a total of 240 trans-cervical artificial inseminations with frozen-thawed semen were performed in Bharat Merino ewes during autumn season either once in the evening (G-I, 10 h after onset of estrus, n = 100) or twice (G-II, 14 h and 22 h after onset of estrus, n = 140) i.e. once in the morning and again in the evening.

Results: The pregnancy rate (proportion of pregnant ewes confirmed by ultrasonography at day 40) and lambing rate (proportion of ewes lambed) were higher in G-II as compared to G-I (26.4 vs 20% and 19.3 vs 10%, respectively). The difference in lambing rates was statistically (P < 0.05) significant. The depth of insemination within cervico-uterine tract had no significant effect on pregnancy and lambing rates.

Conclusions: The results indicate that lambing rate in sheep following TCAI with frozen-thawed semen was significantly influenced by time of inseminations. Two inseminations after 14 and 22 h of onset of estrus enhanced the lambing rates of Bharat Merino sheep as compare to single insemination after 10 h of onset of estrus. The TCAI technique with frozen-thawed ram semen is promising and may serve as a valuable tool for genetic improvement of sheep breeds. Research efforts are going on worldwide to overcome the poor fertility following TCAI with frozen-thawed semen.

Keywords: Frozen semen, Ewe, Non-invasive intrauterine insemination, Lambing

Background

Artificial insemination (AI) can serve as a powerful tool to the sheep owners for making rapid genetic progress of their flock. The AI is expected to play a crucial role for large-scale production and multiplication of fecundity gene carrier germplasm via introgression of fecundity gene into the non-prolific sheep breeds. The AI in sheep is mostly performed using fresh semen, since fertility following AI with frozen semen is limited by the inability of frozen-thawed sperm to transit the cervix. Even when very large number of motile spermatozoa is deposited in the cervix, fertility is lower for frozen semen than for fresh semen [1]. Therefore, AI with frozen-thawed semen is not as widespread in sheep as it is in other domestic species. However, to get maximum benefits through the use of AI, frozen-thawed semen is a prerequisite because instead of acceptable fertility, the short shelf life of fresh semen coupled with a limitation on the number of insemination doses achievable per unit time restricts the widespread use of individual sires.

---
* Correspondence: davendraror@gmail.com
Division of Animal Physiology and Biochemistry, Central Sheep and Wool Research Institute, Avikanagar, 304501 via Jaipur, Rajasthan, India

In all domestic species AI with frozen-thawed semen gives acceptable pregnancy rates only when semen is deposited intrauterine. In sheep, the cervix is a formidable barrier to penetration for trans-cervical intrauterine deposition of frozen-thawed semen due to its caudally facing, eccentric series of four to five funnel-like rings [2-4]. At present satisfactory and reliable lambing results with frozen-thawed semen are only obtainable by circumventing the cervical barrier through laparoscope aided intrauterine AI [5-7]. But the cost, invasive nature and need for professional labor limits its wider use [8]. The development of trans-cervical artificial insemination (TCAI) technique has got immense potential for using frozen-thawed semen because of its non-invasive nature and as an alternative technique for invasive laparoscope-aided intrauterine AI [9]. The pregnancy and lambing rates achieved through TCAI with frozen semen is not still acceptable and there is need to improve lambing rates following TCAI. Besides breed, frozen-thawed semen quality, sperm numbers, ewe condition and time of insemination, the frequency of insemination to meet out the availability of sperm near ovulation for successful fertilization, and catheter penetration (site of semen deposition) are important factors which may influence success rate of TCAI.

The time of insemination is related to the time of ovulation. Ovulation occurs around the end of estrus. Insemination should be performed at a time sufficiently before ovulation so that by the time of ovulation a large population of sperm is established in the ampulla, the site of fertilization. In general, with twice daily inspections for heat detection, if animal showing estrus in the morning it should be inseminated in the evening (10 h after onset of estrus), and if animal showing estrus in the evening it should be inseminated in the morning (14 h after onset of estrus). However, time of insemination needs to be adjusted according to whether the semen is fresh or frozen-thawed, since the former should have a fertilizing life in the female reproductive tract well in excess of 24 h, while for the latter the fertilizing life may not be more than 12 h. Therefore, with frozen-thawed semen two inseminations 8 h apart may play role in providing large population of live sperm (owing to the short life span of frozen-thawed sperm) at the site of fertilization by the time of ovulation. Keeping this in view a fertility trial was conducted in Bharat Merino ewes to determine the influence of time of insemination (10 h Vs 14 & 22 h after onset of estrus) and depth of insemination within cervico-uterine tract on pregnancy and lambing rates following TCAI with frozen-thawed ram semen.

Results and discussion

The results are presented in Table 1. The time of insemination had significant (p < 0.05) effect on lambing rates with higher values in G-II as compare to G-I. However, time of insemination had no significant effect on pregnancy rates. The depth of insemination within cervico-uterine tract i.e. cervical penetration had no significant effect on pregnancy and lambing rates. However, both pregnancy and lambing rates were slightly higher in case of full cervical penetration of insemination needle.

The success rate attained in this study in terms of pregnancy and lambing rates following TCAI of ewes with frozen-thawed semen for one oestrus cycle is at par with other reports [10-16]. The variation in the results may be due to the differences in breed, season, feeding, management, freezing and thawing procedures, insemination dose (number of spermatozoa per straw), equipment used for TCAI, level of penetration of catheter into the cervix, time and frequency of insemination in relation to ovulation and onset of estrus whether natural or synchronized. Moreover, the process of manipulating an insemination catheter through the cervix has been linked to reductions in pregnancy and lambing rates. It has been suggested that cervical trauma and vaginal/cervical stimulation caused by the catheterization may activate pathways that are associated with uterine immune defenses and that interrupt pregnancy at its early stages [17,18,14]. It is believed that frozen-thawed sperm are less motile and lack stamina to transverse the highly viscous cervical mucus, because of ultrastructural, biochemical and functional changes undergone by a large sperm population but phagocytosis of the sperm by leukocytes is also considered a cause of the reduced fertility [19]. Ultrastructural changes mainly affect sperm membranes because during the frozen-thawed process there is a redistribution of lipids that alters lipid-lipid and lipid-protein relations, which are necessary for the normal function of sperm membranes [20].

The higher lambing rate of the ewes inseminated 14 and 22 h after onset of estrus as compare to ewes inseminated 10 h after onset of estrus indicates that the time and/or frequency of insemination of frozen-thawed semen is a critical determinant of fertility, due primarily to the significantly reduced lifespan of frozen-thawed semen in the female reproductive tract. However, Salamon [21] reported no difference between the mean lambing results for frequency of inseminations i.e. single (44.4%, 59/133) and double insemination (46.2%, 61/132) with frozen semen. The effect of double insemination depends upon the time when the first insemination is performed in relation to ovulation. So, the effect of the double insemination is more reliable when the first insemination is carried out at the beginning of estrus than when it takes place in the middle or at the end of estrus. The beneficial effect of two insemination at 14 and 22 h after onset of estrus in the present study may be because insemination is artificial, natural sexual behavior is absent and the male and female

Table 1 Effect of timing and depth of insemination within cervico-uterine tract on pregnancy and lambing rates of Bharat Merino ewes inseminated with frozen-thawed semen

| Timing of insemination after onset of estrus | Depth of insemination within cervico-uterine tract | | | | | | | | | Overall | | |
| | Full (uterine) | | | Mid (3rd to last cervical fold) | | | Partial (os to 2nd cervical fold) | | | | | |
	n	Pregnancy rate (%)	Lambing rate (%)	n	Pregnancy rate (%)	Lambing rate (%)	n	Pregnancy rate (%)	Lambing rate (%)	n	Pregnancy rate (%)	Lambing rate (%)
14 & 22 h	89	31.5	20.2	23	17.4	17.4	28	17.9	17.9	140	26.4	19.3[a]
10 h	73	21.9	11.0	12	16.7	8.3	15	13.3	6.7	100	20	10[b]
Overall	162	27.2	16.0	35	17.1	14.3	43	16.3	14.0	240	23.3	15.4

Values with different superscripts in same column are significantly (P < 0.05) different.

reproductive events have to be carefully synchronized to minimize the gamete waiting time before fertilization occurs. Moreover, two insemination may prevent from the possibility that insemination is practiced too early in relation with ovulation as seems in the G-I where insemination was done at 10 h after onset of estrus.

In the present study the depth of cervical penetration i.e. site of semen deposition did not influence statistically (p < 0.05) pregnancy and lambing rates, however, these are slightly more in case of uterine deposition as compared with cervical deposition. Taking into consideration the anatomy of the sheep cervix, a highly complex fibrous structure with many folds that obliterate the lumen, it is understandable. In partial cervical penetration, a part of semen or the whole quantity can flow back into the vagina where the acidic pH is unfavorable condition for the sperm. In mid cervical penetration, a large number of sperm practically come to a dead end in the rings of the cervix and cannot get into the body of the uterus. In full cervical penetration, semen enters directly into the body of the uterus. The depth of cervical penetration is affected by the anatomy of cervical lumen. Cervices with a less convoluted lumen are more penetrable [22]. The eccentric folds are the ones most difficult to pass and they influence the design of the tip of the AI needle [23]. Generally, the second fold is more misaligned than the first and third, which originates a narrow cervical lumen [3,24]. The depth from the os to the first 2 folds of the cervix has been shown to be the major factor limiting the cervical penetrability [25,26]. Several authors have reported that low fertility obtained when semen is applied via the cervix, is associated with low penetration. The high fertility rate on intrauterine deposition of semen has also been reported by several workers [10,12,27-30]. However, recently it has also been reported that the degree of catheter penetration into the cervix had no correlation with fertility of ewes inseminated with frozen-thawed semen [16,31].

The linear-array real-time ultrasound scanning by the transabdominal route is a reliable method for (early) pregnancy diagnosis in sheep with accuracy of more than 90% [32,33]. The difference in pregnancy and lambing rates in the present study may be due to fetal mortality or unobserved abortions that took place after day 40 of pregnancy. Dixon et al. [34] reported that approximately 19.9% of the ewes experienced late embryonic loss, fetal loss, or both; and 21.2% of the embryos or fetuses were lost from d 25 to term. The proportions of fetuses lost were associated with breed type, concentrations of progesterone, estradiol and vascular endothelial growth factor. Cryoinjuries of spermatozoa associated with freezing and thawing, and manipulations associated with TCAI technique can all contribute to fetal mortality, probably due to inadequate vascular growth in the endometrium and perhaps conceptus. Langford et al. [35] and Olesen [36] also reported increased embryonic mortality between day 18 and term when frozen-thawed semen was used. The contribution of false positive results to the differences between pregnancy and lambing rates may not be ignored.

Conclusions

The results of this study indicate that lambing rate in sheep following TCAI with frozen-thawed semen was significantly influenced by time of inseminations. Two inseminations after 14 and 22 h of onset of estrus enhanced the lambing rates of Bharat Merino sheep as compared to single insemination after 10 h of onset of estrus. The TCAI technique with frozen-thawed ram semen is promising and may serve as a valuable tool for genetic improvement of sheep breeds. Research efforts are going on worldwide to overcome the poor fertility following TCAI with frozen-thawed semen. Improvement in the freezing protocol and the use of exogenous cervical dilatators in sheep have been investigated but a much better understanding of sheep cervical physiology and the mechanism of natural cervical dilatation at estrus is required to facilitate TCAI for sheep with frozen-thawed semen to achieve acceptable fertility.

Methods

The study was conducted with the approval of Institute Animal Ethics Committee (IAEC), which is in the semi-arid tropical area of the India at 75°-28′E longitude, 26°-26′N latitude and at an altitude of 320 m above mean sea level. The climate of this place is typically hot with yearly minimum and maximum temperature of 4° and 46°C, respectively. The annual rainfall ranges from 400 to 700 mm with an erratic distribution throughout the year.

A total of 240 TCAI with frozen-thawed semen were performed in 2–4 year old Bharat Merino ewes having body weight of 35–42 kg, during autumn season when major breeding activities commence at the institute farm. Bharat Merino sheep has been evolved at Central Sheep and Wool Research Institute, Avikanagar to cater the indigenous demand of fine wool for woollen sector by crossbreeding native ewes (Nali, Chokla, Malpura and Jaisalmeri) with exotic Rambouillet/Soviet Merino rams and stabilizing the population at 75% exotic inheritance [37]. Twelve adult Bharat Merino rams of 3–4 years of age and 60–75 kg body weights were used as semen donors. The rams and ewes were maintained under the semi-intensive management system before and after inseminations. All the animals were allowed for 8–10 h daily grazing on natural vegetation interspersed with seasonal shrubs, grasses and forbs (*Achyranthes aspera, Commelina forskalaei, Eleusine aegypticae* and *Sorghum helepense*). In addition to grazing, they were provided a concentrate mixture of 300 g/day. After grazing, the animals were housed in a chain-link fence enclosure having asbestos sheet roof open from all the sides (Figure 1).

The semen of individual rams was frozen well in advance and kept cryopreserved till the day of insemination. On the day of freezing, ejaculates were obtained from donor rams in quick succession by artificial vagina after mounting on the restrained estrus ewe secured in the service crate. The semen samples were evaluated for volume, consistency, wave motion (0–5 scale), concentration (photometrically) and percentage of motile spermatozoa (0-100%). Ejaculates having thick consistency, rapid wave motion, 90% initial motility and more than 3×10^9 spermatozoa per ml were immediately diluted with a TEST-yolk-glycerol extender [38] at 25°C to a final concentration of 1×10^9 spermatozoa per ml. The extended semen samples from individual ram were frozen in a Planer R-204 programmable cell freezer under controlled cooling and freezing rates [39]. Briefly, diluted samples were aspirated into 0.5 ml size French plastic straws (IMV Technologies, L' Aigle, France), sealed with polyvinyl alcohol powder, submerged in water kept in a rectangular glass tray at 25°C and after drying straws were loaded vertically in the programmable cell freezer pre-cooled to 25°C. Controlled-rate cooling of straws was initiated in the cell freezer at the linear rate of 0.15°C per min from 25 ° to 5°C followed by a holding time of 2 h at 5°C and freezing from 5° to –125°C at the rate of 25°C per min and then plunged into liquid nitrogen for storage until required. Thawing of individual straws was done at 60°C for 10 second in a water bath just prior to TCAI. Post-thaw attributes of spermatozoa were objectively assessed by a computer-assisted semen analysis technique (Hamilton-Thorn HTM-S version 7.2 Y motility analyzer) straw per straw. Single straw (0.25 ml French straw having 250 million sperm) of frozen-thawed semen with more than 50% post-thaw motility was immediately used for TCAI of one ewe.

Estrus was detected in ewes twice daily (07:00 and 19:00) by parading adult rams, fitted with an apron. Ewes

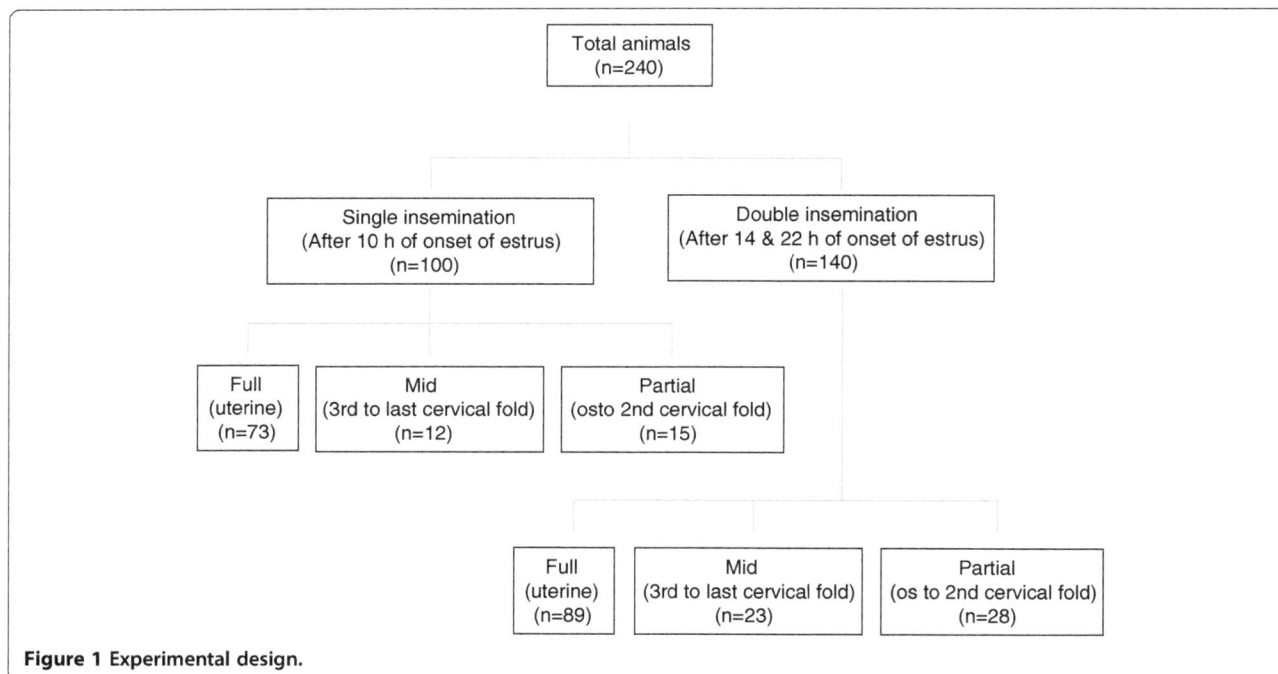

Figure 1 Experimental design.

were inseminated for one cycle either once, only in the morning (G-I, single insemination, n = 100) or twice, once in the morning and again in the evening (G-II, double inseminations, n = 140). Ewes exhibiting estrus in the morning along with ewes in estrus on the previous evening were brought daily to the semenology laboratory for insemination. Ewes exhibited estrus on previous evening were kept in G-II and inseminated twice 8 h apart i.e. between 09:00 to 10:00 as well as 17:00 to 18:00 (14 h and 22 h after onset of estrus). Ewes exhibited estrus same day in the morning were kept in G-I and inseminated once between 17:00 to 18:00 (10 h after onset of estrus). The inseminations were made using individual ram semen allotted to the particular ewe as per the breeding plan (improvement by selective breeding) of Bharat Merino sheep.

The equipment and cradle used for TCAI were fabricated as described earlier by Naqvi et al. [12]. The ewes were restrained in the cradle in dorsal decumbency with the hind quarter elevated so that the vulva was positioned at an angle of 80 to 90° from the ground. The insemination was performed throughout the study by the same inseminator using the procedure of Naqvi et al. [12]. In brief, the speculum with plunger lubricated with small amount of medical gel was introduced into the vagina and external os was located after removal of the plunger with the aid of the light. The needle was gently introduced into cervix and manipulated through the cervical folds into the deepest location possible. A maximum time limit of 5 minutes was set to achieve full penetration of the cervix. If full penetration had not been achieved semen was deposited in the cervix. The depth of cervical penetration was scored as: (i) os to second fold: partial, (ii) third to last fold: mid and (iii) needle inserted to its full length: full (uterine deposition).

The inseminated ewes were closely observed for return in estrus over two-cycle period and until lambing. At 40 days of gestation, all the inseminated ewes were examined for the presence of fetuses by using a real time Desktop Veterinary Ultrasound Scanner system equipped with convex array 5.0 MHz/40R/60D transducer (Model SA-600 V, MEDISON Company, Limited. Korea). Data on pregnancy rate (proportion of pregnant ewes confirmed by ultrasonography) and lambing rate (proportion of ewes lambed) were recorded. The data were statistically analyzed by chi square test as per Snedecor and Cochran [40] to determine the effect of time of inseminations and depth of insemination within cervico-uterine tract on pregnancy and lambing rates.

Competing interests
The authors declare that they have no competing interests.

Authors' contributions
DK carried out freezing, thawing and post-thaw evaluation of semen, made acquisition of data, analysis and interpretation of data and drafted the manuscript. SMKN designed the experiment, carried out TCAI, and given final

approval of the version to be published. Both authors read and approved the final manuscript.

Acknowledgements
The authors are thankful to the Director of the Institute for providing all the research facilities and Mr Munir Ahmed and Mr N.L. Gouttam for rendering technical assistance during this study.

References
1. Maxwell WMC, Hewitt LJ: A comparison of vaginal, cervical and intrauterine insemination of sheep. *J Agricul Sci* 1986, 106:191–193.
2. Halbert GW, Dobson H, Walton JS, Buckrell BC: The structure of the cervical canal of the ewes. *Theriogenology* 1990, 33:977–992.
3. Naqvi SMK, Pandey GK, Guatam KK, Joshi A, Geethalaxmi V, Mittal JP: Evaluation of gross anatomical features of cervix of tropical sheep using cervical silicone moulds. *Anim Reprod Sci* 2005, 85:337–344.
4. Anel L, Alvarez M, Martinez-Pastor F, Garcia-Macias V, Anel E, Paz P: Improvement strategies in ovine artificial insemination. *Reprod Domestic Anim* 2006, 41:30–42.
5. Salamon S, Maxwell WMC: Frozen storage of ram semen. II. Causes of low fertility after cervical insemination and methods of improvement. *Anim Reprod Sci* 1995, 46:89–96.
6. Naqvi SMK, Joshi A, Mathur AK, Bag S, Mittal JP: Intrauterine artificial insemination of Malpura ewes in natural estrus with frozen ram semen. *Ind J Anim Sci* 1997, 67:982–983.
7. Fukui Y, Kohno H, Togari T, Hiwasa M, Okabe K: Fertility after artificial insemination using a soybean-based semen extender in sheep. *J Reprod Develop* 2008, 54:286–289.
8. Aisen EG, Alvarez HL, Venutrino A, Garde JJ: Effect of trehalose and EDTA on cryoprotective action of ram semen diluents. *Theriogenology* 2000, 53:1053–1061.
9. Buckrell BC, Buschbeck C, Gartlet CJ, Kroetsch T, McCutcheon W, Martin J, Penner WK, Walton JS: A breeding trial using a transcervical technique for artificial insemination in sheep. *Proc 12th Int Cong Anim Reprod* 1992, 3:1531–1533.
10. Windsor DP, Szell AZ, Buschbeck C, Edward AY, Milton JTB, Buckrell BC: Transcervical artificial insemination of Australian Merino ewes with frozen-thawed semen. *Theriogenology* 1994, 42:147–157.
11. Buckrell BC, Buschbeck C, Gartley CJ, Kroetsch T, McCutcheon W, Martin J, Penner WK, Walton JS: Further development of a transcervical technique for artificial insemination in sheep using previously frozen semen. *Theriogenology* 1994, 42:601–611.
12. Naqvi SMK, Joshi A, Bag S, Pareek SR, Mittal JP: Cervical penetration and transcervical AI of tropical sheep (Malpura) at natural estrus using frozen-thawed semen. Technical Note. *Small Rum Res* 1998, 29:329–333.
13. Cappai P, Sanna SR, Branca A, Fraghi A, Bomboi G: Comparison of laparoscopic and transcervical insemination with frozen semen in Sarda dairy ewes. *Anim Sci* 1998, 66:369–373.
14. Wulster-Radcliffe MC, Wang S, Lewis GS: Transcervical artificial insemination in sheep: effect of new transcervical artificial insemination instrument and traversing the cervix on pregnancy and lambing rates. *Theriogenology* 2004, 62:990–1002.
15. Gunduz MC, Turna O, Cirit U, Ucmak M, Tek C, Sabuncu A, Bacinoglu S: Lambing rates and litter size following carazolol administration prior to insemination in Kivircik ewes. *Anim Reprod Sci* 2010, 118:32–36.
16. Alvarez M, Chamorro CA, Kaabi M, Anel-Lopez L, Boixo JC, Anel E, Anel L, de Paz P: Design and "in vivo" evaluation of two adapted catheters for intrauterine transcervical insemination in sheep. *Anim Reprod Sci* 2012, 131:153–159.
17. Campbell JW, Harvey TG, McDonald MF, Sparksman RI: Transcervical insemination in sheep: an anatomical and histological evaluation. *Theriogenology* 1996, 45:1535–1544.
18. Wulster-Radcliffe MC, Lewis GS: Development of a new transcervical artificial insemination method for sheep: effects of a new transcervical artificial insemination catheter and traversing the cervix on semen quality and fertility. *Theriogenology* 2002, 58:1361–1371.
19. Salamon S, Maxwell WMC: Storage of ram semen. *Anim Reprod Sci* 2000, 62:77–111.

20. Park JE, Graham JK: Effects of cryopreservation procedures on sperm membranes. *Theriogenology* 1992, **38**:209–222.

21. Salamon S: Fertility following deposition of equal numbers of frozen-thawed ram spermatozoa by single and double insemination. *Aust J Agri Res* 1977, **28**:477–479.

22. Kershaw C, Khalid M, McGowan M, Ingram K, Leethongdee S, Wax G, Scaramuzzi R: The anatomy of the sheep cervix and its influence on the transcervical passage of an inseminating pipette into the uterine lumen. *Theriogenology* 2005, **64**:1225–1235.

23. Halbert GW, Dobson H, Walton JS, Buckrell BC: A technique for transcervical intrauterine insemination of ewes. *Theriogenology* 1990, **33**:993–1010.

24. Kaabi M, Alvarez M, Anel E, Chamorro CA, Boixo JC, de Paz P, Anel L: Influence of breed and age on morphometry and depth of inseminating catheter penetration in the ewe cervix: a postmortem study. *Theriogenology* 2006, **66**:1876–1883.

25. Eppleston J, Maxwell WMC: Recent attempts to improve the fertility of frozen ram semen inseminated into the cervix. *Wool Tech Sheep Breed* 1993, **41**:291–302.

26. Eppleston J, Salamon S, Moore NW, Evans G: The depth of cervical insemination and site of intrauterine insemination and their relationship to the fertility of frozen-thawed ram semen. *Anim Reprod Sci* 1994, **36**:211–225.

27. Paulenz H, Soderquist L, Adnoy T, Nordsroga A, Berg KA: Effect of vaginal and cervical deposition of semen on the fertility of sheep inseminated with frozen-thawed semen. *Vet Rec* 2005, **156**:372–375.

28. Szabados T, Gergatz E, Vitinger E, Zi T, Gyoker E: Lambing rate as a function of artificial insemination depth in ewe lambs, primiparous and multiparous ewes. *Acta Agraria Kaposvariensis* 2005, **9**:41–49.

29. Leethongdee S: Development of trans-cervical artificial Insemination in sheep with special reference to anatomy of cervix. *Suranaree J Sci Technol* 2010, **17**:57–69.

30. Taqueda GS, Azevedo HC, Santos EM, Matos JE, Bittencourt RF, Bicudo SD: Influence of anatomical and technical aspects on fertility rate based on sheep transcervical Artificial insemination performance. *ARS Veterinaria* 2011, **27**:127–133.

31. Richardson L, Hanrahan JP, Donovan A, Martí JI, Fair S, Evans AC, Lonergan P: Effect of site of deposition on the fertility of sheep inseminated with frozen-thawed semen. *Anim Reprod Sci* 2012, **131**:160–164.

32. Kahn W, Achtzehn J, Kahn B, Richter A, Schulz J, Wolf M: Sonography of pregnancy in sheep. II Accuracy of transrectal and transcutaneous pregnancy diagnosis. *Dtsch Tierarztl Wochenschr* 1993, **100**:29–31.

33. Yotov S: Diagnostics of early pregnancy in Stara Zagora dairy sheep breed. *Bulg J Vet Med* 2005, **8**:41–45.

34. Dixon AB, Knights M, Winkler JL, Marsh DJ, Pate JL, Wilson ME, Dailey RA, Seidel G, Inskeep EK: Patterns of late embryonic and fetal mortality and association with several factors in sheep. *J Anim Sci* 2007, **85**:1274–1284.

35. Langford GA, Marcus GJ, Hackett IJ, Ainsworth L, Wolynetz MS, Peters HF: A comparison of fresh and frozen semen in the insemination of confined sheep. *Canad J Anim Sci* 1979, **59**:685–691.

36. Olesen I: Effects of cervical insemination with frozen semen on fertility and litter size of Norwegian sheep. *Livestock Prod Sci* 1993, **37**:169–184.

37. Singh VK, Arora AL, Mehta BS, Mishra AK, Kumar S, Prince LLL: Bharat Merino. In *New Strains of Sheep Evolved*. Avikanagar (India): CSWRI; 2006:21.

38. Schmehl MK, Anderson SP, Vazques IA, Graham EF: The effect of dialysis of extended ram semen prior to freezing on post-thaw survival and fertility. *Cryobiology* 1986, **23**:406–414.

39. Joshi A, Kumar D, Naqvi SMK, Maurya VP: Effect of controlled-rate cooling and freezing on motion characteristics and acrosomal integrity of cryopreserved ram spermatozoa. *Cell Pres Tech* 2008, **6**:277–284.

40. Snedecor GW, Cochran WG: *Statistical Methods*. 8th edition. Ames, Iowa: Iowa State University Press; 1989.

2

Dietary Conjugated Linoleic Acid (CLA) increases milk yield without losing body weight in lactating sows

Sung-Hoon Lee[1*], Young-Kuk Joo[1], Jin-Woo Lee[1], Young-Joo Ha[1], Joon-Mo Yeo[2] and Wan-Young Kim[2]

Abstract

This study was conducted to evaluate the effects of dietary conjugated linoleic acid (CLA) on the performance of lactating sows and piglets as well as the immunity of piglets suckling from sows fed CLA. Eighteen multiparous Duroc sows with an average body weight (BW) of 232.0 ± 6.38 kg were randomly selected and assigned to two dietary treatments (n = 9 for each treatment), control (no CLA addition) and 1% CLA supplementation. For the control diet, CLA was replaced with soybean oil. Experimental diets were fed to sows during a 28-day lactation period. Litter size for each sow was standardized to nine piglets by cross-fostering within 24 hours after birth. Sow milk and blood samples were taken from sows and piglets after 21 and 27 days of lactation, respectively. Loss of BW was significantly ($p < 0.05$) higher in sows fed control diet compared to sows fed CLA diet. Piglet weights at weaning and weight gain during suckling were significantly ($p < 0.05$) higher in sows fed CLA compared to sows fed control diet. Serum non-esterified fatty acid (NEFA) and urea nitrogen concentrations were significantly ($p < 0.05$) lower in sows fed CLA than in sows fed soybean oil. IgG concentrations of the groups supplemented with CLA increased by 49% in sow serum ($p < 0.0001$), 23% in milk ($p < 0.05$), and 35% in piglet serum ($p < 0.05$) compared with the control group. Sows fed CLA showed an increase of 10% in milk yield compared with sows fed soybean oil ($p < 0.05$), even though there was no difference in daily feed intake between the treatments. Milk fat content was significantly ($p < 0.05$) lower in sows fed CLA than in sows fed soybean oil. Solid-not-fat yield was significantly ($p < 0.05$) higher in sows supplemented with CLA than in sows fed control diet and also protein-to-fat ratio in milk was significantly ($p < 0.05$) higher in sows fed CLA compared with the control group. The results show that CLA supplementation to sows increased milk yield without losing BW during lactation, whereas soybean oil supplementation resulted in severe BW loss.

Keywords: Conjugated linoleic acid, Milk yield, Body weight, Sows, Piglets

Background

The growth rate of suckling piglets is determined by the amount of milk produced by sows [1]. Greater milk production in sows increases pig weaning weights as well as viability of offspring, and pigs with heavier weaning weights grow more rapidly at post-weaning [2,3]. It has been well documented that sows mobilize sufficient energy from their body tissue stores for milk production [4-6]. Deprivation of milk from sows has been shown to

reduce body weight (BW) of sows during lactation [7]. Further, several studies have reported that a BW loss between 10 to 12% during lactation reduced reproductive performance in the subsequent parity [8,9]. Thus, it is important to minimize BW loss in sows during lactation as well as maintain both maximal growth of piglets and subsequent reproductive performance.

To solve this problem, it was intended to feed lactating sows a diet containing conjugated linoleic acid (CLA), which is composed of an isomeric mixture of linoleic acid containing conjugated double bonds, predominantly *cis*-9, *trans*-11 CLA (c9, t11) and *trans*-10, *cis*-12 CLA (t10, c12), produced from polyunsaturated fatty acids by ruminal bacteria during biohydrogenation [10]. CLA has

---
* Correspondence: hoonlee@korea.kr
[1]Livestock Experiment Station, Gyeongsangnamdo Livestock Promotion Research Institute, 251 Cheonghyun-ro, Sinan-myeon, Sancheong 666-962, Republic of Korea
Full list of author information is available at the end of the article

been extensively studied due to its beneficial effects on humans and animals [11,12]. Consumption of CLA by sows during lactation has been found to lower backfat thickness loss as well as increase weaning weight in piglets [13,14]. Higher BW at weaning is closely associated with higher milk yield [6,15]. In contrast, Harrell et al. [16] and Peng et al. [17] showed that inclusion of CLA did not affect piglet weights at weaning. As these results are contradictory, further research is needed to elucidate the effects of dietary CLA on milk yield and body weight changes in lactating sows. Regarding immunity in piglets, Corino et al. [13,18] found that increasing levels of CLA fed to sows or weaned piglets markedly increased the immunoglobulin G (IgG) concentrations of piglets during suckling and post-weaning. Therefore, transfer of CLA from sows to milk could reduce mortality in piglets by increasing immunity.

The objective of this study was to determine the effects of dietary CLA on the performance of lactating sows and piglets as well as blood and milk compositions. It further examined the effects of CLA on IgG concentrations in sera from sows and piglets, and milk.

Methods

The animal use and care protocol was approved by the Institutional Animal Care and Use Committee of the Gyeongsangnamdo Livestock Promotion Research Institute, Korea.

Animal and experimental diets

Eighteen multiparous Duroc sows in their 2nd to 5th parities with an average body weight (BW) of 232.0 ± 6.38 kg were randomly selected and assigned to two dietary treatments (n = 9 for each treatment), control (0% CLA) and 1% CLA addition. CLA was replaced with soybean oil in the control diet. The sows were moved into farrowing rooms after 108 days of gestation and were housed individually in crates (2.4 m × 1.7 m) with slatted floors.

Experimental diets were provided in the form of flour and were formulated to meet the recommended amounts of crude protein and digestible energy (DE) as required by the NRC [19]. The two diets were isoenergetic and isonitrogenous. Ingredients and chemical compositions of the experimental diets are shown in Table 1. Diets were provided from 110 days of gestation until weaning (28 days postpartum). Sows were fed twice daily at approximately 07:00 and 17:00 h and had *ad libitum* access to water. The diets were restricted to 2.0 kg/day for each animal and were administered 24–48 hours prior to farrowing in order to prevent excessive gut fill from obstructing the farrowing process *per se* as well as to minimize any problem with mastitis-metritis-agalactia. After farrowing, sows were initially fed 1.5 kg of their treatment diet twice daily (08:00 and 16:30 h), and this was increased daily by 0.5 kg until 7

Table 1 Ingredients and chemical compositions of experimental diets

	Sow diets	
	CON[1]	CLA[2]
Ingredients, %		
Corn	45.60	45.60
Wheat	8.00	8.00
Wheat bran	5.00	5.00
Rice bran	12.25	12.25
Soybean meal	16.00	16.00
Rape seed meal	5.00	5.00
Palm oil meal	1.50	1.50
Cotton seed meal	1.00	1.00
Soybean oil[3]	1.00	0.00
Conjugated linoleic acid oil[4]	0.00	1.00
Monocalcium phosphate	0.10	0.10
Limestone	1.67	1.67
Tallow	2.00	2.00
Sodium chloride	0.60	0.60
Vitamin and mineral premix[5]	0.20	0.20
L-lysine HCl	0.08	0.08
Total	100.00	100.00
Chemical composition		
Moisture, %	12.28	12.27
Crude protein, %	17.48	17.10
Crude fat, %	9.38	9.84
Crude fiber, %	4.58	4.60
Crude ash, %	5.53	5.56
Calcium, %	50.75	50.63
Phosphorus, %	0.93	0.90
Lysine, %	0.53	0.51
Nitrogen-free extracts, %	0.86	0.85
Digestible energy, kcal/kg[6]	3,393	3,393

[1]CON = diets supplemented with soybean oil.
[2]CLA = diets replaced soybean oil with CLA.
[3]Soybean oil, purchased from Cheiljedang (Seoul, Korea), contains C16:0 10.92%; C18:0 4.20%; C18:1*cis*-9 24.04%; C18:2*n*-6 54.38%; and C18:3*n*-36.46%.
[4]Conjugated linoleic acid oil, purchased from HK Biotech (Jinju, Korea), contains C16:0 6.57%; C18:0 2.45%; C18:1*cis*-9 10.01%; C18:2*n*-6 1.76%; CLA *cis*-9, *trans*-11 33.91%; and CLA *trans*-10, *cis*-12 41.47%.
[5]Vitamin-mineralpremix provides per kg of feed: vitamin A 10,000IU; vitamin D$_3$ 2,000IU; vitamin E 44 mg; vitamin K$_3$ 2 mg; vitamin B$_1$ 1.3 mg; vitamin B$_2$ 4.0 mg; vitamin B$_6$ 1.3 mg; vitamin B$_{12}$ 0.015 mg; pantothenic acid 12 mg; nicotinic acid 20 mg; biotin 0.2 mg; folic acid 1.3 mg; iron 80 mg; copper 5 mg; cobalt 0.30 mg; zinc 50 mg; manganese 20 mg; iodine 0.14 mg.
[6]Calculated value.

days postpartum. Thereafter, sows had free access to their diets until weaning.

To evaluate sow performance, litter sizes were adjusted (nine piglets per sow) by cross-fostering piglets within 24 hours after birth. Piglets had no access to creep feed. At birth, the litters were subjected to normal management procedures, including cutting of teeth and tails, ear notching, and iron shots. Males were not castrated.

All weaned piglets were mixed across treatments and then moved to pens (10 piglets per pen, arranged based on similar BW) in an environmentally controlled room. Piglets were raised on the same commercial diet (PuKyung Pig Farmers Agricultural Cooperative Feed Mill, Gimhae, Korea) until 70 days of age. Piglets after weaning were separated into pre-starter and starter phases, respectively. Pre-starter diet consisting of 19% crude protein, 1.3% lysine, 3,500 kcal/kg DE, 0.76% calcium, and 0.62% phosphorus was given from 29–49 days, followed by starter diet consisting of 17.5% crude protein, 1.0% lysine, 3,350 kcal/kg DE, 0.67% calcium, and 0.57% phosphorus until 70 days of age.

Measurements and sampling

The BWs of sows were recorded at farrowing and weaning (28 days), and differences in BW were calculated. For piglets, birth and weaning weights were also recorded, and daily BW gain was calculated. After weaning, individual BWs of piglets were recorded at 70 days of age. Average daily gain was calculated during the 42-day post-weaning period and total period (0 to 70 days of age). Backfat thickness of sows was measured by ultrasound at farrowing and weaning using a Renco Lean Meater® (Renco Corporation, Minneapolis, USA). Measurements were taken 65 mm from the midline at the last rib. Feed intake during lactation was recorded daily. The number of pigs that died during the lactation period was counted, and mortality (%) was calculated. Weaning-to-estrus interval was determined by monitoring estrus from 3 to 10 days after weaning.

On day 21 of lactation, milk samples were collected. Litters were separated from sows for 1 hour prior to milking, after which approximately 50 mL of milk was obtained after intramuscular injection of 20 IU of oxytocin (Komioxytocin inj.; Komipharm International Co., Ltd., Siheung, Korea). Milk samples were then frozen immediately at −80°C for milk composition and IgG analysis. Blood samples taken from the jugular vein were collected from sows prior to feeding in the morning on the day before weaning. Three randomly selected piglets per sow were then subjected to blood collection from the anterior vena cava, after which blood samples were pooled within the same litter. Before collecting blood samples, suckling piglets were not segregated from sows. Serum was separated by centrifugation (3,000 × g for 15 min at 4°C) and frozen immediately at −80°C until analyses.

Milk production by sows was measured on day 21 of lactation by a modified weigh-suckle-weigh (WSW) method of Speer and Cox [15]. Briefly, litters were separated from their dam for 1 hour. Piglets were placed in a pen under a heat lamp during separation. Litters then were weighed to obtain their pre-suckling BW, returned to their mothers, allowed to suckle until the end of vigorous synchronized suckling by the litter, and then immediately collected and weighed to obtain their post-suckling BW. This procedure was repeated hourly until a minimum of three consistent measurements of hourly milk yield were obtained. Hourly milk yields were a measurement of BW gain, as litter milk intake was based on the difference between pre- and post-suckling litter BWs. Mean hourly milk yield multiplied by 24 was used to estimate daily milk yield. Suckling frequency was not controlled on the other days of lactation.

Sample analyses

Compositional analysis of the experimental diets was carried out according to the procedures of AOAC [20]. Fatty acids of the two oil sources (soybean oil and CLA oil) used in this study were analyzed by the one-step procedure described by Sukhija and Palmquist [21]. Serum glucose content was determined using an enzymatic kit (Glucose Hexokinase kit, Bayer, US). Serum total protein and urea nitrogen contents were determined using an auto analyzer (model 704, Hitachi). Serum total cholesterol, triacylglyceride, and non-esterified fatty acid (NEFA) levels were determined by enzymatic spectrophotometric assay (Boehringer Mannheim, Germany). Serum total lipid content was determined by colorimetric assay (Hitachi 7180, Japan). Serum LDL- and HDL-cholesterol levels were determined by enzymatic colorimetric assay (Roche, Germany). IgG concentrations of serum and milk were determined by the radial immunodiffusion method of Mancini et al. [22] using a commercial kit (Bethyl Laboratories Inc., Montgomery, TX). Serum thyroxine (T4) and triiodothyronine (T3) levels were measured using radioimmunoassay (RIA) kits (ICN Pharmaceuticals, Inc., Costa Mesa, CA). Analyses for T4 and T3 levels were performed in duplicate. Milk composition (%) was analyzed using Milkoscan FT 120 (FOSS Electric, Korea).

Statistical analysis

Data were analyzed by t-test for a completely randomized design using the GLM procedure of SAS [23]. Least squares means were calculated for each independent variable. Individual sows and their litters were used as the experimental unit. In weaned piglets, gender effect was ignored and thus not included in the model. Differences were considered significant at $p < 0.05$.

Results

Sow performance

Results on BW, backfat thickness, total feed intake, and weaning-to-estrus interval of sows during lactation are shown in Table 2. Loss of BW during the lactation period was significantly ($p < 0.05$) higher in sows fed control diet compared to sows fed CLA diet. Sows fed

Table 2 Body weight, backfat thickness, total feed intake, and weaning-to-estrus interval in lactating sows fed diets supplemented with CLA

| | Sow diets | | SED[3] | p-value |
	CON[1]	CLA[2]		
No. of sows	9	9		
Parity	3.08	3.09	0.57	0.9896
Body weight, kg				
24 h postpartum	232.19	231.82	13.16	0.9781
Weaning (day 28)	222.16	231.52	11.83	0.4401
Difference	−10.03	−0.31	4.26	0.0364
Backfat thickness, mm				
24 h postpartum	15.56	15.33	1.37	0.8727
Weaning (day 28)	14.30	14.97	1.35	0.6295
Difference	−1.26	−0.36	0.85	0.3095
Total feed intake, kg	181.36	184.16	2.20	0.2179
Weaning-to-estrus interval, day	4.22	4.78	0.46	0.2536

[1]CON = diets supplemented with soybean oil.
[2]CLA = diets replaced soybean oil with CLA.
[3]SED = standard error of difference.

CLA diet showed little change in BW during lactation. No difference in backfat thickness, total feed intake, or weaning-to-estrus interval was observed between the treatments during the lactation period.

Performances of litters and piglets

Performances of litters and piglets during the 28-day lactation period are presented in Table 3. Dietary CLA had no effect on the number of weaned piglets or piglet mortality during lactation. Litter weight at weaning ($p = 0.0722$) and daily litter weight gain ($p = 0.096$) tended to be higher in sows fed CLA diet compared with sows fed control diet. Furthermore, piglet weight at weaning was significantly ($p < 0.05$) increased by sow consumption of CLA diet, but this was not evident after 42 days of weaning. Piglet weight gain was significantly ($p < 0.05$) higher in sows fed CLA diet during the suckling period, whereas it was unaffected from weaning until 70 days of age as well as birth to 70 days of age.

Blood metabolites, thyroid hormones, and IgG concentrations

Results on blood metabolites, thyroid hormones, and IgG concentrations in blood and milk are shown in Table 4. Blood metabolite levels of sows were unaffected, whereas NEFA and urea nitrogen concentrations were significantly ($p < 0.05$) reduced by CLA diet. There was no significant difference in T3 or T4 concentration between the treatments. IgG concentrations were significantly higher in sera of sows ($p < 0.01$) and piglets ($p < 0.05$) fed CLA diet

Table 3 Performances of litters and piglets in lactating sows fed diets supplemented with CLA

| | Sow diets | | SED[3] | p-value |
	CON[1]	CLA[2]		
Litter size, no. of piglets				
After cross-fostering (day 0)	9	9		
At weaning (day 28)	8.71	8.63	0.26	0.7377
Pre-weaning mortality, %	3.17	4.17	2.90	0.7377
Litter weight, kg				
After cross-fostering (day 0)	14.45	14.87	1.11	0.7097
At weaning (day 28)	62.88	69.14	3.20	0.0722
Gain (day 0 to 28), kg/d	1.73	1.94	0.12	0.0960
Piglet weight, kg				
After cross-fostering (day 0)	1.60	1.65	0.12	0.6965
At weaning (day 28)	7.21	8.03	0.33	0.0278
42 days post-weaning (day 70)	24.54	25.98	1.03	0.1841
Piglet weight gain, g/day				
Day 0 to 28	199.94	227.75	11.21	0.0276
Day 28 to 70	412.46	431.51	23.07	0.4239
Day 0 to 70	327.67	347.57	14.25	0.1859

[1]CON = diets supplemented with soybean oil.
[2]CLA = diets replaced soybean oil with CLA.
[3]SED = standard error of difference.

as well as milk ($p < 0.05$) of sows fed CLA diet compared to sows fed control diet.

Milk yield and composition

Milk yield and composition in sows fed CLA diet during lactation are shown in Table 5. Sows fed CLA diet showed a significantly ($p < 0.05$) higher milk yield than those fed control diet, even though there was no significant difference in daily lactational feed intake between the treatments. Fat content was significantly ($p < 0.05$) lower in sows fed CLA diet than in those fed control diet. However, there were no differences in other milk components by dietary CLA consumption during lactation. Except for lactose and solid-not-fat yields, there were no differences in yields of other components upon dietary CLA consumption. Yield of lactose tended ($p = 0.0532$) to increase in sows fed CLA diet compared with sows fed control diet, and solid-not-fat yield was significantly ($p < 0.05$) higher in sows supplemented with CLA compared to those fed control diet. Further, protein-to-fat ratio was significantly ($p < 0.05$) higher in sows fed CLA diet compared with sows fed control diet.

Discussion

In the present study, sows fed CLA showed little change in BW during the lactation period. This result is in disagreement with previous studies that have reported no dietary CLA effect on BW change in sows [13,16,24]. In

Table 4 Serum concentrations of metabolites and thyroid hormone in lactating sows fed control diet or CLA diet, and serum and milk IgG concentrations of sows and their progeny

| | Sow diets | | SED[3] | p-value |
	CON[1]	CLA[2]		
Sow serum characteristics				
Glucose, mg/dL	72.42	76.73	5.46	0.4586
Total protein, g/dL	7.19	7.11	0.17	0.6358
Total lipid, mg/dL	225.50	191.50	24.13	0.2085
Triacylglycerol, mg/dL	19.33	21.36	3.07	0.5151
NEFA, µEq/L	306.50	170.44	64.77	0.0484
Total cholesterol, mg/dL	97.50	99.91	6.58	0.7180
LDL-cholesterol, mg/dL	50.33	49.40	4.08	0.8213
HDL-cholesterol, mg/dL	49.75	47.46	4.47	0.6130
HDL/Total cholesterol	0.51	0.49	0.04	0.6007
LDL/HDL cholesterol	1.03	1.08	0.16	0.7708
Urea nitrogen, mg/dL	13.49	11.60	0.66	0.0134
Thyroxine (T4), µg/dL	2.88	3.01	0.18	0.4661
Triiodothyronine (T3), ng/mL	0.61	0.67	0.05	0.1729
T3:T4, %	2.15	2.24	0.16	0.5722
IgG, mg/dL				
Sow serum	596.42	889.33	41.54	<0.0001
Milk	42.16	52.06	4.11	0.0468
Piglet serum	187.78	253.88	26.48	0.0189

[1]CON = diets supplemented with soybean oil.
[2]CLA = diets replaced soybean oil with CLA.
[3]SED = standard error of difference.

Table 5 Daily lactational feed intake, milk yield, and composition in lactating sows fed control diet or CLA diet

| | Sow diets | | SED[3] | p-value |
	CON[1]	CLA[2]		
Lactational feed intake, kg/day	6.48	6.58	0.08	0.2117
Milk yield at 21 days of lactation, kg/day	8.54	9.49	0.39	0.0288
Fat				
%	8.44	5.92	1.05	0.0308
Yield, kg/day	0.72	0.56	0.10	0.1158
Protein				
%	5.15	4.80	0.28	0.1838
Yield, kg/day	0.44	0.46	0.02	0.5033
Lactose				
%	5.19	5.44	0.18	0.1855
Yield, kg/day	0.45	0.52	0.03	0.0532
Solids not fat				
%	10.93	10.92	0.24	0.9492
Yield, kg/day	0.93	1.04	0.05	0.0367
Protein: fat	0.64	0.85	0.09	0.0288

[1]CON = diets supplemented with soybean oil.
[2]CLA = diets replaced soybean oil with CLA.
[3]SED = standard error of difference.

contrast, sows fed control diet containing soybean oil (high in linoleic acid (LA); Table 1) at the expense of CLA showed severe loss of BW. It seems that lactating sows that consume a diet containing soybean oil rather than CLA oil more readily mobilize energy from their body reserves to produce milk. This may be related to the acceleration of body fat catabolism. Sanz et al. [25] reported that broiler chickens fed sunflower oil diet, which is also high in LA, had higher specific activities of fat-catabolic enzymes such as carnitine palmitoyltransferase I (CPT I) and L-3-hydroxyacyl-CoA dehydrogenase (L3HOAD). Likewise, Shimomura et al. [26] observed lower fat deposition in rats fed a diet rich in safflower oil compared to those fed tallow. However, Vicente et al. [27] found that fat sources had no effect on BW of lactating sows, which contradicts the results of the present study.

In our study, there was no significant difference in backfat thickness between the treatments. In support of our result, Harrell et al. [16] previously reported no difference in backfat thickness, whereas Cordero et al. [14] showed a less loss of backfat thickness in sows fed CLA

diet. Conversely, Park et al. [28] reported that backfat thickness became thinner at a higher level of CLA supplementation or longer feeding time in lactating sows. Meanwhile, in grower-finisher pigs, many researchers reported that CLA treatment reduced backfat thickness [29-33]. These contradictory results on CLA feeding may be due to the different physiological state of swine. Therefore, the backfat thickness of sows fed CLA during lactation must be further investigated.

In the present study, BW changes during lactation did not affect the subsequent weaning-to-estrus interval. Thus, it seems that BW loss was not sufficiently low enough to reduce subsequent reproductive performance. On the other hand, Reese et al. [34] demonstrated that little relationship existed between sow weight loss during lactation and the interval between weaning and first estrus. Total feed intake of sows was not influenced by dietary inclusion of CLA, in accordance with previous studies [13,14,16,24].

Litter size at weaning and mortality of suckling piglets were not affected by CLA supplementation. This result is in agreement with those of Cordero et al. [14], who observed no effect of dietary CLA on the number of weaned piglets or piglet mortality. Reduction of litter size at weaning and mortality during the suckling period may be dependent on numerous factors, such as milk quality of sows, suckling intensity, disease, accidents, environmental conditions, and unknown factors [35].

Although litter weight was maintained after cross-fostering piglets, weaning litter weight as well as litter weight gain increased in the CLA dietary group (Table 3), reflecting increased milk yield by sows [6,36]. Moreover, weaning litter weight and weight gain were drastically elevated in piglets from sows fed CLA compared to those from sows fed soybean oil, even though piglet weights at birth were similar between the treatments. These results corroborate the published data of Corino et al. [13] and Cordero et al. [14], who reported that piglets from sows fed a diet supplemented with 0.5% or 1% CLA during lactation were markedly heavier than piglets from control sows. Cabrera et al. [37] further demonstrated a link between higher weaning weight and reduced time to reach finishing weight in pigs, which implies market pigs (BW = 110 kg) can be produced more economically. However, maternal CLA background had no effect on piglet weights or weight gain throughout the 42-day post-weaning period, which is in agreement with the observations of Bontempo et al. [24] and Corino et al. [13]. On the other hand, Bee [38] observed that, irrespective of starter diet, pigs reared by sows fed CLA during lactation showed greater feed intake, weight gain, and final weights after weaning than pigs reared by sows fed LA diet. These discrepancies might be due to the higher level of CLA (2%) used by Bee [38].

It was previously reported that CLA reduces body fat content [39]. In the present study, serum NEFA concentration was drastically reduced in sows fed CLA compared with sows fed soybean oil, suggesting the reduction of body fat mobilization. Further, a lower NEFA concentration is consistent with higher BW of sows, as shown in Table 2. In contrast, Bontempo et al. [24] and Corino et al. [13] reported that 0.5% CLA supplementation to sows during lactation had no effect on NEFA levels. Corino et al. [18] also observed no difference in NEFA concentration with increasing amount of CLA, whereas triacylglycerol levels markedly decreased in rabbits. On the other hand, Ostrowska et al. [40] found that dietary CLA treatment significantly increased plasma NEFA and triacylglycerol levels in pigs, with no effect on plasma glucose and insulin levels. In addition, Gutgesell et al. [41] revealed that lactating rats fed CLA diets showed a greater concentration of NEFA in plasma compared to rats fed diets containing sunflower oil.

As shown in this study, dietary CLA had no effect on total protein, total lipid, or triacylglycerol concentration, which is in accordance with the results of Stangl [42,43]. The glucose concentration also did not differ between the treatments. This result corroborates data from Bontempo et al. [24] and Corino et al. [13], who reported that the serum glucose concentration of lactating sows remained unchanged after CLA treatment. Ramsay et al. [44] also found that CLA supplementation up to 2% had

no effect on the serum glucose concentration of growing pigs. On the contrary, Stangl [42] found that rats fed 5% CLA exhibited a higher concentration of glucose than control rats.

In the present study, CLA had no effect on total, LDL-, or HDL-cholesterol concentration, which leads to no difference in the HDL-to-total cholesterol or LDL-to-HDL-cholesterol ratio. There is little information on the effects of dietary CLA on the serum cholesterol profile of lactating sows. Stangl [42] found that rats fed 5% CLA showed significantly reduced total, LDL-, and HDL-cholesterol levels, whereas rats fed less than 5% CLA showed no differences. Mele et al. [45] also reported that dietary intake of 0.8 g of cis-9, trans-11 CLA per day in humans tended to reduce the plasma LDL-cholesterol level. Stangl [43] also reported that the serum cholesterol level of growing rats fed 3% CLA under conditions of enhanced fat mobilization remained unchanged. Moreover, Stangl et al. [46] observed no difference in the serum concentration of total, LDL-, or HDL-cholesterol in adult female pigs fed CLA at a dietary level of 1%, which is in accordance with data from this study. On the other hand, Corino et al. [18] found that total cholesterol level in rabbits was markedly reduced by consumption of 0.25 or 0.5% CLA. Nunes et al. [47] also reported that dogs fed 0.5% CLA for nine months showed a 34% reduction in the total cholesterol level, as well as 28% reduction in the levels of LDL and non-HDL-cholesterol. Therefore, CLA can have differential effects depending on its level, dominant isomer, animal species used, and physiological status.

The present data show that the serum urea nitrogen concentration was remarkably higher in sows fed soybean oil compared to those fed CLA. This result may be due to higher catabolism of body protein during lactation in sows fed soybean oil compared to those fed CLA.

The present study examined the thyroid hormones T4 and T3 as an indicator of mammary gland growth and development [48]. In the present study, there was no difference in the T4 or T3 concentration between the treatments. This result is in agreement with data from Stangl et al. [46] and Bontempo et al. [24]. In contrast, Corino et al. [13] found that sows fed CLA until weaning had a high concentration of serum thyroxine, resulting in heavier piglets via increased milk yield. Beckett et al. [49] found that conversion of T4 to T3, the biologically active form of thyroid hormone, is controlled by selenium-containing enzymes. Consequently, the present study again confirmed no relationship between CLA and T3.

As in other species, IgG is the most representative antibody in pig serum, and its concentration is an index of antibody production against antigenic stimuli not processed by T-cells. In the present study, the IgG concentrations of sows fed CLA increased by 49% in sow serum, 23% in milk, and 35% in piglet serum compared with levels of control groups. Furthermore, piglets from sows fed CLA diet also showed increased serum IgG concentrations. These results support data from Bontempo et al. [24] and Corino et al. [13]. Corino et al. [18] also found that weaned piglets fed increasing levels of CLA showed elevated IgG concentrations, suggesting a positive correlation between CLA intake and serum IgG. Moreover, Peng et al. [17] reported that consumption of 0.5 and 1% CLA by lactating sows increased CLA contents in milk and plasma as well as backfat and muscle in their suckling piglets, reflecting maternal CLA transport. However, this study did not analyze the fatty acid profiles of serum and milk. Thus, in the present study, the elevation of IgG concentrations in piglets from sows fed CLA could be attributed to maternal IgG transfer directly to piglets through milk CLA.

Whether or not dietary CLA affects milk yield in lactating sows remains unknown, as milk yield fluctuates depending on breed, parity, suckling intensity, lactational day, and health conditions. Indeed, as this study indirectly measured milk yield using the weigh-suckle-weigh technique [15] based on daily changes in litter BW of sows, there could be differences between real and estimated figures for milk yield. Although there was no difference in daily lactational feed intake between groups, milk yield increased by almost 10% in sows fed CLA compared with sows fed soybean oil. In past studies using lactating cows [50] and lactating ewes [51,52], rumen-protected or unprotected CLA supplementation increased milk yield. These results are similar to those of the present study, even though it was compared to the milk yield of ruminants other than monogastrics.

On the other hand, CLA reduced milk fat content by about 30%, whereas milk protein, lactose, and solid-not-fat contents were unaffected (Table 5). This result is similar to the data of Harrell et al. [16], who reported that sows fed 1% CLA diet showed 36% reduced milk fat content, whereas milk protein or ash content was unaffected. Moreover, Cordero et al. [14] observed a 14% reduction in crude fat content in milk from sows fed 1% CLA diet compared to those fed control diet. Poulos et al. [53] also reported that 0.5% CLA supplementation from day 40 of gestation until weaning reduced milk fat by 17%. Griinari et al. [54] demonstrated that milk fat depression can occur by either a shortage of precursors for milk fat synthesis or by direct inhibition of milk fat synthesis. Despite the fact that the milk fat content was reduced by CLA supplementation, milk fat yield did not show any significant difference between the treatments. This might be attributable to the offset of milk fat reduction by increased milk yield.

Until now, there has been no report on lactose in lactating sows. In the present study, increased lactose yield due to dietary CLA appeared to be associated with higher circulation of blood glucose to mammary glands for lactose synthesis. Further, although the blood glucose concentration was not significantly different between the treatments (Table 4), a slight increase in blood glucose in sows fed CLA resulted in elevated synthesis of lactose.

In the present study, lower milk fat content due to dietary CLA increased the protein-to-fat ratio. This higher protein-to-fat ratio in sows fed CLA can increase weight at weaning in suckling piglets. Nam and Aherne [55] found that increasing ratios of lysine to DE in weanling piglets linearly increased average daily gain and feed efficiency.

Conclusion

In conclusion, CLA supplementation to sows during lactation resulted in little change in BW despite increased milk yield, suggesting less mobilization of body stores. Piglets from sows fed CLA showed higher weights at weaning as well as weight gain, but there was no difference in piglet growth after weaning regardless of maternal CLA background. Sows fed CLA also showed reduced serum NEFA and urea nitrogen concentrations as a result of lesser body expenditure, whereas levels of other blood metabolites and thyroid hormone were unaltered. CLA supplementation to sows during lactation increased IgG concentrations not only in sera of sows and piglets but also in milk, implying that CLA may improve the health status and growth of piglets. However, milk produced by sows fed CLA contained lower milk fat content, resulting in a higher protein-to-fat ratio. Further, yield of solid-not-fat in response to dietary CLA consumption remarkably increased due to higher lactose yield.

Implications

These results show that CLA supplementation to sows increased milk yield without losing BW during lactation, which might positively affect subsequent reproduction and piglet growth. Moreover, soybean oil supplementation to lactating sows resulted in severe BW loss, which might be helpful in reducing BW of obese sows.

Competing interests
The authors declare that they have no competing interests.

Authors' contributions
SHL designed the experiment, wrote the manuscript, and performed the statistical analysis. JM and WY revised the manuscript. All authors were responsible for interpreting the data, added scientific comments, and approved the final version of the manuscript.

Acknowledgements
This study was supported by the research fund of the Gyeongsangnamdo Provincial Government, Republic of Korea. The authors extend thanks to the staff of Livestock Experiment Station (Sancheong), especially Man-Dal Lee for his enthusiastic animal care and technical assistance as well as veterinarians Hee-Sook Kwon and Do-KyungYoun for the collection of blood.

Author details
[1]Livestock Experiment Station, Gyeongsangnamdo Livestock Promotion Research Institute, 251 Cheonghyun-ro, Sinan-myeon, Sancheong 666-962, Republic of Korea. [2]Department of Beef & Dairy Science, Korea National College of Agriculture and Fisheries, 212 Hyohaeng-ro, Bongdam-eup, Hwaseong 445-760, Republic of Korea.

References

1. Boyd RD, Kensinger RS. In *Metabolic precursors for milk synthesis*, Page 71 in The Lactating Sow. Edited by Verstegen MWA, Moughun PJ, Schrama JW. Wageningen, the Netherlands: Wageningen Press; 1998.
2. Mahan DC, Lepine AJ: **Effect of weaning weight and associated nursery feeding programs on subsequent performance to 105 kilograms of body weight.** *J Anim Sci* 1991, 69:1370–1378.
3. Azain MJ: *Nutrition of the young pig, use of liquid diets*, Pages 1–14 in Proc. 13th Annu. Raleigh, NC: Canolina Swine Nutr. Conf. Raleigh, NC. Canolina Feed Ind. Assoc; 1997.
4. King RH, Williams IH: **The effect of nutrition on the reproductive performance of first-litter sows. 2. Protein and energy intakes during lactation.** *Anim Prod* 1984, 38:249–256.
5. King RH, Dunkin AC: **The effect of nutrition on the reproductive performance of first-litter sows. 4. The relative effects of energy and protein intakes during lactation on the performance of sows and their piglets.** *Anim Prod* 1986, 43:319–325.
6. Noblet J, Etienne M: **Estimation of sow milk nutrient output.** *J Anim Sci* 1989, 67:3352–3359.
7. Quesnel H, Etienne M, Père M-C: **Influence of litter size on metabolic status and reproductive axis in primiparous sows.** *J Anim Sci* 2007, 85:118–128.
8. Clowes EJ, Aherne FX, Foxcroft GR, Baracos VE: **Selective protein loss in lactating sows is associated with reduced litter growth and ovarian function.** *J Anim Sci* 2003, 81:753–764.
9. Thaker MYC, Bilkei G: **Lactation weight loss influences subsequent reproductive performance of sows.** *Anim Reprod* 2005, 88:309–318.
10. Kepler CR, Hirons KP, McNeill JJ, Tove SB: **Intermediates and products of the biohydrogenation of linoleic acid by *Butyrivibrio fibrisolvens*.** *J Biol Chem* 1966, 241:1350–1354.
11. Belury MA: **Dietary conjugated linoleic acid in health: Physiological effects and mechanisms of action.** *Annu Rev Nutr* 2002, 22:505–524.
12. Pariza MW: **Perspective on the safety and effectiveness of conjugated linoleic acid.** *Am J Clin Nutr* 2004, 79:1132S–1136S.
13. Corino C, Pastorelli G, Rosi F, Bontempo V, Rossi R: **Effect of dietary conjugated linoleic acid supplementation in sows on performance and immunoglobulin concentration in piglets.** *J Anim Sci* 2009, 87:2299–2305.
14. Cordero G, Isabel B, Morales J, Menoyo D, Pineiro C, Daza A, Lopez-Bote CJ: **Conjugated linoleic acid (CLA) during last week of gestation and lactation alters colostrums and milk fat composition and performance of reproductive sows.** *Anim Feed Sci Technol* 2011, 168:232–240.
15. Speer VC, Cox DF: **Estimating milk yield of sows.** *J Anim Sci* 1984, 59:1281–1285.
16. Harrell RJ, Phillips O, Boyd RD, Dwyer DA, Bauman DE: *Effects of conjugated linoleic acid on milk composition and baby pig growth in lactating sows*, Annu Swine Report 2002. USA: NC State University; 2002.
17. Peng Y, Ren F, Yin JD, Fang Q, Li FN, Li DF: **Transfer of conjugated linoleic acid from sows to their offspring and its impact on the fatty acid profiles of plasma, muscle, and subcutaneous fat in piglets.** *J Anim Sci* 2010, 88:1741–1751.
18. Corino C, Bontempo V, Sciannimanico D: **Effects of dietary conjugated linoleic acid on some aspecific immune parameters and acute phase protein in weaned piglets.** *Can J Anim Sci* 2002, 82:115–117.
19. NRC: *Nutrient Requirements of Swine*. 10 revth edition. Washington, DC: Natl. Acad. Press; 1998.
20. AOAC: *Official Methods of Analysis (18th Ed)*. Washington DC: Association of Official Analytical Chemists; 2005.
21. Sukhija PS, Palmquist DL: **Rapid method for determination of total fatty acid content and composition of feedstuffs and feces.** *J Agric Food Chem* 1988, 36:1202–1206.
22. Mancini G, Carbonara AO, Heremans JF: **Immunological quantitation of antigens by single radial immunodiffusion.** *Immunochemistry* 1965, 2:235–243.

23. SAS Institute: *Statistics*. Cary, NC: SAS/STAT user's guide, SAS Institute; 2004.
24. Bontempo V, Sciannimanico D, Pastorelli G, Rossi R, Rosi F, Corino C: **Dietary conjugated linoleic acid positively affects immunologic variables in lactating sows and piglets.** *J Nutr* 2004, 134:817–824.
25. Sanz M, Lopez-Bote CJ, Menoyo D, Bautista JM: **Abdominal fat deposition and fatty acid synthesis are lower and beta-oxidation is higher in broiler chickens fed diets containing unsaturated rather than saturated fat.** *J Nutr* 2000, 130:3034–3037.
26. Shimomura Y, Tamura T, Suzuki M: **Less body-fat accumulation in rats fed a safflower oil diet than in rats fed a beef tallow diet.** *J Nutr* 1990, 120:1291–1296.
27. Vicente JG, Isabel B, Cordero G, Lopez-Bote CJ: **Fatty acid profile of the sow diet alters fat metabolism and fatty acid composition in weanling pigs.** *Anim Feed Sci Technol* 2013, 181:45–53.
28. Park JC, Kim YH, Jung HJ, Moon HK, Kwon OS, Lee BD: **Effects of dietary supplementation of conjugated linoleic acid (CLA) on piglets' growth and reproductive performance in sows.** *Asian Aust J Anim Sci* 2005, 18:249–254.
29. Dugan MER, Alhus JL, Schaefer AL, Kramer JKG: **The effect of linoleic conjugated acid on fat to lean repartitioning and feed conversion in pigs.** *Can J Anim Sci* 1997, 77:723–725.
30. Ostrowska E, Muralitharan M, Cross RF, Bauman DE, Dunshea FR: **Dietary conjugated linoleic acids increase lean tissue and decrease fat deposition in growing pigs.** *J Nutr* 1999, 129:2037–2042.
31. Thiel-Cooper RL, Parrish FC Jr, Sparks JC, Wiegand BR, Ewan RC: **Conjugated linoleic acid changes swine performance and carcass composition.** *J Anim Sci* 2001, 79:1821–1828.
32. Wiegand BR, Sparks JC, Parrish FC Jr, Zimmerman DR: **Duration of feeding conjugated linoleic acid influences growth performance, carcass traits, and meat quality of finishing barrows.** *J Anim Sci* 2002, 80:637–643.
33. Barnes KM, Winslow NR, Shelton AG, Hlusko KC, Azain MJ: **Effect of dietary conjugated linoleic acid on marbling and intramuscular adipocytes in pork.** *J Anim Sci* 2012, 90:1142–1149.
34. Reese DE, Peo ER Jr, Lewis AJ: **Relationship of lactation energy intake and occurrence of postweaning estrus to body and backfat composition in sows.** *J Anim Sci* 1984, 58:1236–1244.
35. Kirkden RD, Broom DM, Andersen IL: **Piglet mortality: Management solutions.** *J Anim Sci* 2013, 91:3361–3389.
36. McNamara JP, Pettigrew JE: **Protein and fat utilization in lactating sows: I. Effects on milk production and body composition.** *J Anim Sci* 2002, 80:2442–2451.
37. Cabrera RA, Boyd RD, Jungst SB, Wilson ER, Johnston ME, Vignes JL, Odle J: **Impact of lactation length and piglet weaning weight on long-term growth and viability of progeny.** *J Anim Sci* 2010, 88:2265–2276.
38. Bee G: **Dietary conjugated linoleic acid consumption during pregnancy and lactation influences growth and tissue composition in weaned pigs.** *J Nutr* 2000, 130:2981–2989.
39. West DB, Delany JP, Camet PM, Blohm F, Truett AA, Scimeca J: **Effects of conjugated linoleic acid on body fat and energy metabolism in the mouse.** *Am J Physiol* 1998, 44:R667–R672.
40. Ostrowska E, Cross RF, Muralitharan M, Bauman DE, Dunshea FR: **Effects of dietary fat and conjugated linoleic acid on plasma metabolite concentrations and metabolic responses to homeostatic signals in pigs.** *Br J Nutr* 2002, 88:625–634.
41. Gutgesell A, Ringseis R, Eder K: **Dietary conjugated linoleic acid down-regulates fatty acid transporters in the mammary glands of lactating rats.** *J Dairy Sci* 2009, 92:1169–1173.
42. Stangl GI: **High dietary levels of a conjugated linoleic acid mixture alter hepatic glycerophospholipid class profile and cholesterol-carrying serum lipoproteins of rats.** *J Nutr Biochem* 2000, 11:184–191.
43. Stangl GI: **Conjugated linoleic acids exhibit a strong fat-to-lean partitioning effect, reduce serum VLDL lipids and redistribute tissue lipids in food-restricted rats.** *J Nutr* 2000, 130:1140–1146.
44. Ramsay TG, Evock-Clover CM, Steele NC, Azain MJ: **Dietary conjugated linoleic acid alters fatty acid composition of pig skeletal muscle and fat.** *J Anim Sci* 2001, 79:2152–2161.
45. Mele MC, Cannelli G, Carta G, Cordeddu L, Melis MP, Murru E, Stanton C, Banni S: **Metabolism of c9, t11-conjugated linoleic acid (CLA) in humans.** *PLEFA* 2013, 89:115–119.
46. Stangl GI, Müller H, Kirchgessner M: **Conjugated linoleic acid effects on circulating hormones, metabolites and lipoproteins, and its proportion in**

fasting serum and erythrocyte membranes of swine. *Eur J Nutr* 1999, **38**:271–277.

47. Nunes EA, Bonatto SJ, de Oliveira HHP, Rivera NLM, Maiorka A, Krabbe EL, Tanhoffer RA, Fernandes LC: **The effect of dietary supplementation with 9-*cis*:12-*trans* and 10-*trans*:12-*cis* conjugated linoleic acid (CLA) for nine months on serum cholesterol, lymphocyte proliferation and polymorphonuclear cells function in Beagle dogs.** *Res Vet Sci* 2008, **84**:62–67.

48. Tucker HA: **Physiological control of mammary growth, lactogenesis, and lactation.** *J Dairy Sci* 1981, **64**:1403–1421.

49. Beckett GJ, Nicol F, Rae PW, Beach S, Guo Y, Arthur JR: **Effects of combined iodine and selenium deficiency on thyroid hormone metabolism in rats.** *Am J Clin Nutr Suppl* 1993, **57**:2405–2435.

50. Bernal-Santos G, Perfield JW II, Barbano DM, Bauman DE, Overton TE: **Production responses of dairy cows to dietary supplementation with conjugated linoleic acid (CLA) during the transition period and early lactation.** *J Dairy Sci* 2003, **86**:3218–3228.

51. Lock AL, Teles BM, Perfield JW II, Bauman DE, Sinclair LA: **A conjugated linoleic acid supplement containing *trans*-10, *cis*-12 reduces milk fat synthesis in lactating sheep.** *J Dairy Sci* 2006, **89**:1525–1532.

52. Husvéth F, Galamb E, Gaál T, Dublecz K, Wágner L, Pál L: **Milk production, milk composition, liver lipid contents and C18 fatty acid composition of milk and liver lipids in Awassi ewes fed a diet supplemented with protected *cis*-9, *trans*-11 and *trans*-10, *cis*-12 conjugated linoleic acid (CLA) isomers.** *Small Rum Res* 2010, **94**:25–31.

53. Poulos SP, Azain MJ, Hausman GJ: **Conjugated linoleic acid (CLA) during gestation and lactation does not alter sow performance or body weight gain and adiposity in progeny.** *Anim Res* 2004, **53**:275–288.

54. Griinari JM, Chouinard PY, Bauman DE: *Trans fatty acid hypothesis of milk fat depression revised.* Ithaca, NY: Proc. Cornell Nutr. Conf; 1997:208–216.

55. Nam DS, Aherne FX: **The effects of lysine: energy ratio on the performance of weanling pigs.** *J Anim Sci* 1994, **72**:1247–1256.

3

Identification of proteins involved in the pancreatic exocrine by exogenous ghrelin administration in Sprague-Dawley rats

Kyung-Hoon Lee[1], Tao Wang[1], Yong-Cheng Jin[2], Sang-Bum Lee[1], Jin-Ju Oh[1], Jin-Hee Hwang[1], Ji-Na Lim[1], Jae-Sung Lee[1] and Hong-Gu Lee[1]*

Abstract

The aims of study were to investigate the effects of intraperitoneal (i.p.) infusion of ghrelin on pancreatic α-amylase outputs and the responses of pancreatic proteins to ghrelin that may relate to the pancreatic exocrine. Six male Sprague-Dawley rats (300 g) were randomly divided into two groups, a control group (C, n = 3) and a treatment group (T, 10.0μg/kg BW, n = 3). Blood samples were collected from rat caudal vein once time after one hour injection. The concentrations of plasma ghrelin, cholecystokinin (CCK) and alfa-amylase activity were evaluated by enzyme immunoassay (EIA) kit. Two-dimensional gel electrophoresis (2-DE) analysis was conducted to separate the proteins in pancreas tissue. Results showed that the i.p. infusion of ghrelin at doses of 10.0 μg/kg body weight (BW) increased the plasma ghrelin concentrations (p = 0.07) and elevated the plasma CCK level significantly (p < 0.05). Although there was no statistically significant, the α-amylase activity tended to increase. The proteomics analysis indicated that some pancreatic proteins with various functions were up- or down- regulated compared with control group. In conclusion, ghrelin may have role in the pancreatic exocrine, but the signaling pathway was still not clear. Therefore, much more functional studies focus on these found proteins are needed in the near future.

Keywords: Alfa-amylase activity, Cholecystokinin, Ghrelin, Pancreatic exocrine, Sprague-Dawley rats, Two-dimensional gel electrophoresis

Background

Ghrelin is a 28-amino-acid peptide isolated from the rat stomach in 1999 which is mainly produced by X/A-like cells in the oxyntic mucosa [1]. It was also found in other parts like hypothalamus, pituitary gland, lung, kidney [2-4], duodenum, ileum, colon and pancreas [5]. The ghrelin receptors (GHS-R) have been detected in many central and peripheral tissues and pancreatic α-cells, β-cells, exocrine cells. Ghrelin shows a number of actions at the gastrointestinal tract level. It is well known as a stronger activator of growth hormone (GH) through central nervous system (CNS) and modulation of food intake. And it also exhibits lots of other biological activities including energy expenditure, stimulation of lactotroph and corticotroph secretion,

influence on sleep and behavior or modulation of heart rate and blood pressure ([6-9]; Assakawa et al. [10]).

Recent studies demonstrated that ghrelin could modulate exocrine secretions as well as pancreatic endocrine secretions. There were two opposite theories. First, Zhang and coworkers demonstrated that intravenous administration of ghrelin to the rats produced inhibition of enzyme secretion, and that this inhibitory effect of ghrelin on pancreatic exocrine secretion is indirect and may be exerted at the level of intrapancreatic neurons [11]. In contrast, other studies revealed that central as well as peripheral administration of ghrelin significantly increased pancreatic fluid and protein output, though the activation of vagal centers in the brainstem and stimulation of vagal efferent nerves [12,13]. It is generally believed that, in the intestinal phase, stimulation of pancreatic enzyme secretion depends, in the main part, on the neuronal mechanism involved in the CCK release and activation of cholinergic vago-vagal enteropancreatic reflex. However, these physiological roles of ghrelin in the

---

* Correspondence: hglee66@konkuk.ac.kr
[1]Department of Animal Science and Technology, College of Animal Bioscience & Technology, Konkuk University, 120 Neungdong-ro, Seoul, Gwangjin-gu 143-701, South Korea
Full list of author information is available at the end of the article

modulation of exocrine pancreatic functions are still unclear (Gherlaldoni et al. [14]).

Therefore, this study was conducted to determine the effects of exogenous ghrelin, given intraperitoneal (i.p.) on plasma CCK level and α-amylase output in the sprague-dawley rats. Meanwhile, the responses of pancreatic proteins to ghrelin were also analyzed using the 2-DE system.

Methods

Materials

Rat ghrelin peptide was obtained from Bachem (Bubendorf, Switzerland). Rat plasma ghrelin and CCK enzyme immunoassay kit were purchased from Phoenix Pharmaceuticals (Mountain View, CA, USA). The α-amylase activity analyzed using an EnzyChrom™ a-Amylase Assay Kit (ECAM-100).

Animals

Sprague-Dawley rats (300 g) (Samtaco, Osan, Korea) were used for all experiments. Animals were housed at one animal per cage on a 12/12 h light cycle (lights on at 8 am) and given access to food and water ad libitum. Ghrelin (0.1, 1.0, 10.0 ug/kg, respectively) were injected intraperitoneally in rats. One hour after injection, blood samples (1 ml) were collected from rat caudal vein once time into heparined tube and immediately centrifuged (3,000 rpm/min, 15 min) to obtain the plasma. Aliquots of plasma were stored at -80°C till analyzed. Then rats were anaesthetized via intramuscular injection of zoletil (Vetbutal) at a dose of 15.0 mg/300 g (BW). And the rat pancreas tissues were collected. All experimental procedures were in accordance with the "Guidelines for the Care and Use of Experimental Animals of Pusan National University".

Determination of plasma ghrelin, CCK concentration and α-amylase activity

The plasma ghrelin and CCK concentration were separately determined by enzyme immunoassay kit (Phoenix Pharmaceuticals, Inc., Burlingame, CA, USA). The plasma α-amylase activity analyzed using an EnzyChrom™ a-Amylase Assay Kit (ECAM-100, BioAssay Systems, Hayward, CA, USA). All the operations were done followed the kit manual.

Pancreas sample preparation and 2-DE analysis

Pancreatic tissues were collected and then pulverized into powder under liquid nitrogen and stored at -80°C until use. Tissues (0.5 g) were homogenized in 1ml lysis solution containing 7 M urea, 2 M thiourea, 4% chaps, 40 mM Tris, 65 mM DTT, 0.5% IPG buffer and 1X protease inhibitor (GE Healthcare, Piscataway, New Jersey, USA). This mixture was stirred every 5 min for 30 min and then centrifuged at 14,000 rpm for 30 min at 20°C. The supernatant was then stored in aliquots at -80°C until analysis.

2-DE was performed of pooling pancreatic tissue samples from three animals in each group. Briefly, protein samples were diluted into isoelectric focusing (IEF) buffer containing 6 M urea, 2 M thiourea, 1% 3-((3-Cholamidopropyl) dimethylammonium)-1-propanesulfonate (CHAPS), 0.002% bromophenol blue, 0.5% phamalyte (pH 3-10NL) and 65 mM dithiothreitol (DTT). Then 100 μg protein smaples of control or treatment was loaded on Immobiline DryStrip gels (pH 3-10NL, 18 cm, GE Healthcare) for rehydration for 12 h at 20°C. The IEF procedures were performed using an IEF electrophoresis unit (GE Healthcare) following the manufacturer's protocols. The following voltage program was used after the 12 h rehydration: linear ramp from 500 to 1000 V over 2 h, then a constant voltage of 8000 V for 7 h to give a total of 56,000 V h. After focusing, gel strips were equilibrated in a solution containing 50 mM Tris/HCl (pH 8.8), 6 M urea, 2% SDS, 30% glycerol, 0.002% w/v bromophenol blue and DTT for 15 min, followed by incubation in the same solution but replacing DTT with 135 mM iodoacetamide (IAA) for another 15 min. After that the equilibrated strips were inserted into sodium dodecyl sulfate polyacrylamide gel electrophoresis (SDS-PAGE) gels (18 cm, 12%). SDS-PAGE was performed using an Ettan DALT 2-D gel system (GE Healthcare). Upon completion, gels were stained using a PluseOne Silver Staining Kit (GE Healthcare). The silverstained gels were scanned using an Umax scanner (PowerLook 2100XL, UMAX Technologies, Inc., Dallas, TX, USA). Scanned gel images were processed by Proteomweaver™ 2-D Analysis Software (Definiens AG, Munich, Germany).

Statistical analysis

Data are presented as mean ± standard error of the mean (SEM). The group mean values were compared with an independent sample t-test (SPSS 14.0, Chicago, IL, USA). A p-value <0.05 was considered to be statistically significant.

Results and discussion

The i.p. infusion of ghrelin at doses of 10.0 μg/kg body weight (BW) increased the plasma ghrelin concentrations ($p = 0.07$) and elevated the plasma CCK level significantly ($p < 0.05$). Although there was no statistically significant, the α-amylase activity tended to increase (Table 1). These data indicated that ghrelin may have role in pancreatic exocrine secretion through the stimulation of CCK release and

Table 1 Relationship between plasma G.I. hormones and α-amylase activity

Items	Ghrelin (ng/ml)	CCK (ng/ml)	α-amylase (U/L)
Control[1]	0.202 ± 0.152[2]*	5.745 ± 2.428	3791.799 ± 208.308
Treatment[1]	3.223 ± 2.140	11.306 ± 3.937	4210.847 ± 262.825

[1]Control = vehicle/500 uLSaline, Treatment = Ghrelin (10.0 ug/kg)/500 uLSaline.
[2]Data are mean ± SEM and *means significantly different, p < 0.05 (Student's paired t-test).

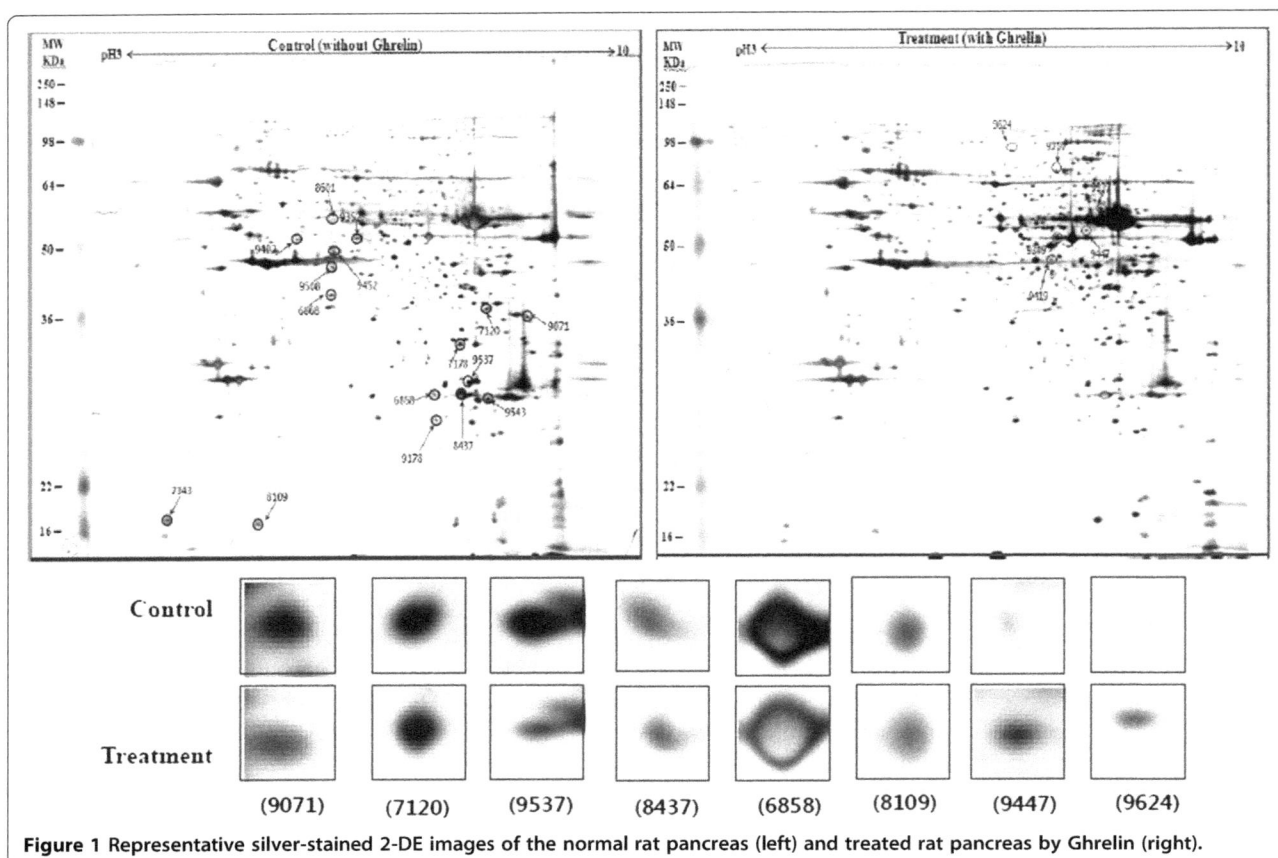

Figure 1 Representative silver-stained 2-DE images of the normal rat pancreas (left) and treated rat pancreas by Ghrelin (right).

activation of dorsal vagal nerve. CCK is one of the major gut hormones which released from duodenal mucosa I cells. It can stimulate pancreatic exocrine secretion via activation of CCK1 receptors and entero-pancreatic vago-vagal reflex [15-19]. Whether CCK plays a role in the exocrine secretion response to ghrelin is unknown.

The response of pancreatic proteins to ghrelin administration were analyzed by 2-DE in order to get a better understanding of the mechanisms involved in the ghrelin-exocrine secretion through CCK signal. Eight spots ranging from 6 to 200 kDa were detected differently expressed in the pancreatic protein map (Figure 1). One

Table 2 Identification of differentially altered protein spots in rat pancreas by ESI/Q-TOF MS

Spot	Protein name	Peptide match	Protein score	Mass (bp)	Expression (treatment/control)
Cell growth and proliferation					
8109	Eukaryotic translation initiation factor 5A-1	295	98.23	16821.4	Down
Gluconeogenesis					
9071	Malate dehydrogenase, mitochondrial precursor	40	100.29	35660.8	Down
De nove purine biosynthesis					
9447	Similar to adenylosuccinate lyase	67	214.27	54817	Up
Oxidative stress					
6858	PRx IV	44	104.23	30988.1	Down
7120	Thiosulfate sulfurtransferase	392	154.28	33385.8	Down
9624	Sarcosine dehydrogenase, mitochondrial precursor	115	238.23	101578.9	Only Treatment
Others					
8437	Ubiquinol-cytochrome c reductase iron-sulfur subunit, mitochondrial precursor	55	100.32	29427.2	Down
9537	Carbonic anhydrase 3	297	148.26	29412.7	Down

spot was up-regulated and six were down-regulated compared with control group. One spot was found only expressed in treatment group. These proteins were classified according to their functions (Table 2). Three down-regulated proteins eukaryotic translation initiation factor 5A-1 (eIF5A), peroxiredoxin IV (PRx IV) and ubiquinol-cytochrome c reductase (UQCR) were particular interested. eIF5A is involved in biosynthesis of pancreatic enzyme by increasing plasma CCK level [20]. And this data was different with ours. Judging by our reports, ghrelin may affect on pancreatic exocrine secretion via other factors as well as releasing CCK. Also, Yannick demonstrated that UQCR and PRx IV down-regulated biosynthesis of insulin related to diabetes [21]. Synthetically, when the effects of exogenous ghrelin, given intraperitoneal (i.p.) on pancreatic exocrine secretion, we have to consider other factor like growth hormone related to insulin synthesis as involving the CCK release and activation of cholinergic vago-vagal enteropancreatic reflex.

Conclusions

In conclusion, our result suggest a role of ghrelin on pancreatic exocrine, but the protein concerning with mechanism functional study was uncompleted. And some following functional study of proteins will be done in the near future.

Competing interests
The authors declare that they have no competing interests.

Authors' contributions
HGL conceived and designed the Experiments. KHL executed the experiment and analyzed the samples. TW revised the manuscript. All authors interpreted the data, critically revised the manuscript for important intellectual contents and approved the final version.

Acknowledgments
This research was supported by Bioindustry Technology Development Program (313020041SB010) for Ministry of Agriculture, Food and Rural Affairs, Republic of Korea.

Author details
[1]Department of Animal Science and Technology, College of Animal Bioscience & Technology, Konkuk University, 120 Neungdong-ro, Seoul, Gwangjin-gu 143-701, South Korea. [2]Department of Animal Science, College of Animal Science and Veterinary Medicine, Jilin University, Changchun 130062, P. R. China.

References

1. Kojima M, Hasoda H, Date Y, Nakazato M, Matsou H, Kangawa K: Ghrelin is a growth-hormone-releasing acylated peptide from stomach. *Nature* 1999, **402:**656–660.
2. Muccioli G, Tschop M, Papotti M, Deghenghi R, Heiman M, Ghigo E: Neuroendocrine and peripheral activities of ghrelin: implications in metabolism and obesity. *Eur J Pharmacol* 2002, **440**(2–3):235–254.
3. Korbonits M, Kojima M, Kangawa K, Grossman AB: Presence of ghrelin in normal and adenomatous human pituitary. *Endocrine* 2001, **14:**101–104.
4. Volante M, Fulcheri E, Allia E, Cerrato M, Pucci A, Papotti M: Ghrelin expression in fetal infant and adult human lung. *J Histochem Cytochem* 2002, **50:**1013–1021.
5. Date Y, Kojima M, Hosoda H, Sawaguchi A, Mondal MS, Suganuma T, Matsukura S, Kangawa K, Nakazato M: Ghrelin, a novel growth hormone-releasing acylated peptide, is synthetized in a distinct endocrine cell type in the gastrointestinal tracts of rats and humans. *Endocrinology* 2000, **141:**4255–4261.
6. Ghigo E, Arvat E, Giordano R, Broglio F, Gianotti L, Maccario M, Bisi G, Graziani A, Papotti M, Muccioli G, Deghenghi R, Camanni F: Biologic activities of growth hormone secretagouges in humans. *Endocrine* 2001, **14:**87–93.
7. Tschop M, Smiley DL, Heiman ML: Ghrelin induces adiposity in rodents. *Nature* 2000, **407:**908–913.
8. Nakazato M, Murakami N, Date Y, Kojima M, Matsuo H, Kangawa K, Matsukura S: A role for ghrelin in the central regulation of feeding. *Nature* 2001, **409:**194–198.
9. Tolle V, Bassant MH, Zizzari P, Poindessous-Jazat F, Tomasetto C, Epelbaum J, Bluet-Pajot MT: Ultradian rhythmicity of ghrelin secretion in relation with GH, feeding behavior, and sleep–wake patterns in rats. *Endocrinology* 2002, **143:**1353–1361.
10. Asakawa A, Inui A, Kaga T, Yuzuriha H, Nagata T, Fujimiya M, Katsuura G, Makino S, Fujino MA, Kasuga M: A role of ghrelin in neuroendocrine and behavioral responses to stress in mice. *Neuroendocrinology* 2001, **74:**143–147.
11. Broglio F, Prodam F, Me E, Riganti F, Lucatello B, Granata R, Benso A, Muccioli G, Ghigo E: Ghrelin: endocrine, metabolic and cardiovascular actions. *J Endocrinol Invest* 2005, **28**(5 suppl):23–25.
12. Sato N, Kanai S, Takano S, Kurosawa M, Funakoshi A, Miyasaka K: Central administration of ghrelin stimulates pancreatic protein exocrine secretion via the vagus in conscious rats. *Jpn J Physiol* 2003, **53**(6):443–449.
13. Li Y, Wu X, Zhao Y, Chen S, Owyang C: Ghrelin acts on the dorsal vagal complex to stimulate pancreatic protein secretion. *Am J Physiol Liver Physiol* 2006, **290**(6):G1350–G1358.
14. Ghelardoni S, Carnicelli V, Frascarelli S, Ronca-Testoni S, Zucchi R: Ghrelin tissue distribution: comparison between gene and protein expression. *J Endocrinol Invest* 2006, **29**(2):115–121.
15. Brzozowska I, Konturek PC, Brzozowski T, Konturek SJ, Kwiecien S, Pajdo R, Drozdowicz D, Pawlik M, Ptak A, Hahn EG: Role of prostaglandins, nitric oxide sensory nerves and gastrin in acceleration of ulcer healing by melatonin and its precursor; L-tryptophan. *J Pineal Res* 2002, **32:**149–162.
16. Jahovic N, Sener G, Cevik H, Ersoy Y, Arbak S, Yeğen BC: Amelioration of metotrexate-induced enteritis by melatonin in rats. *Cell Biochem Function* 2004, **22:**169–178.
17. Jaworek J, Nawrot K, Konturek SJ, Leja-Szpak A, Thor P, Pawlik WW: Melatonin and its precursor; L-tryptophan: influence on pancreatic amylase secretion in vivo and in vitro. *J Pineal Res* 2004, **36:**156–164.
18. Leja-Szpak A, Jaworek J, Nawrot-Porabka K, Palonek M, Mitis-Musioł M, Dembiński A, Konturek SJ, Pawlik WW: Modulation of pancreatic enzyme secretion by melatonin and its precursor; L-tryptophan. Role of CCK and afferent nerves. *J Physiol* 2004, **55:**441–451.
19. Li Y, Owyang C: High affinity CCK-A receptors on the vagus nerve mediate CCK-stimulated pancreatic secretion in the rat. *Am J Physiol* 1997, **273:**G679–G685.
20. Bragado MJ, Tashiro M, Williams JA: Regulation of the initiation of pancreatic digestive enzyme protein synthesis by cholecystokinin in rat pancreas in vivo. *Gastroenterology* 2000, **119**(6):1731–1739.
21. Brunner Y, Schvartz D, Priego-Capote F, Couté Y, Sanchez JC: Glucotoxicity and pancreatic proteomics. *J Proteomics* 2009, **71**(6):576–591.

Effects of sires with different weight gain potentials and varying planes of nutrition on growth of growing-finishing pigs

Duck-Min Ha[1], Dae-Yun Jung[1], Man Jong Park[1], Byung-Chul Park[2] and C Young Lee[1*]

Abstract

The present study was performed to investigate the effects of two groups of sires with 'medium' and 'high' weight gain potentials (M-sires and H-sires, respectively) on growth of their progenies on varying planes of nutrition during the growing-finishing period. The ADG of the M-sires' progeny was greater ($P < 0.05$) than that of the H-sires' progeny (0.51 vs. 0.47 kg) during a 26- to 29-d early grower phase beginning from 55 d of age, but the opposite was true (0.66 vs. 0.72 kg) during the latter grower phase. Overall grower-phase ADG was greatest on the high plane of nutrition (H plane) followed by the medium (M) and low (L) planes (0.65, 0.61, and 0.51 kg, respectively; $P < 0.05$) in the M-sires' progeny, whereas in the H-sires' progeny, ADG was greater on the H and M planes vs. L plane (0.63, 0.62, and 0.54 kg, respectively). The ADG of pigs on the M or H plane during the grower phase and switched to the H plane thereafter (M-to-H or H-to-H planes) was greater than that of pigs on the L-to-L planes (0.99 vs. 0.78 kg) during the early finisher phase in the M-sires' progeny ($P < 0.01$). However, in the H-sires' progeny, ADG of pigs on the L-to-L planes did not differ from that of pigs on the M-to-M or H-to-M planes (0.94 vs. 0.96 kg). Results suggest that the H-to-H or H-to-M planes and M-to-M or M-to-L planes are optimal for maximal growth of the M- and H-sires' progenies, respectively.

Keywords: Sire, Progeny, Weight gain, Plane of nutrition, Growing-finishing pig

Background

The rate of weight gain or lean gain is a most important economic trait in pig production [1-3]. Breeding pigs have thus been selected for greater ADG/lean gain and less backfat thickness for decades in the Western countries, partly because most consumers prefer lean pork, partly because lean gain is much more efficient energetically than fat gain [4,5]. Given the genetic background, growth of pigs is determined primarily by nutrition as well as environmental stressors including the presence/absence or severity of diseases and heat stress [6-9].

Backfat thickness, a negative indicator of lean gain, which is known to be moderately correlated with the rate of weight gain [5], decreased through the mid-2000s as a result of selection of breeding pigs for greater lean gain in Korea [10]. However, backfat thickness rebounded by

approximately 2 mm during the latter 2000s due primarily to an allegation that lean pigs are less resistant to diseases, especially to the post-weaning multi-systemic wasting syndrome (PMWS) prevalent during that period, and also are less preferred by domestic consumers [11]. This resulted in a comparable increase in the backfat thickness of market pigs with a lag time of approximately 2 years [12]. Currently, the average backfat thickness of market pigs in Korea is 21.3 mm with an average liveweight of 114.5 kg [13]. This is much greater than the estimated 17 mm or less at the same liveweight in the USA [14] where the average liveweight of market pigs was 125.4 kg in 2013 [15]. These data suggest that overall growth potential of growing-finishing pigs in Korea is lower than that in the USA. As an initial step to see the influences of genetic growth potential on growth efficiency and thereby suggesting optimal feeding programs for pigs with different growth potentials, effects of sires with a high weight gain potential vs. a medium potential on growth of their

* Correspondence: cylee@gntech.ac.kr
[1]Regional Animal Industry Center, Gyeongnam National University of Science and Technology, Jinju 660-758, Korea
Full list of author information is available at the end of the article

progenies during the growing-finishing period were investigated in the present study.

Methods

Breeding

Four Duroc boars of JSR Genetics Ltd. (East Yorkshire, UK) origin, which were judged to have a high weight gain potential (H-sires) based on their breeding indices, were purchased from Darby Genetics Inc. (Anseong, Korea) for the present experiment in the fall of 2011. The other four boars with a medium weight gain potential (M-sires), which were suitable as sires under the carcass grading standard of 2011 [16], were selected from boars available in a commercial farm where the present experiment was performed. The H-sires and M-sires were mated to more than 20 Yorkshire × Landrace 'M-dams' of JSR origin per sire group by artificial insemination in mid- February, 2012. The progeny used for the present experiment were born on June 7 and received cross-fostering within the litters born to each sire group, iron injection, and other routine cares for suckling pigs during 21 d of lactation.

Experimental animals and diets

Weanling pigs were pooled by sex in each progeny from the respective group of sires, randomly distributed to the nursery pens (34 animals per pen), and fed the 3-phase nursery diets as previously described [17]. At 55 d of age, three pens of females and three pens of castrated males in each of the two progeny groups were selected and moved to grower pens by the pen unit for the present experiment beginning from August 1, 2012. Each pen in each progeny group × sex combination was placed on a high, medium, or low plane of nutrition during the grower phase (Table 1) for 55 to 57 d, after which the pigs were subjected to one of the two dietary treatments in each progeny group. In the M-sires' progeny, pigs on the high or medium plane of nutrition during the grower phase were fed the high (H)- and medium (M)-plane finisher diets during the finisher phases 1 and 2 up to marketing, respectively, whereas those on the low (L) plane of nutrition during the grower phase were fed the L-plane finisher diet during both finisher phases 1 and 2 (Table 2). In the progeny of the

H-sires, pigs on the H or M plane of nutrition during the grower phase were fed the M- and L-plane finisher diets during the finisher phases 1 and 2, respectively; those on the L plane of nutrition during the grower phase was fed the L-plane finisher diet during both finisher phases 1 and 2.

A total of 120 pigs in 12 pens (10 animals per pen), which had been selected randomly at the outset of the experiment, were weighed at the beginning as well as at the ends of grower phases 1 and 2 and finisher phases 1 and 2. The animals were transported to a local abattoir on the last day of the finisher phase 2 and slaughtered the following day after overnight lairage. Backfat thickness of the carcass was adjusted to a liveweight of 115 kg as described previously [18,19]. The number of days required to reach the 115-kg liveweight was estimated as follows: age in days at the end of the finisher phase 2 + (115 kg - liveweight of the finisher phase 2 in kg)/ADG during the finisher phase 2.

The experimental protocol conformed to the guidelines of the Institutional Animal Care and Use Committee (IACUC) at Gyeongnam National University of Science and Technology. The animals used in the present study were treated humanely throughout the study and did not receive any prolonged constraint.

Statistical analysis

All data were analyzed using SAS (SAS Inst. Inc., Cary, NC, USA). Individual animal was the experimental unit in all analyses. Data for the grower phase were analyzed using the GLM procedure as a completely randomize design with a 2 (sire type) × 2 (sex) × 3 (plane of nutrition) factorial arrangement of treatments. Finisher phase results of the two progeny groups born to their respective sire groups were analyzed separately using the ANOVA procedure for estimation of P-values for the fixed errors and the their interactions as well as using the GLM procedure for estimation of the least squares means and comparisons between means by the PDIFF option. Effects of the fixed errors and their interactions and differences between two means were considered to be significant when the corresponding P-value was less than 0.05.

Table 1 Declared minimum nutritional values of the commercial diets used in the present study (as-fed basis)

Item	Grower phase 1[1]			Grower phase 2[2]			Finisher[3]		
	H[4]	M[5]	L[6]	H	M	L	H[7]	M	L
Crude protein, %	18.2	17.5	15.5	16.6	16.0	15.6	13.5	13.8	13.4
Lysine, %	1.1	1.0	0.8	1.0	0.8	0.8	0.7	0.8	0.7
Crude fat, %	6.0	4.0	3.0	5.0	3.0	3.0	3.0	3.0	2.5
DE, Mcal/kg	3.45	3.35	3.30	3.45	3.30	3.30	3.24	3.30	3.25

[1]-[3]Fed from 16–29 kg, 29–48 kg, and 48–114 kg of body weight, respectively, on average.
[4]-[6]High-, medium, and low planes of nutrition, respectively.
[7]It was informed from the formulator of the experimental diets that actual nutrients densities of the finisher H were higher than those of the finisher M and also comparable to those of the phase 2 grower M.

Table 2 Experimental design: dietary regimens for the pigs born from two groups of sires with medium and high weight gain potentials

Sire group	Plane of nutrition (PN)	Grower				PN	Finisher			
		Phase 1		Phase 2			Phase 1		Phase 2	
		Diet[1]	Days	Diet[1]	Days		Diet[1]	Days[2]	Diet[1]	Days[2]
Medium weight gain potential	High (H)	H	26	H	29	H	H	28 or 29	M	38 to 42
	Medium (M)	M	27	M	29					
	Low (L)	L	29	L	28	L	L	35 or 36	L	45 or 46
High weight gain potential	H	H	26	H	29	M	M	30 or 31	L	37 or 38
	M	M	27	M	29					
	L	L	29	L	28	L	L	37	L	38

[1]See Table 1 for the nutritional composition of the diets for the high (H)-, medium (M)-, and low (L)-planes of nutrition during each phase.
[2]See Tables 4 and 5 for more details.

Results

Grower phase

Growth performance of the animals during the grower phase is shown in Table 3. The effect of sex was not significant in any of the variables ($P > 0.05$), and thus the results on sex within each sire group's progeny × plane of nutrition as well P-values for sex-associated interactions are not shown in this table for brevity. The ADG during the phase 1 was greater in the M-sires' progeny than in the H-sires' progeny (0.51 vs. 0.47 kg; SE = 0.01 kg; $P <$ 0.01). Moreover, ADG during the phase 1 was greater ($P < 0.01$) in animals on the H or M plane of nutrition (H or M group) than in those on the L plane of nutrition (L group; 0.54, 0.51, and 0.42 kg in the H, M, and L groups, respectively; SE = 0.02 kg). In addition, phase 1 ADG was greater in the H and M groups than in the L group in the progeny of the M-sires whereas in the

progeny of the H-sires, ADG was greater only in the H group vs. the L group.

The ADG during the phase 2 was greater in the H-sires' progeny vs. the M-sires' progeny (0.72 vs. 0.66 kg; SE = 0.01 kg; $P < 0.01$), which was opposite to the result of the phase 1. On the other hand, phase 2 ADG was greater ($P < 0.01$) in the H and M groups vs. the L group (0.74, 0.71, and 0.61 kg for the H, M, and L groups, respectively; SE = 0.02 kg) as in the phase 1. The ADG for the entire grower phase did not differ between the two progeny groups. Within the M-sires' progeny, overall ADG was greater in the H vs. M group ($P < 0.05$) and also in the M vs. L group (0.66, 0.61, and 0.49 kg for the H, M, and L groups, respectively; $P < 0.01$). However, in the H-sires' progeny, overall ADG was greater in the H and M group vs. the L group, with no difference between the H and M groups (0.63, 0.62, and 0.54 kg for the H, M, and L groups, respectively).

Table 3 Growth performance of the pigs born from two groups of sires with high and medium weight gain potentials, respectively, on varying planes of nutrition during the grower phase beginning from August 1 through September 26, 2012

Item	Medium potential			High potential			SEM	Sex			P-value			
	H[1]	M[2]	L[3]	H[1]	M[2]	L[3]		B[4]	G[4]	SEM	Sr[4]	PN[4]	Sr × PN	S[4]
Initial wt, kg	15.2	15.8	16.0	15.7	16.4	14.3	0.44	15.5	15.6	0.25	0.63	0.11	0.02	0.81
Days on feed														
Phase (P) 1	26	27	29	26	27	29								
P 2	29	29	28	29	29	28								
Final wt, kg														
P 1	30.7	30.5	27.6	28.7	29.1	27.0	0.8	28.8	29.0	0.5	0.05	<0.01	0.67	0.77
P 2	51.6	49.7	44.0	49.9	50.4	45.1	1.2	48.8	48.1	0.7	0.96	<0.01	0.46	0.48
ADG														
P 1	0.59	0.54	0.40	0.50	0.47	0.44	0.02	0.49	0.49	0.01	<0.01	<0.01	<0.01	0.87
P 2	0.72	0.66	0.59	0.76	0.77	0.64	0.02	0.71	0.67	0.01	<0.01	<0.01	0.27	0.08
Overall	0.66	0.61	0.49	0.63	0.62	0.54	0.02	0.60	0.58	0.01	0.48	<0.01	0.13	0.26

[1], [2], ^{3}H, high plane of nutrition; M, medium plane of nutrition; L, low plane of nutrition. Data are least squares means of 20 animals. Average ADFI for both H, M, and L groups were 1.07, 1.06, and 0.84 kg and 1.78, 1.72, and 2.02 kg during the phases 1 and 2, respectively.
[4]B, barrow; G, gilt; Sr, sire; PN, plane of nutrition; S, sex.

Finisher phase

Results for the M-sires' and H-sires' progenies during the finisher phase, which included the effects of the grower diet, finisher diet, sex, and their interactions, are shown in Tables 4 and 5, respectively. The effect of the grower × finisher interaction was significant in all variables due to the unbalanced and partially cross-ward transition of the planes of nutrition at the turn of the finisher phase, but no difference was detected between the two sub-groups of animals fed the H- and M-plane grower diets within the group of animals fed the same H-plane (Table 4) or M-plane (Table 5) finisher diet. For these reasons, results for the finisher phase were limited to comparison of growth performance of the animals on the M- or H-plane finisher diet vs. the L-plane finisher unless the effect of sex was significant.

In the progeny of the M-sires (Table 4), ADG during the finisher phase 1 was greater (P < 0.01) on the H-plane finisher diet vs. the L-plane finisher (0.99 vs. 0.78 kg). However, ADG during the finisher phase 2 was greater in the group on the L-plane finisher diet during both phases 1 and 2 (L group) than in the other group on the H- and M-plane finishers during the phases 1 and 2, respectively (H group; 0.96 vs. 0.85 kg). Consequently, ADG during

the entire finisher period did not differ between the L and H groups (0.91 vs. 0.88 kg). The ADG was not influenced by sex during either phase or the entire finishing period. The estimated number of days required to reach 115 kg of liveweight in the L group was greater than that in the H group (194.3 vs. 183.9 d), which was largely attributable to the former having 6.6 kg lower BW at the beginning of the finisher phase. However, the number of days required to reach the 115-kg liveweight was not influenced by sex. Carcass weight and dressing percentage did not differ between the dietary groups or sexes. Backfat thickness at slaughter adjusted for the 115-kg liveweight was greater (P < 0.01) in the L group than in the H group (26.1 vs. 23.4 mm; SE = 0.7 mm) as well as in barrows vs. gilts (25.6 vs. 23.8 mm). Furthermore, backfat thickness was greater in barrows than in gilts in the H group (P < 0.01), but in the L group, it did not differ between the two sexes.

In those pigs born from the H-sires, ADG during phase 1 did not differ between the L group and the M group fed the M- and L-plane finisher diets during the phases 1 and 2, respectively (Table 5). However, during the phase 2 when all pigs were fed the L-plane finisher, the L group exhibited a greater ADG than the M group (1.05 vs. 0.88 kg; P < 0.01). Likewise, ADG during the

Table 4 Effects of the plane of nutrition on growth of finishing pigs born from the sires with a medium weight gain potential

Item	HP Gr[1] HP F[1,2] B[1]	G[1]	MP Gr[1] HP F[1,2] B	G	LP Gr[1] LP F[1,3] B	G	SEM	Gr	F	S[1]	Gr × F	F × S
Initial wt, kg	48.4	52.3	45.7	53.6	43.8	44.2	1.8	<0.01	<0.01	<0.01	<0.01	0.09
Days on feed												
Phase (P) 1	28	29	28	28	35	36						
P 2	41	39	42	38	46	45						
Final wt, kg												
P 1	77.1	78.9	73.4	82.7	71.8	72.1	2.2	<0.01	<0.01	0.03	<0.01	0.15
P 2	113.4	111.2	109.8	113.2	116.7	114.2	2.4	0.26	0.12	0.84	<0.01	0.42
ADG												
P 1	1.01	0.92	0.99	1.04	0.80	0.77	0.04	<0.01	<0.01	0.58	<0.01	0.65
P 2	0.88	0.85	0.86	0.82	0.98	0.93	0.04	<0.01	<0.01	0.15	<0.01	1.00
Overall	0.94	0.88	0.91	0.90	0.90	0.86	0.03	0.56	0.28	0.11	<0.01	0.71
Days to 115 kg	184.3	182.1	190.5	178.5	192.5	196.2	3.3	<0.01	<0.01	0.11	<0.01	0.08
Carcass wt, kg	84.7	85.0	83.4	84.9	88.4	84.9	2.0	0.42	0.21	0.66	<0.01	0.22
Dressing, %	74.6	75.9	75.9	75.0	75.7	74.4	0.8	0.90	0.70	0.61	<0.01	0.22
BFT[4], mm	25.1	22.0	25.8	20.5	25.8	26.4	1.2	0.03	0.01	0.01	<0.01	0.03

[1] HP, high plane; MP, medium plane; LP, low plane; Gr, grower; F, finisher; B, barrow; G, gilt; S, sex. Data are least squares means of 10 animals in each plane of nutrition × sex.
[2] Provided with the HP and MP finisher diets during the finisher phases 1 and 2, respectively (see Table 1). The average ADFI during the phases 1 and 2 were 2.64 and 3.14 kg, respectively.
[3] Provided with the LP finisher during both phases 1 and 2 (see Table 1). The average ADFI during the phases 1 and 2 for pigs born from both groups of sires with the medium and high (see Table 5) weight gain potentials were 2.89 and 4.10 kg, respectively.
[4] Average of backfat thickness measurements between the 11 and 12th ribs and at the last rib adjusted for a 115-kg liveweight.

Table 5 Effects of the plane of nutrition on growth of finishing pigs born from the sires with a high weight gain potential

Item	HP Gr[1]		MP Gr[1]		LP Gr[1]		SEM	P-value				
	MP F[1), 2)]		MP F		LP F[1), 3)]			Gr	F	S[1]	Gr × F	F × S
	B[1]	G[1]	B	G	B	G						
Initial wt, kg	52.8	47.0	53.7	47.0	45.8	44.5	1.9	0.01	<0.01	<0.01	<0.01	0.10
Days on feed												
Phase (P) 1	31	31	30	31	37	37						
P 2	38	38	37	37	38	38						
Final wt, kg												
P 1	82.2	78.3	82.4	76.1	81.8	78.0	2.8	0.95	0.96	0.05	<0.01	0.80
P 2	119.3	111.0	111.4	109.9	121.5	118.3	3.3	0.02	0.01	0.11	<0.01	1.00
ADG												
P 1	0.95	1.01	0.95	0.95	0.97	0.91	0.04	0.60	0.50	0.94	<0.01	0.18
P 2	0.97	0.86	0.81	0.91	1.05	1.06	0.04	<0.01	<0.01	0.87	<0.01	0.79
Overall	0.96	0.93	0.88	0.93	1.01	0.98	0.03	<0.01	<0.01	0.95	<0.01	0.35
Days to 115 kg	178.8	186.8	178.2	179.7	180.7	184.5	4.1	0.58	0.72	0.07	<0.01	0.38
Carcass wt, kg	89.3	81.8	84.8	85.2	87.7	86.9	2.1	0.48	0.23	0.13	<0.01	0.51
Dressing, %	74.8	73.7	74.5	75.4	73.9	73.4	0.8	0.33	0.23	0.78	<0.01	0.76
BFT[4], mm	21.7	23.3	23.0	25.9	26.5	26.0	1.1	<0.01	<0.01	0.31	<0.01	0.10

[1]HP, high plane; MP, medium plane; LP, low plane; Gr, grower; F, finisher; B, barrow; G, gilt; S, sex. Data are least squares means of 10 animals in each plane of nutrition × sex.
[2]Provided with the MP and LP finisher diets during the finisher phases 1 and 2, respectively (see Table 1). The average ADFI during the phases 1 and 2 were 2.40 kg and 3.96 kg, respectively.
[3]Provided with the LP finisher during both phases 1 and 2 (see Table 1). The average ADFI during the phases 1 and 2 for the pigs born from both groups of sires with the medium (see Table 4) and high weight gain potentials were 2.89 and 4.10 kg, respectively.
[4]Average of backfat thickness measurements between the 11 and 12th ribs and at the last rib adjusted for a 115-kg liveweight.

entire finisher phase was greater in the L group vs. M group (1.00 vs. 0.93 kg; P < 0.01). However, ADG was not influenced by sex during either finisher phase. The estimated number of days required to reach the 115-kg liveweight was not influenced by either sex or plane of nutrition (181.1 vs. 182. 6 for the M and L groups, respectively), even though BW at the beginning of the finisher phase was 5.0 kg less in the L group vs. the M group. Carcass weight and dressing percentage were not influenced by the dietary treatment or sex. Backfat thickness adjusted for the 115-kg liveweight was greater (P < 0.01) in the L group than in the M group (26.3 vs. 23.5 mm; P < 0.01).

Overall, the number of days required to reach the 115-kg liveweight and the liveweight-adjusted backfat thickness of the H group of the M-sires' progeny on the H plane during the entire growing-finishing period did not differ from those of the M group of the H-sires' progeny (183. 2 vs. 185.6 d and 23.7 vs. 24.4 mm, respectively) when the raw data of Tables 4 and 5 were pooled and further analyzed. In those pigs on the L plane throughout the growing-finishing period, the number of days required to reach the 115-kg liveweight was less (P < 0.01) in the H-sires' progeny than in the M-sires' progeny (182.6 vs. 194.0 d; SE = 2.7 d), whereas the adjusted backfat

thickness did not differ between the two groups (26.3 and 26.1 mm in the former and latter, respectively).

Discussion

Main results of the present study are as follows. The M-sires' progeny grew faster than the H-sires' progeny during the early grower phase whereas during the latter grower phase, the H-sires' progeny grew faster. The M group of the H-sires' progeny gained weight as well as the H group of the M-sires' progeny throughout the growing-finishing period. Moreover, the L group of the H-sires' progeny grew faster than that of the M-sires' progeny during the entire growing-finishing period. Collectively, these results suggest that the H plane of nutrition is necessary to elicit maximal weight gain in the M-sires' progeny during the grower phase beginning from early August whereas the M plane is sufficient for the H-sires' progeny. Moreover, during the early finisher phase, the L plane is probably sufficient for maximal weight gain for the H-sires' progeny whereas the H or M plane is necessary for the M-sires' progeny.

Unexpectedly, both L groups of the M- and H-sires' progenies exhibited greater ADG during the latter finisher phase and greater backfat thickness at slaughter than the H and M groups of the respective progeny groups. It was also evident that feed intake was greater in the L group vs. the M or H group during the latter finisher phase, although only the average intake of each diet was measured due to practical limitations of the commercial farm where the present experiment was performed. These responses of the L group, which are consistent with the typical phenomena of the animals undergoing compensatory growth [20], suggest that the accelerated growth of the L group during the latter finisher phase resulted from compensatory growth. In this regard, the relatively long lag period between growth retardation and compensatory growth in the L group seems to be related to the long period of inadequate nutrition during the early developmental stage when the growth retardation occurred [20,21]. It was not clear, however, whether or not the apparently high ADG of the M and H groups during the early finisher phase (0.95 to 1.00 kg) also resulted from compensatory growth. Further, the daily high ambient temperature during this period frequently exceeded 30°C (data not shown), which is far above 26 to 28°C of the threshold temperature causing the heat stress in growing-finishing pigs [22,23]. Accordingly, it also remains unknown how much of the observed growth rates of the whole animals were influenced by a presumptive heat stress during the grower phase. Further studies are therefore warranted to determine the effects of the heat stress and plane of nutrition during the hot summer as well as their interactions on subsequent growth on varying planes of nutrition during the fall.

According to the grading standard for pig carcasses, the upper and lower limits of carcass weight for the 1^+ grade are 83.5 and 93.5 kg, respectively in Korea [24], which are equal to 109.9 and 123.0 kg by the liveweight, respectively, at a dressing percentage of 76%. In addition, backfat thickness should fall within a range between 17.5 and 24.5 mm to be eligible for the 1^+ grade by Korean standards, which means that the optimum backfat thickness of market pigs is 21.5 mm if a standard deviation of 3 mm [25] is deducted from the upper limit of the 1^+ grade as a safety margin. As such, the average backfat thickness of both M- and H-sires' progenies at the market weight, which was optimal on the M or H plane (23.5 mm), but not on the L plane (over 26 mm), under the previous carcass grading standard (26.5 mm maximum to be eligible for the 1^+A grade; [16]), is now too high by the current standard. Evidently, sires leaner than those used in the present study are now needed not only to meet the current grading standard calling for leaner carcasses but also to improve feed efficiency.

Conclusions

The present results suggest that the H-to-H or H-to-M planes and M-to-M or M-to-L planes are optimal for maximal growth of the M- and H-sires' progenies, respectively.

Competing interests
The authors declare that they have no competing interests.

Authors' contributions
MJP, BCP, and CYL designed the experiment. DMH, DYH, MJP, and CYL performed the feeding trial. BCP and CYL analyzed the results. All authors read and approved the final manuscript.

Acknowledgments
This work was supported in parts by Taewon Farm and Gyeongnam National University of Science and Technology. We authors thank Dr. Il-Joo Lee for the assistance and advice regarding pig breeding.

Author details
[1]Regional Animal Industry Center, Gyeongnam National University of Science and Technology, Jinju 660-758, Korea. [2]Sunjin Co., Ltd, 517-3 Doonchon-dong, Kangdong-gu, Seoul 134-060, Korea.

References
1. Pettigrew JE, Esnaola MA: Swine nutrition and pork quality: a review. *J Anim Sci* 2001, **79**((Suppl):E31–E342.
2. NSNG: *National Swine Nutrition Guide.* Ames, IA, USA: U.S. Pork Center of Excellence (USPCE), Iowa State University; 2010.
3. NRC: *Nutrient Requirements of Swine.* 11th edition. Washington, D.C., USA: National Academy Press; 2012.
4. Silence MN: **Technologies for the control of fat and lean deposition in livestock.** *Vet J* 2004, **167**:242–257.
5. van Wijk HJ, Arts DJG, Matthews JO, Webster M, Ducro BJ, Knol EF: **Genetic parameters for carcass composition and pork quality estimated in a commercial production chain.** *J Anim Sci* 2005, **83**:324–333.
6. McFarland DC: **Nutrition and growth.** In *Biology of Growth of Domestic Animals.* Edited by Scanes CG. Ames, IA, USA: Iowa State Press, A Blackwell Publishing Company; 2003:249–262.
7. Apple JK, Maxwell CV, Brown DC, Friesen KG, Musser RE, Johnson ZB, Armstrong TA: **Effects of dietary lysine and energy density on performance and carcass characteristics of finishing pigs fed ractopamine.** *J Anim Sci* 2004, **82**:3277–3287.
8. Park BC, Lee CY: **Feasibility of increasing the slaughter weight of finishing pigs.** *J Anim Sci Technol* 2011, **53**:211–222.
9. Schinckel AP, Einstein ME, Jungst S, Matthews JO, Fields B, Booher C, Dreadin T, Fralick C, Tabor S, Sosnicki A, Wilson E, Boyd RD: **The impact of feeding diets of high or low energy concentration on carcass measurements and weight of primal and subprimal lean cuts.** *Asian Aust J Anim Sci* 2012, **25**:531–540.
10. KAIA: *Korea Animal Improvement Association.* Statistics; 2011.
11. Oh SH, See MT: **Pork preference for consumers in China, Japan and South Korea.** *Asian Aust J Anim Sci* 2012, **25**:143–150.
12. Seo JT: **Changes in carcass characteristics and drawbacks accompanying an increase in slaughter weight.** In *10th Symposium of the Korean Pig Industry Research Society "Strategies for Increasing Pork Supply through an Increase of the Market Weight", Jul. 13, 2011,* Proceedings; 2011:35–57.
13. KPPA: *Korea Pork Producers Association.*Statistics; 2014.
14. Gerrard DE, Grant AL: *Principles of Animal Growth and Development,* revised printing. Dubuque, IA, USA: Kendall/Hunt Publishing Company; 2006.
15. NASS: *Agricultural Statistics.* USDA, National Agricultural Statistics Service; 2014.
16. MIFAFF: *Grading Standards for Livestock Products (published in Korean).* Notification No. 2011–46 of the Ministry of Food, Agriculture, Forestry and Fisheries, Republic of Korea; 2011.
17. Ha DM, Jang KS, Won HS, Ha SH, Park MJ, Kim SW, Lee CY: **Effects of creep feed and milk replacer and nursery phase-feeding programs on pre- and post-weaning growth of pigs.** *J Anim Sci Technol* 2011, **53**:333–339.
18. Lee CY, Lee HP, Jeong JH, Baik KH, Jin SK, Lee JH, Sohn SH: **Effects of restricted feeding, low-energy diet, and implantation of trenbolone acetate plus estradiol on growth, carcass traits, and circulating

concentrations of insulin-like growth factor (IGF)-I and IGF-binding protein-3 in finishing barrows. *J Anim Sci* 2002, **80**:84–93.

19. Ha DM, Kim GD, Han JC, Jeong JY, Park MJ, Park BC, Joo ST, Lee CY: **Effects of dietary energy level on growth efficiency and carcass quality traits of finishing pigs.** *J Anim Sci Technol* 2010, **52**:191–198.

20. Lawrence TLJ, Fowler VR, Novakofski JE: *Growth of Farm Animals.* 3rd edition. Wallingford, UK: CABI; 2012.

21. Whang KY, McKeith FK, Kim SW, Easter RA: **Effect of starter feeding program on growth performance and gains of body components from weaning to market weight in swine.** *J Anim Sci* 2000, **78**:2885–2895.

22. Noblet J, Le Dividich J, Van Milgen JV: **Thermal environment and swine nutrition.** In *Swine Nutrition.* 2nd edition. Edited by Lewis AJ, Southern LL. New York: CRC Press; 2001:519–544.

23. Le Bellego L, van Milgen J, Noblet J: **Effect of high temperature and low-protein diets on the performance of growing-finishing pigs.** *J Anim Sci* 2002, **80**:691–701.

24. MAFRA: *Grading Standards for Livestock Products (published in Korean).* Notification No. 2014–4 of the Ministry of Agriculture, Food and Rural Affairs, Republic of Korea; 2014.

25. Park MJ, Park BC, Ha DM, Kim JB, Jang KS, Lee DH, Kim GT, Jin SK, Lee CY: **Effects of increasing market weight of finishing pigs on backfat thickness, incidence of the 'caky-fatty' belly, carcass grade, and carcass quality traits.** *J Anim Sci Technol* 2013, **55**:195–202.

5

Assessing the association of single nucleotide polymorphisms in thyroglobulin gene with age of puberty in bulls

María Elena Fernández[1,3†], Daniel Estanislao Goszczynski[1,3†], Alberto José Prando[2], Pilar Peral-García[1], Andrés Baldo[2], Guillermo Giovambattista[1*] and Juan Pedro Liron[1*]

Abstract

Puberty is a stage of sexual development determined by the interaction of many loci and environmental factors. Identification of genes contributing to genetic variation in this character can assist with selection for early pubertal bulls, improving genetic progress in livestock breeding. Thyroid hormones play an important role in sexual development and spermatogenic function. The objective of this study was to evaluate the association between single nucleotide polymorphisms (SNPs) located in thyroglobulin(TG) gene with age of puberty in Angus bulls. Four SNPs were genotyped in 273 animals using SEQUENOM technology and the association between markers and puberty age was analyzed. Results showed a significant association ($P < 0.05$) between these markers and puberty age estimated at a sperm concentration of 50 million and a progressive motility of 10%. This is the first report of an association of TG polymorphisms with age of puberty in bulls, and results suggest the importance of thyroidal regulation in bovine sexual development and arrival to puberty.

Keywords: Bovine, Puberty, Polymorphism, Thyroglobulin, Association study

Background

Puberty is a stage of sexual development determined by the interaction of many loci hierarchically arranged in networks, and environmental factors (Ojeda et al., [1]). Puberty in cattle is an important target for genetic improvement so early prediction using genetic markers is a goal for livestock breeding. The identification of new genes and/or mutations contributing to genetic variation in puberty can assist with the selection for early pubertal bulls, reducing the generation interval and increasing fertility and genetic progress (Johnston et al., [2], Fortes et al., [3]).

Thyroid hormones (THs) exert a broad range of effects on metabolism, growth, homeostasis control, and other biological processes (Warner and Mittag, [4], Mullur et al., [5]), and show connection with nearly every biological endocrine system. Several studies provide evidence to confirm the role of THs in sexual differentiation and gonadal development (Jannini et al., [6], Mendis-Handagama and Siril Ariyaratne, [7], Flood et al., [8], Duarte-Guterman et al., [9]). For example, it has been discovered that deiodinases [the enzymes responsible for ioding thyroglobulin (TG) to obtain the active forms triiodothyronine (T3) and thyroxine (T4)] and thyroid receptors (TRs, encoded by *trα* and *trβ* genes) are present within gonadal tissues, suggesting that THs must have an action on these organs (Wagner et al., [10]). The presence of TH machinery in testicular tissues implies that TH axis must regulate aspects of testicular functioning. Indeed, it has been shown that hypo- and hyperthyroid males exhibit testes and sperm dysfunction (Krassas et. al., [11]). In addition, THs are involved in the regulation of androgen receptors(ARs) expression in testicular tissues through thyroid response elements (TREs) located in the promoter of *ar* genes(Flood et al., [8]). Furthermore, THs may also regulate other genes involved in androgen biosynthesis and signaling. For example, THs enhance 5α-reductase expression and activity within the testes, increasing 5α-dihydrotestosterone concentrations (Ram and Waxman, [12], Duarte-Guterman et al., [9]). On the other hand, it has

* Correspondence: ggiovam@fcv.unlp.edu.ar; juanpedroliron@gmail.com
†Equal contributors
[1]Instituto de Genética Veterinaria (IGEVET), CCT La Plata – CONICET - Facultad de Ciencias Veterinarias, Universidad Nacional de La Plata, Calle 60 y 118 s/n, La Plata B1900AVW, CC 296, Argentina
Full list of author information is available at the end of the article

been reported that TH-related transcription factors influence the expression of *sox9*, which induces differentiation of the bipotential cells in the testes into Sertoli cells (Zhou et al., [13]). *Sox9* stimulates the nuclear receptor steroidogenic factor 1 (*sf1*), which is primarily expressed in Leydig cells and plays an important role in sexual differentiation (Zhao et al., [14]). *Dax1* gene works in parallel with *sf1* to regulate testicular differentiation and can regulate TH-related gene expression (Sugawara et. al., [15]).

Beyond the fact that THs have considerable effects on the hypothalamic-pituitary-gonadal axis (HPG) and sexual development, some studies examined the involvement of androgens in TH synthesis and metabolism. For example, gonadotropin-releasing hormone (GnRH) interferes with the hypothalamic-pituitary-thyroid axis (HPT), increasing thyroid stimulating hormone (TSH) secretion. Moreover, ARs were identified in the thyroid gland of different vertebrate species, suggesting that the androgen axis directly regulates TH synthesis and metabolism (Pelletier et al., [16]). Other works have demonstrated the potential of ARs to regulate TH axis showing that TRs transcriptional levels and distribution within testes are responsive to androgen fluctuations through the presence of androgen response elements (AREs) in the promoter regions of TH-related genes. These evidences demonstrate the existence of a considerable cross-regulation between both axes.

Considering the existing evidence on the biological connection between metabolic status, regulated to a large extent by thyroid hormones, and sexual development, and taking into account the fact that THs play a role in the regulation of sexual function, we decided to evaluate the possible associations between SNPs located in the 3' flanking region of TG gene with the age of puberty in bull calves.

Methods

In order to reach this objective, DNA was extracted from blood samples belonging to 273 Angus bulls using Wizard Genomic kit, following manufacturer instructions (Promega, Madison, WI, USA). Four SNPs (rs378215592, rs110406764, rs109662686, rs109057985) in TG previously reported by Hou et al., [17] were genotyped using SEQUENOM platform by GeneSeek Inc. genotyping services (Lincoln, NE, USA). This technology is based on primer-extension reaction that generates allele-specific products with distinct masses, which are then detected through MALDI-TOF mass spectrometry (www.sequenom.com/). Detailed information of the studied SNPs is presented in Table 1. Animal samples, phenotypic measurements and estimation of puberty ages used in this work were reported in Liron et al., [18]. The estimated puberty ages were: i. age at 28 cm of scrotal circumference (SC28),

Table 1 Information of single nucleotide polymorphisms (SNP) used in this work

SNP	Gene-Chromosome	Gene region	Position (UMD 3.1)	Allele change
rs378215592	TG- Ch14	3' UTR	9281431	T/C
rs110406764	TG- Ch14	3' UTR	9281469	G/A
rs109662686	TG- Ch14	3' UTR	9281507	A/G
rs109057985	TG- Ch14	3' UTR	9281510	T/G

and ii. age at sperm concentration 50 million and percentage of progressive motility 10% (C50 - M10). In order to evaluate the linkage disequilibrium (LD) between the three studied SNPs, the haplotypes for each individual were constructed using Phase algorithm (Li and Stephens, [19]).

The association between haplotype markers of TG gene and the estimated puberty ages was analyzed utilizing MIXED procedure implemented in SAS 9.0 software (SAS Inst. Inc.). The linear mixed model used to analyze the association between puberty age and genotypes was the following:

$$Y_{ijkl} = \mu + S_i + G_j + B_k + O_l + e_{ijkl}$$

Where Y_{ijkl} = phenotypic observation of the I bull, μ = the overall mean, S_i = the fixed effect of ith year, G_j = the fixed effect of jth genotype, B_k = the fixed effect of kth herd, O_l = random effect of lth sire, and e_{ijkl} = random error.

Results and discussion

After analyzing the genotyping results, one SNP (rs109662686) was removed given that it exhibited a call rate lower than 57%. The LD analysis indicated that the three remaining SNPs were completely linked (r^2 = 1). Only two (TGT and CAG) of the eight possible haplotypes were found, with gene frequencies of 0.81 and 0.19 for TGT and CAG haplotypes, respectively. Genotype frequencies values were 0.64 for homozygote TGT, 0.03 for homozygote CAG and 0.33 for heterozygote bulls. The obtained haplotypes for the 273 bulls were tested for association with phenotypic data for the two estimated puberty ages mentioned above. The association analysis showed a significant association (P < 0.05) between the haplotype markers and puberty age estimated at C50 and M10. Homozygote TGT exhibited a mean +/- S.E. age at C50 and M10 of 289.74 ± 8.13 days, while homozygote CAG showed 347.67 ± 22.51 days, resulting in a difference of 57.93 days of age. Heterozygote animals showed a mean +/- S.E. age of 299.65 ± 8.72 days of age. No significant association was found between both haplotypes and age at puberty estimated at SC28 (TGT/TGT = 277.95 ± 23.85 days, TGT/CAG = 281.09 ± 25.82 days and CAG/CAG = 292.00 ± 17.76 days, P > 0.05).

The results obtained here constitute the first report of an association of TG gene polymorphisms with age of puberty in bulls and could be explained by the vast amount of works in which the modulatory influence of THs on male reproduction is demonstrated (Flood et al., [8], Duarte-Guterman et al., [9]). The potential of THs in the modulation of male reproductive functions during or preceding puberty was determined to be of such importance at the point that any alteration in their expression and/or concentration has profound effects on male reproduction (Krassas et al., [11], Weber et al., [20]). Despite the indirect regulation of sexual maturation by THs through their known roles in development, metabolism, hormonal regulation and other physiological processes, evidence indicates that THs have direct effects on sexual development, reproductive function and associated molecular mechanisms and pathways. Although the specific mechanisms underlying this regulation are not completely established, these direct effects could be exerted mainly through the presence of TH machinery in gonadal tissues. For example, TRs are widely distributed and expressed in different compartments of the testis in mammalian species, which suggests a direct regulatory role for THs in male gonadal development and function (Jannini et al., [6], Kumar et al., [21], Wagner et al., [10]). Studies have also identified deiodinases in the testes of vertebrate species, whose role within testicular functioning in mammalian species has been reviewed by Wagner et al., [10]. Another evidence for the involvement of THs in puberty is the presence of TREs in androgen receptors in the testis and in GnRH and luteinizing hormone receptor (LHR) promoter region (Tsai-Morris et al., [22]), which demonstrates that THs can directly regulates androgen biosynthesis. Hyper- and hypo-thyroidic conditions alter GnRH concentrations in mammals, consequently affecting LH and FSH production and secretion (Chiao et al., [23]). Furthermore, several studies have shown a fall in circulating testosterone levels in hypothyroid humans (Kumar et al, [21]). As we said before, this regulation is bidirectional and there area lot of studies describing the considerable cross-regulation existing between HPT and HPG axes in vertebrates. Interestingly, Fortes et al., [24] detected SERPINA7 gene on chromosome X associated with percentage of normal sperm (PNS) in Brahman bulls. This gene codes for thyroxine-binding globulin (TBG), the major TH transport protein in serum. This evidence reinforces the hypothesis that THs play an important role in bovine male fertility.

Despite the broad range of effects in vertebrates and the connections in nearly every biological endocrine system, a lot of studies provide enough evidence to confirm the role of THs in sexual differentiation and gonadal development in mammalian and non-mammalian species. We can affirm that THs influence steroidogenesis and spermatogenesis and as described above there is extensive evidence that links TG gene to testicular development, existing a direct crosstalk between HPG and thyroid hormones axis (Wagner et al., [10], Nonneman et al., [25]).

Conclusions

In conclusion, we detected an association between TG polymorphisms and age at puberty at C50 and M10 in male Angus cattle. Our results could contribute to the investigation on regulation of bovine puberty and fertility. THs are essential for normal growth, sexual development and reproductive function, and normal thyroid activity seems to be a requisite for an adequate male reproductive function. However, the knowledge about the interaction between both endocrine axis, HTP and HTG, is still rudimentary and needs further investigation.

Competing interests
None of authors of this paper has financial, personal or other relationship with other people or organizations that could inappropriately influence the content of this paper.

Authors' contribution
MEF and DEG participated in the genetic and statistical studies, drafted the manuscript. AP conceptualized the study and collected the phenotypic data. PPG participated in the design of the study. AB conceptualized the study and collected the phenotypic data. GG and JPL conceptualized and supervised the whole study, analyzed the data, performed the statistical analysis, drafted the manuscript. All authors read and approved the final version of the manuscript.

Author details
[1]Instituto de Genética Veterinaria (IGEVET), CCT La Plata – CONICET - Facultad de Ciencias Veterinarias, Universidad Nacional de La Plata, Calle 60 y 118 s/n, La Plata B1900AVW, CC 296, Argentina. [2]Departamento de Producción Animal, Facultad de Ciencias Veterinarias, Universidad Nacional de La Plata, La Plata, Argentina. [3]Fellow of the Consejo Nacional de Investigaciones Científicas y Técnicas (CONICET), La Plata, Argentina.

References
1. Ojeda SR, Dubay C, Lomniczi A, Kaidar G, Matagne V, Sandau US, Dissen GA: Gene networks and the neuroendocrine regulation of puberty. *Mol Cell Endocrinol* 2010, 324:3–11.
2. Johnston DJ, Barwick SA, Corbet NJ, Fordyce G, Holroyd RG, Williams PJ, Burrow HM: Genetics of heifer puberty in two tropical beef genotypes in northern Australia and associations with heifer and steer-production traits. *Anim Prod Sci* 2009, 49:399–412.
3. Fortes MRS, Lehnert SA, Bolormaa S, Reich C, Fordyce G, Corbet NJ, Whan V, Hawken RJ, Reverter A: Finding genes for economically important traits: Brahman cattle puberty. *Anim Prod Sci* 2012, 52:143–150.
4. Warner A, Mittag J: Thyroid hormone and the central control of homeostasis. *J Mol Endocrinol* 2012, 49:29–35.
5. Mullur R, Liu YY, Brent GA: Thyroid hormone regulation of metabolism. *Physiol Rev* 2014, 94(Suppl2):355–382.
6. Jannini EA, Crescenzi A, Rucci N, Screponi E, Carosa E, de Matteis A, Macchia E, d'Amati G, D'Armiento M: Ontogenetic pattern of thyroid hormone receptor expression in the human testis. *J Clin Endocrinol Metab* 2000, 85:3453–3457.
7. Mendis-Handagama SM, Siril Ariyaratne HB: Leydig cells, thyroid hormones and steroidogenesis. *Indian J Exp Biol* 2005, 43(Suppl11):939–962.
8. Flood DE, Fernandino JI, Langlois VS: Thyroid hormones in male reproductive development: evidence for direct crosstalk between the androgen and thyroid hormone axes. *Gen Comp Endocrinol* 2013, 192:2–14.

9.   Duarte-Guterman P, Navarro-Martín L, Trudeau VL: **Mechanisms of crosstalk between endocrine systems: Regulation of sex steroid hormone synthesis and action by thyroid hormones.** *Gen Comp Endocrinol* 2014, doi:10.1016/j. ygcen.2014.03.015.

10. Wagner MS, Wajner SM, Maia AL: **Is there a role for thyroid hormone on spermatogenesis?** *Microsc Res Tech* 2009, **72**:796–808.

11. Krassas GE, Papadopoulou F, Tziomalos K, Zeginiadou T, Pontikides N: **Hypothyroidism has an adverse effect on human spermatogenesis: A prospective, controlled study.** *Thyroid* 2008, **18**(Suppl12):1255–1259.

12. Ram PA, Waxman DJ: **Pretranslational control by thyroid hormone of rat liver steroid 5 alpha-reductase and comparison to the thyroid dependence of two growth hormone-regulated CYP2C mRNAs.** *J Biol Chem* 1990, **265**(31):19223–19229.

13. Zhou RJ, Bonneaud N, Yua CX, Barbara PD, Boizet B, Tibor S, Scherer G, Roeder RG, Poulat F, Berta P: **SOX9 interacts with a component of the human thyroid hormone receptor-associated protein complex.** *Nucleic Acids Res* 2002, **30**(Suppl 14):3245–3252.

14. Zhao L, Bakke M, Krimkevich Y, Cushman LJ, Parlow AF, Camper SA, Parker KL: **Steroidogenic factor 1 (SF1) is essential for pituitary gonadotrope function.** *Development* 2001, **128**(Suppl 2):147–154.

15. Sugawara T, Kiriakidou M, McAllister JM, Holt JA, Arakane F, Strauss JF: **Regulation of expression of the steroidogenic acute regulatory protein (StAR) gene: a central role for steroidogenic factor 1.** *Steroids* 1997, **62**:5–9.

16. Pelletier G: **Localization of androgen and estrogen receptors in rat and primate tissues.** *Histol Histopathol* 2000, **15**:1261–1270.

17. Hou GY, Yuan ZR, Zhou HL, Zhang LP, Li JY, Gao X, Wang DJ, Gao HJ, Xu SZ: **Association of thyroglobulin gene variants with carcass and meat quality traits in beef cattle.** *Mol Biol Rep* 2011, **38**:4705–4708.

18. Lirón JP, Prando A, Fernández ME, Ripoli MV, Rogberg-Muñoz A, Goszczynski DE, Posik DM, Peral-García P, Baldo A, Giovambattista G: **Association between GNRHR, LHR and IGF1 polymorphisms and timing of puberty in male Angus cattle.** *BMC Genet* 2012, **5**:13–26.

19. Li N, Stephens M: **Modelling linkage disequilibrium, and identifying recombination hotspots using SNP data.** *Genetics* 2003, **165**:221–233.

20. Weber G, Vigone MC, Stroppa L, Chiumello G: **Thyroid function and puberty.** *J Pediatr Endocrinol Metab* 2003, **16**(Suppl 2):253–257.

21. Kumar A, Shekhar S, Dhole B: **Thyroid and male reproduction.** *Indian J Endocr Metab* 2014, **18**(Suppl 1):23–31.

22. Tsai-Morris CH, Xie XZ, Buczko E, Dufau ML: **Promoter and regulatory regions of the rat luteinizing hormone receptor gene.** *J Biol Chem* 1993, **268**(Suppl 6):4447–4452.

23. Chiao YC, Lee HY, Wang SW, Hwang JJ, Chien CH, Huang SW, Lu CC, Chen JJ, Tsai SC, Wang PS: **Regulation of thyroid hormones on the production of testosterone in rats.** *J Cell Biochem* 1999, **73**:554–562.

24. Fortes MRS, Reverter A, Hawken RJ, Bolormaa S, Lehnert SA: **Candidate Genes Associated with Hormone Levels of Inhibin, Luteinising Hormone, and Insulin-Like Growth Factor 1, Testicular Development, and Sperm Quality in Brahman Bulls.** *Biol Reprod* 2012, **87**(Suppl 3):58.

25. Nonneman D, Rohrer GA, Wise TH, Lunstra DD, Ford JJ: **A variant of porcine thyroxine-binding globulin has reduced affinity for thyroxine and is associated with testis size.** *Biol Reprod* 2005, **72**:214–220.

6

Fatty acid analysis as a tool to infer the diet in Illinois river otters (*Lontra canadensis*)

Damian Satterthwaite-Phillips[1], Jan Novakofski[2] and Nohra Mateus-Pinilla[1*]

Abstract

Fatty acids (FA) have recently been used in several studies to infer the diet in a number of species. While these studies have been largely successful, most have dealt with predators that have a fairly specialized diet. In this paper, we used FA analysis as a tool to infer the diet of the nearctic river otter (*Lontra canadensis*). The river otter is an opportunistic predator known to subsist on a wide variety of prey including, fishes, crayfish, molluscs, reptiles and amphibians, among others. We analyzed the principle components of 60 FA from otters and 25 potential prey species in Illinois, USA. Prey species came from 4 major taxonomic divisions: fishes, crayfish, molluscs and amphibians. Within each division, most, but not all, species had significantly different profiles. Using quantitative FA signature analysis, our results suggest that, by mass, fish species are the most significant component of Illinois River otters' diet ($37.7 \pm 1.0\%$). Molluscs ranked second ($32.0 \pm 0.8\%$), followed by amphibians ($27.3 \pm 4.3\%$), and finally, crayfish ($3.0 \pm 0.6\%$). Our analysis indicates that molluscs make up a larger portion of the otter diet than previously reported. Throughout much of the Midwest there have been numerous otter reintroduction efforts, many of which appear to be successful. In regions where mollusc species are endangered, these data are essential for management agencies to better understand the potential impact of otters on these species. Our analysis further suggests that quantitative FA signature analysis can be used to infer diet even when prey species are diverse, to the extent that their FA profiles differ. Better understanding of the otter's metabolism of FA would improve inferences of diet from FA analysis.

Keywords: River otter (*Lontra canadensis*), Inferring diet, Quantitative fatty acid signature analysis (QFASA), Predation, Gas chromatography, Principal components analysis (PCA)

Background

In Illinois, and throughout the Midwestern region of the United States, river otter (*Lontra canadensis*) reintroduction programs have led to recent increase in numbers of river otters and their geographical distribution [1,2]. In regions where otters have been present for longer periods of time [3-5], the species' diet varies considerably with geographic region. However, little is known about how this diet varies in regions where otters have been recently reintroduced. Furthermore, information about diets have generally been obtained by examining gut or fecal contents, which relies primarily on the identification of hard or bony tissues. These methods underestimate the contribution of dietary items without such hard tissues,

such as molluscs and insect larvae [6,7]. In regions where prey species may be endangered, wildlife management agencies need to better understand the potential impact of otter predation on these species to better protect them.

To compensate for the limitations in earlier methods, diet can also be inferred from fatty acid analysis. For example, quantitative fatty acid signature analysis (QFASA [8]) is a technique capable of providing high resolution in differentiating diet species. Unlike fecal and gut-content analyses, tissue fatty acids (FA) reflect diet over a longer time span, and a predator's FA profile is a proportionate representation of its prey. QFASA has been used to determine the diet of marine mammals, including cetacean [9-11] and pinniped species [8,12,13], and is based on: (1) variable ability to metabolize and store FAs depending on chain length and saturation, and (2) concentration of unusual FA up the food chain. Experiments with harbor seals in Prince William Sound,

---

* Correspondence: nohram@illinois.edu
[1]Illinois Natural History Survey, University of Illinois Urbana-Champaign, 1816 S. Oak Street, Champaign, IL 61820, USA
Full list of author information is available at the end of the article

Alaska demonstrated that QFASA was able to determine the specific fish species in the diet of seals, and reflect differences in diet with age and fine-scale habitat [13]. Because different FAs may be metabolized differently, the FA profile of the predator's diet may not match the FA profile of the predator itself. Nevertheless, experimental studies indicate that QFASA has a high rate (88%) of correctly inferring diets when diets were known [14].

Our primary objective in this paper was to determine how well differences in fatty acid signatures could be used to identify different taxonomic groups among prey species that river otters are known or suspected to consume. We hypothesized that differences in fatty acid profiles would allow for the identification of large taxonomic groups (fishes, crayfish, amphibians, and molluscs), and possibly lower-level taxonomic distinctions of animals being consumed by otters. We demonstrate that the different taxa considered have significantly different FA profiles, and provide the otter diet as inferred by QFASA.

Methods

Species and tissues analyzed

All animals included in this analysis were from Illinois, USA. The Illinois Department of Natural Resources provided us with river otters obtained from incidental deaths including road kills and unintended trapping. Otter carcasses were stored frozen until transport to our laboratory. All otter carcasses were gathered between 21 Oct and 15 Mar, with 12 (26%) obtained in the fall (Oct, Nov), 26 (57%) in the winter (Dec-Feb), and 8 (17%) in the spring (Mar). Further details on the otters sampled are available in Carpenter et al. [15]. Crayfish samples were collected Feb-Mar, and fish, amphibian, and mollusc samples between Feb-May.

River otter adipose tissue was dissected from two deposits: dorsal subcutaneous adipose from the ventral side of the base of the tail, approximately 5 cm posterior of the anus ($n = 46$), and from the footpads of the forefeet ($n = 19$). Lipid was extracted separately from each the two adipose deposits. A smaller number of samples were analyzed from the footpads, as these were intended only to verify the assumption that this deposit would be cold-adapted as suggested in Käkelä and Hyvärinen [16].

Species that were potential prey for otters, including 15 fish species, 5 mollusc species, 3 amphibian species, and 2 species of crayfish (Table 1) were obtained opportunistically from colleagues at the Illinois Natural History Survey, in conjunction with their individual research agendas. For all prey species, whole-body homogenate of the entire specimen, excluding shells for molluscs, was used for lipid extraction.

Table 1 River otter and its candidate prey species evaluated in the study, Illinois, USA

Taxonomic division	Scientific name
Carnivores (predator species)	Nearctic River Otter *Lontra canadensis* ($n = 46$; 46 tail, and 19 footpad samples)
Fishes (prey)	
Clupeiformes	Gizzard Shad *Dorosoma cepedianum* ($n = 6$)
Cypriniformes	Asian Carp[a] *Hypophthalmichthys nobilis* ($n = 5$)
	Hornyhead Chub *Nocomis biguttatus* ($n = 3$)
	Creek Chub *Semotilus atromaculatus* ($n = 7$)
Perciformes	Greenside Darter *Etheostoma blennioides* ($n = 6$)
	Bluegill *Lepomis macrochirus* ($n = 10$)
	Redear Sunfish *Lepomis microlophus* ($n = 9$)
	Smallmouth Bass *Micropterus dolomieu* ($n = 7$)
	Largemouth Bass *Micropterus salmoides* ($n = 6$)
	White Crappie *Pomoxis annularis* ($n = 1$)
	Black Crappie *Pomoxis nigromaculatus* ($n = 4$)
Siluriformes	Black Bullhead *Ameiurus melas* ($n = 8$)
	Blue Catfish *Ictalurus furcatus* ($n = 8$)
	Channel Catfish *Ictalurus punctatus* ($n = 8$)
	Brindled Madtom *Noturus miurus* ($n = 2$)
Molluscs (prey)	Threeridge *Amblema plicata* ($n = 2$)
	Asian Clam *Corbicula fluminea*[a] ($n = 10$)
	Wabash Pigtoe *Fusconaia flava* ($n = 10$)
	Fat Mucket *Lampsilis siliquoidea* ($n = 10$)
	Round Pigtoie *Pleurobema sintoxia* ($n = 5$)
Amphibians (prey)	Cricket Frog *Acris crepitans* ($n = 10$)
	Western Chorus Frog *Pseudacris triseriata* ($n = 1$)
	American Bullfrog *Rana catesbeiana* [*Lithobates catesbeianus*] ($n = 1$)
Crayfish (prey)	Northern Clearwater Crayfish *Orconectes propinquus* ($n = 10$)
	Virile Crayfish *Orconectes virilis* ($n = 4$)

[a]Invasive species.

Gas–liquid chromatography

Lipids were extracted from tissue or homogenates with choloroform:methanol (2:1 by volume) by mixing in a Polytron for 30 seconds [17]. The chloroform phase containing lipid was removed, the extraction repeated twice, and chloroform evaporated under a stream of nitrogen at room temperature. FA methyl esters were prepared as described by AOCS official method Ce 2–66 [18]. Subsequently, FA methyl esters were analyzed using a gas chromatograph (Hewlett Packard 5890 series II) with a DB-wax capillary column (30 m × 0.25 mm × 0. 25 μm film coating, Agilent Technologies, Santa Clara, CA). The column was under a constant pressure at 1.30 kg/cm^2 using helium as the carrier gas. Temperature of the injector and of the flame-ionization detector was held constant at 250°C and 260°C, respectively. The oven was operated at 170°C for 2 min (programmed temperature to increase 2°C /min up to 240°C and then held constant for 10 min). Chromatographs from FA methyl esters were integrated using Agilent Chemstation software for gas chromatographs systems (Version B.01.02, Agilent Technologies, Inc.®). We identified FA methyl esters by comparing retention times of known standards (GLC 461A, Nu-check-prep, Elysian, MN). Some FA peaks did not correspond to the 30 fatty acids present in this standard. We estimated the total number of fatty acids (60) based on the number previously reported for several species [16], and assigned identity of unknown peaks by running the k-means clustering method in R (version 2.15.0) [19,20] to account for variation in retention times between runs. Centroids for unidentified peaks were randomly initialized, with a uniform distribution between the minimum and maximum observed retention-time values. We repeated the process of initializing centroids and grouping peaks by the k-means algorithm 1000 times. For each run, variance explained by the clustering ($SS_{between\ clusters}/SS_{total}$) was recorded and the run that explained the greatest amount of the variance was used to determine the best-fit assignment of unknown peaks.

Principal components analysis

Principal components analysis (PCA) is an analytical method that compresses multiple variables into a more tractable number of linearly uncorrelated variables, while simultaneously maximizing the amount of variance explained by the new, compressed and lower-dimensional set of variables.

The FA profile of each individual animal was comprised of 60 different FA. The proportion of each individual FA was treated as a variable and PCA was conducted on otters and prey. An analysis was performed on otter adipose deposits to determine if fatty acid profiles were different for subcutaneous and footpad adipose. Individual PCA were also performed on all samples within each taxonomic group (crayfish, frogs, fishes, and molluscs) and on species within a given taxonomic group to determine if PCA could be used to identify individual species on the basis of their fatty acid profiles. All analyses were conducted using the FactoMineR library for R (described in [21]).

For each PCA, we performed an analysis of variance (ANOVA) on the first several principal components to determine significant differences of mean contributions to each component. For the PCA including all samples, the factors were the large taxonomic divisions (otter, mollusc, fish, crayfish, frog). For PCA within each subdivision, the factors were individual species, and for the

PCA on otter adipose only, the factors were the two deposits—tail and footpad. The p values for all ANOVA were corrected using Tukey's Honest Significant Differences.

Diet inference—quantitiative fatty acid signature analysis

We estimated the contribution of each taxonomic group to the otter diet using the quantitative fatty acid signature analysis method described in Iverson et al. [8]. Means and standard errors for each proportion are based on 500 bootstrapped samples.

Results and discussion

In both otters and their potential prey species, including, fishes, frogs, crayfish, and molluscs, 16:0, 16:1, 18:0, and 18:1 were the predominate FA (Figure 1). This is not surprising, as these four fatty acids predominate in most animals [22]. Across all samples, these four FA comprised a mean of 55.9% (±9.2% sd) of the total FA. Within these four FA, however, molluscs and crayfish differed from the vertebrate animals. Molluscs had higher proportions of 16:0 and 16:1, whereas vertebrate taxa had higher proportions of 18:0 and 18:1. Crayfish have approximately equal proportions of 16 and 18-carbon FA. These four FA were important in PCA because of their predominance in most species. However, the remaining FA were particularly important for identifying taxon-specific variation. For example, considerable variation is evident in the long-chain FA, particularly the 20-carbon FA (20:0 – 20:5n3). Fish, mollusc, and crayfish species all showed higher proportions of omega-3 FA (20:5n3) than frogs or otters; molluscs, fish, and frog species had higher proportions of omega-6 (20:4n6) than otters or crayfish. Longer chain FA (22-C and above) were in low proportions in all taxa other than fishes and molluscs. A summary of the fatty acid signatures for all species is provided in Table 2.

Principal components analysis
Prey species by taxonomic group

The 60-dimensional FA signature of all prey species projected onto the first 2 principal components (Figure 2) provides a quick visual interpretation of overall similarity between fatty acid signatures, as more similar profiles will be projected more closely together. The first PC accounts for 14.9% of the total variation in FA in all potential prey species, and separates the molluscs from the remaining taxonomic groups (ANOVA: $p < 0.001$ for all pairwise differences with molluscs). Mean contributions to the component also separated frogs and crayfish ($p = 0.011$). This component is dominated by relatively high proportions of 20:1, and relatively low proportions of 18:1, 14:0, and 20:3n3 in molluscs. These values can be quantified by how strongly the proportion of each FA

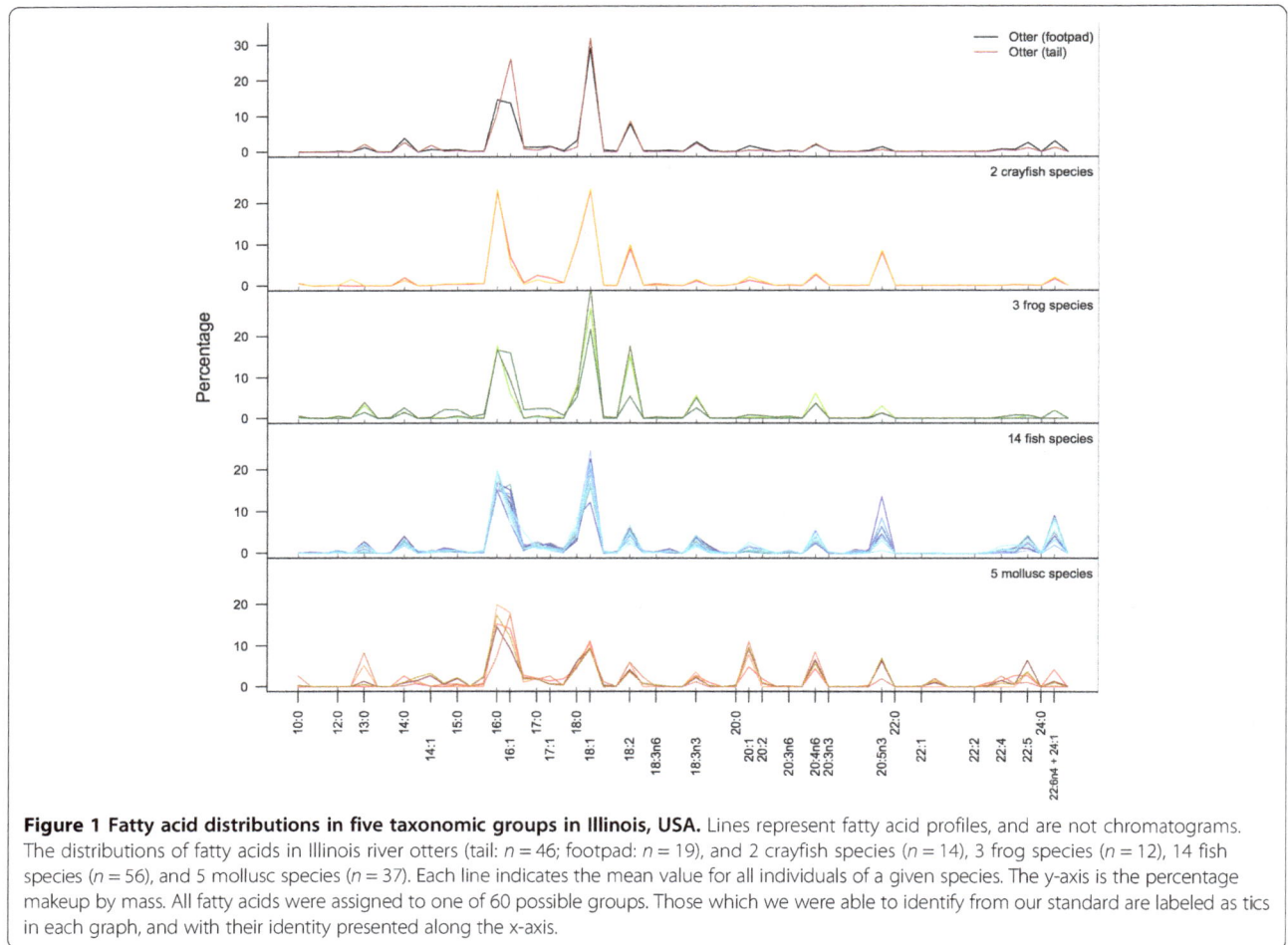

Figure 1 Fatty acid distributions in five taxonomic groups in Illinois, USA. Lines represent fatty acid profiles, and are not chromatograms. The distributions of fatty acids in Illinois river otters (tail: $n = 46$; footpad: $n = 19$), and 2 crayfish species ($n = 14$), 3 frog species ($n = 12$), 14 fish species ($n = 56$), and 5 mollusc species ($n = 37$). Each line indicates the mean value for all individuals of a given species. The y-axis is the percentage makeup by mass. All fatty acids were assigned to one of 60 possible groups. Those which we were able to identify from our standard are labeled as tics in each graph, and with their identity presented along the x-axis.

correlates with the component. Specifically, 20:1 shows a strong positive correlation (0.89) with the first PC, whereas 18:1, 14:0, and 20:3n3 have strong negative correlations (−0.66, −0.62, and −0.53 respectively).

The second PC accounts for an additional 12.9% of the total variation, and separates fish species from frogs and crayfish (both $p < 0.001$), contrasting relatively higher proportions of 17:1 (correlation = 0.60) and 16:1 (0.60) in fishes, and relatively low proportions of 18:0, 18:2, and 10:0 (correlations = −0.64, −0.52, and −0.44 respectively).

Comparing species within a single taxonomic group

Analysis of fatty acid signatures within the 5 mollusc species again yielded apparent clusters (Figure 3). The first PC accounts for 22.1% of the total variation among mollusc FA signatures and separates the invasive Asian clam (*Corbicula fluminea*) from the native species ($p < 0.001$ for all pairwise differences), as well as separating the round pigtoe (*Pleurobema sintoxia*) from the fat mucket (*Lampsilis siliquoidea*; $p = 0.045$). This component is characterized by relatively high

proportions of 10:0, 20:1, and 13:0 in native species (correlations = −0.57, −0.55, −0.46).

The second PC accounts for an additional 18.5% of the total variation in molluscs, and separates the round pigtoe (*Pleurobema sintoxia*) from both the fat mucket and the Wabash pigtoe (*Fusconaia flava*; both $p < 0.001$), as well as further separating the round pigtoe from the Asian clam ($p = 0.019$). This component is characterized by relatively high proportions of 17:1 and 16:1 (correlations = 0.63, 0.60) in *Pleurobema*, and low proportions of 14:1 and 22:1 (correlation = 0.70 for both). The 4th PC (not shown) separates the Wabash pigtoe from the fat mucket ($p = 0.004$). The first 5 PC all failed to separate the three ridge (*Amblema plicata*) from either the round pigtoe or the Wabash pigtoe (all $p > 0.05$).

Other taxonomic divisions had similar results. The two crayfish species differed in their mean contributions to the 3rd PC ($p = 0.011$). Among frogs, both *Acris* and *Pseudacris* differed from *Lithobates* along the 1st PC ($p = 0.003$ and 0.010 respectively), and from one another along the 5th PC ($p < 0.001$). Among the fish species, 87% of the pairwise differences were significant ($p <$

Table 2 Fatty acids of the river otter (*Lontra canadensis*) and several candidate prey species; Illinois, USA

FA	Otter		Crayfish	Frog	Mollusc	Fish[a]			
	Tail	Footpad				Clup.	Cypr.	Perc.	Silur.
10:0	0.0 ± 0.03[b]	0.0 ± 0.02	0.5 ± 0.42	0.2 ± 0.19	0.4 ± 1.47	0.1 ± 0.04	0.1 ± 0.08	0.1 ± 0.06	0.0 ± 0.05
12:0	0.2 ± 0.10	0.1 ± 0.07	0.1 ± 0.10	0.6 ± 0.51	0.0 ± 0.01	0.1 ± 0.02	0.2 ± 0.21	0.3 ± 0.26	0.5 ± 0.57
13:0	1.3 ± 2.09	2.3 ± 3.41	0.0 ± 0.00	3.1 ± 3.51	1.7 ± 3.61	0.3 ± 0.30	1.3 ± 1.73	1.0 ± 1.78	1.1 ± 2.25
14:0	3.8 ± 1.05	2.7 ± 0.99	1.5 ± 0.64	1.6 ± 0.47	1.3 ± 0.85	4.1 ± 0.81	2.5 ± 0.96	2.7 ± 0.64	2.5 ± 1.06
14:1	0.7 ± 0.45	1.9 ± 1.04	0.1 ± 0.14	0.2 ± 0.15	1.6 ± 1.95	0.2 ± 0.03	0.4 ± 0.25	0.4 ± 0.25	0.3 ± 0.18
15:0	0.6 ± 0.25	0.4 ± 0.24	0.4 ± 0.19	0.5 ± 0.55	1.3 ± 1.13	0.8 ± 0.24	0.6 ± 0.15	0.6 ± 0.23	0.6 ± 0.11
16:0	14.7 ± 3.33	11.3 ± 5.91	23.3 ± 8.09	17.7 ± 1.6	15.6 ± 4.08	17.0 ± 3.33	15.9 ± 1.98	17.7 ± 2.57	18.9 ± 1.73
16:1	13.7 ± 4.65	26.1 ± 9.01	5.7 ± 2.94	7.1 ± 3.38	13 ± 4.05	15.2 ± 7.16	10.2 ± 3.29	12.4 ± 3.23	8.9 ± 2.77
17:0	1.3 ± 0.47	0.4 ± 0.35	1.7 ± 0.89	0.5 ± 0.63	1.9 ± 0.44	1.9 ± 0.39	1.7 ± 0.46	1.9 ± 0.62	1.8 ± 0.51
17:1	1.6 ± 0.40	1.4 ± 0.51	1.0 ± 0.92	0.5 ± 0.65	1.0 ± 0.83	2.0 ± 0.47	1.2 ± 0.87	1.5 ± 0.73	0.9 ± 0.29
18:0	3.2 ± 1.60	1.4 ± 0.69	10.7 ± 5.57	7.7 ± 2.24	5.2 ± 1.20	3.0 ± 1.32	6.9 ± 2.59	4.8 ± 1.35	7.7 ± 1.76
18:1	29 ± 4.80	31.9 ± 8.71	23.3 ± 5.33	26.7 ± 3.38	9.8 ± 1.37	22.6 ± 5.28	19.8 ± 5.93	18.2 ± 2.96	20.1 ± 2.64
18:2	7.9 ± 3.30	8.7 ± 3.70	9.7 ± 4.12	14.6 ± 3.25	4.3 ± 0.99	3.3 ± 0.33	5.2 ± 2.63	5.4 ± 1.79	3.2 ± 1.74
18:3n6	0.3 ± 0.10	0.1 ± 0.12	0.2 ± 0.26	0.1 ± 0.17	0.3 ± 0.26	0.4 ± 0.06	0.4 ± 0.13	0.4 ± 0.11	0.2 ± 0.07
18:3n3	2.8 ± 0.89	2.4 ± 1.25	1.4 ± 1.37	5.4 ± 4.83	2.5 ± 0.72	3.9 ± 0.62	2.4 ± 1.29	3.4 ± 1.47	1.8 ± 0.76
20:0	0.1 ± 0.07	0.0 ± 0.03	0.3 ± 0.10	0.1 ± 0.13	0.3 ± 0.20	0.2 ± 0.06	0.2 ± 0.11	0.2 ± 0.09	0.3 ± 0.04
20:1	1.6 ± 0.73	0.4 ± 0.37	1.9 ± 1.54	0.2 ± 0.26	8.8 ± 2.98	0.8 ± 0.20	1.5 ± 0.70	1.0 ± 0.91	1.3 ± 0.77
20:2	0.8 ± 0.24	0.3 ± 0.27	0.9 ± 0.42	0.2 ± 0.19	1.2 ± 1.27	0.7 ± 0.21	0.7 ± 0.41	0.5 ± 0.44	0.5 ± 0.28
20:3n6	0.4 ± 0.10	0.2 ± 0.20	0.1 ± 0.12	0.3 ± 0.26	0.1 ± 0.10	0.3 ± 0.07	0.6 ± 0.42	0.3 ± 0.17	0.4 ± 0.11
20:4n6	2.0 ± 0.8	2.3 ± 1.22	2.9 ± 1.31	5.8 ± 2.33	5.8 ± 1.66	2.5 ± 0.92	4.7 ± 2.10	3.6 ± 1.22	5.0 ± 1.90
20:3n3	0.3 ± 0.09	0.1 ± 0.13	0.1 ± 0.16	0.1 ± 0.14	0.0 ± 0.05	0.1 ± 0.03	0.2 ± 0.11	0.3 ± 0.19	0.2 ± 0.07
20:5n3	1.4 ± 0.63	0.5 ± 0.37	8.4 ± 3.83	2.8 ± 0.83	5.8 ± 2.00	4.6 ± 0.86	7.8 ± 4.49	4.3 ± 2.72	6.1 ± 1.36
22:0	0.0 ± 0.02	0.0 ± 0.02	0.0 ± 0.03	0.0 ± 0.04	0.1 ± 0.54	0.0 ± 0.04	0.0 ± 0.00	0.1 ± 0.06	0.1 ± 0.06
22:1	0.1 ± 0.06	0.0 ± 0.01	0.0 ± 0.02	0.0 ± 0.01	0.2 ± 0.45	0.1 ± 0.05	0.0 ± 0.02	0.1 ± 0.06	0.0 ± 0.05
22:2	0.0 ± 0.01	0.0 ± 0.00	0.0 ± 0.00	0.0 ± 0.00	0.0 ± 0.05	0.0 ± 0.02	0.0 ± 0.00	0.0 ± 0.02	0.0 ± 0.02
22:4	0.7 ± 0.25	0.6 ± 0.51	0.0 ± 0.00	0.1 ± 0.15	1.3 ± 0.69	0.4 ± 0.23	0.4 ± 0.38	0.9 ± 0.65	1.0 ± 0.57
22:5	2.5 ± 0.88	1.1 ± 0.98	0.0 ± 0.13	0.6 ± 0.54	3.7 ± 2.09	1.3 ± 0.25	2.2 ± 1.59	3.3 ± 1.32	3.0 ± 0.57
24:0	0.0 ± 0.01	0.0 ± 0.00	0.0 ± 0.00	0.0 ± 0.00	0.0 ± 0.04	0.0 ± 0.02	0.0 ± 0.00	0.1 ± 0.09	0.1 ± 0.08
22:6n4 + 24:1[c]	3.0 ± 1.11	1.2 ± 1.06	1.9 ± 0.92	1.8 ± 1.27	1.8 ± 1.56	4.3 ± 0.77	6.2 ± 2.59	6.8 ± 2.72	6.4 ± 2.41

[a]Because a relatively large number of fish species were included in this study, they are subdivided here by order: Clupeiformes, Cypriniformes, Perciformes and Siluriformes (see Table 1).
[b]All values given as percentage of total fatty acids by mass ± 1sd.
[c]The column used in gas chromatography did not separate 22:6n4 and 24:1 and therefore they were grouped together.

0.05) on at least one of the first 5 PC—higher PC were not analyzed.

River otter fatty acids by fat deposit

Within our river otter samples, fatty acid profiles differed according to the deposit from which the adipose sample was dissected (either from the base of the tail or the footpad of the forepaw). The 1st PC alone accounts for 33.7% of the total variance, and there is already considerable separation ($p < 0.001$) of the two deposits along this component (Figure 4). This component is primarily characterized by higher proportions of 20:1, 20:0, and 17:0 in the tail (correlations = 0.87, 0.86, 0.79), and higher proportions of 16:1 and 14:1 (−0.84, −0.79) in the footpad. More generally, the footpad deposit is characterized by lower concentrations of saturated fats. With the exception of 13:0, all other saturated fats are positively correlated with this component, indicating higher proportions in the tail (correlations for 10:0, 12:0, 14:0, 15:0, 16:0, 17:0, 18:0, 20:0, 22:0, and 24:0 are 0.34, 0.75, 0.39, 0.62, 0.26, 0.79, 0.57, 0.86, 0.25, and 0.29 respectively). In a study of several species, Käkelä and Hyvärinen [16] demonstrate similarly high concentrations of unsaturated FA in the extremities of cold-adapted species. The FA signature of the footpads is therefore likely to be cold-adapted as well, and thus a biased indicator

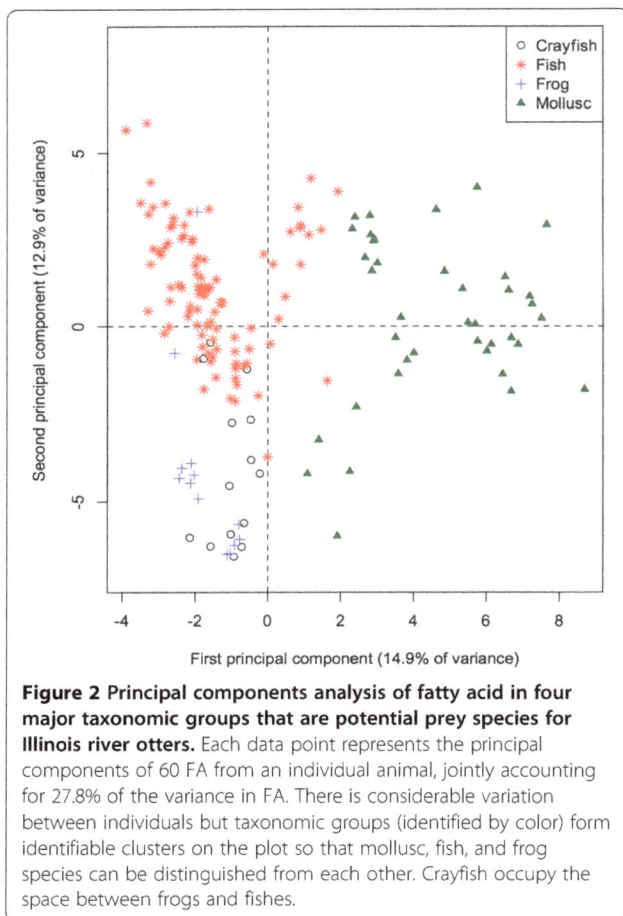

Figure 2 Principal components analysis of fatty acid in four major taxonomic groups that are potential prey species for Illinois river otters. Each data point represents the principal components of 60 FA from an individual animal, jointly accounting for 27.8% of the variance in FA. There is considerable variation between individuals but taxonomic groups (identified by color) form identifiable clusters on the plot so that mollusc, fish, and frog species can be distinguished from each other. Crayfish occupy the space between frogs and fishes.

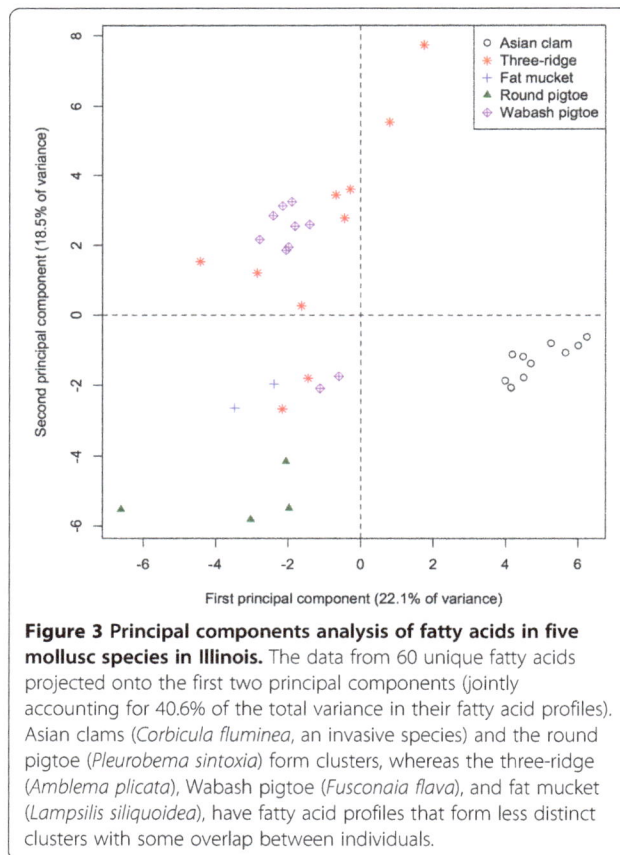

Figure 3 Principal components analysis of fatty acids in five mollusc species in Illinois. The data from 60 unique fatty acids projected onto the first two principal components (jointly accounting for 40.6% of the total variance in their fatty acid profiles). Asian clams (*Corbicula fluminea*, an invasive species) and the round pigtoe (*Pleurobema sintoxia*) form clusters, whereas the three-ridge (*Amblema plicata*), Wabash pigtoe (*Fusconaia flava*), and fat mucket (*Lampsilis siliquoidea*), have fatty acid profiles that form less distinct clusters with some overlap between individuals.

for inferring diet. Thus, when analyzing the otter FA signatures relative to the other species, we included only the adipose tissue from the base of the tail, with the assumption that this depot did not suffer the same bias.

Otter and potential prey combined

Adding the otter data to the PCA of its potential prey species (Figure 5) indicates that the otters have fatty acid signatures that are most similar to fish species. Like fish and frogs, the proportions of the 18-carbon chains are higher than those of 16-carbon chains. Similarly, like fish, and unlike frogs, otters show higher proportions of the long-chain FA (particularly 24:0) than the other taxa.

Estimation of the otter diet

The proportions of each taxon were estimated at 37.7 ± 1.0% (SE) fishes, 32.0 ± 0.8% mollusc, 27.3 ± 4.3% amphibian, and 3.0 ± 0.6% crayfish by mass.

Discussion

Large-scale taxonomic differences were discernible from FA signatures. Species-level differences were largely significant as well, though not universally. Within the

mollusc and fish divisions, some species did not differ significantly on any of the first 5 principle components. Where significance was not demonstrable, it may have been due to small sample sizes, but even with such small sample sizes, the majority of pairwise differences were significant. These results provide further evidence that FA signatures are capable of demonstrating fine-grain differences between species, and potentially diet. Nevertheless, we recognize a number of limitations in interpreting these results. In particular, it is possible that FA signatures for a given species may show significant variation both seasonally and geographically, as both may lead to differences in availability of prey that are sufficient to change the species' FA makeup. In using FA signatures to infer diet, it is necessary to determine the degree to which individual species or taxonomic groups may be discerned from each other on the basis of their FA signatures as we have done here. However, this information alone is insufficient to make more than general inferences about the diet. In order to make more accurate inferences, further information is needed regarding how the predator species metabolizes different FA. The QFASA adjusts for this shortcoming by measuring the distance between the predator (otter) FA profile and that of the diet by using the Kulback-Liebler (KL) distance

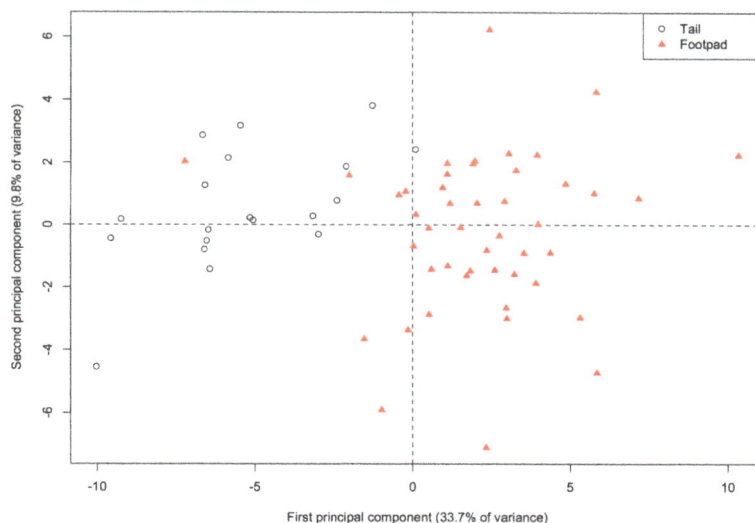

Figure 4 Principal components analysis of fatty acids in Illinois river otters. The data from 60 unique fatty acids projected onto the first two principal components (jointly accounting for 27.8% of the total variance in otter fatty acid profiles). Each data point represents a tissue sample from a single animal. Footpad and tail deposits have largely distinct fatty acid profiles, as there is very little overlap between the two clusters.

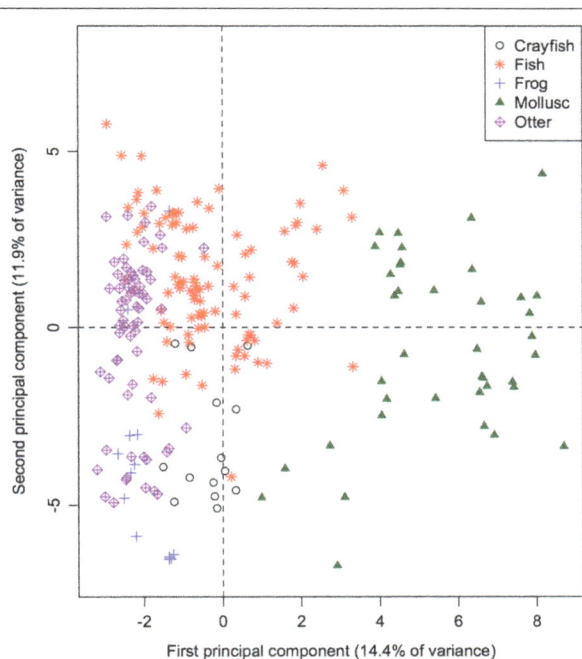

Figure 5 Principal components analysis of 60 fatty acids in Illinois river otters (*Lontra canadensis*) and several of its candidate prey species. Of the candidate prey species considered here, the river otter fatty acid signatures are most similar to those of fish species as indicated by their proximity in the graph. This suggests that fish are the predominant component of the otter diet, though a better understanding of how river otters metabolize different fatty acids is necessary for a fuller inference of the PCA results for diet.

(in [8]), which gives greater weight to rare FA. When comparing results among different distance metrics, including the usual squared error, squared relative error, the squared error distance of logs, and the KL on a controlled diet, KL was shown to perform well and is a natural distance metric for comparing distributions [8].

Throughout the results above, we reported differences only for those FA that we were able to identify from the standard. However, variations also existed in the unidentified FA but since the FA were unidentified, we omitted this information.

Conclusion

We demonstrate that there are taxon-specific differences in fatty acid signatures that allow for the unique identification of taxonomic groups (Figure 2). In previous studies, QFASA has been largely successful at inferring diets in marine mammals. One study on harbor seals even inferred the geographic location where predation had occurred, and the average size of the fish being preyed upon on the basis of unique FA signatures [13]. However, river otters have been reported to prey on a greater variety of species, including, most commonly, a wide variety of fish species, crayfish, and frogs, but have also been reported to eat other reptiles and amphibians—including snakes, salamanders, and turtles—molluscs, insects and insect larvae, and occasionally birds and other mammals. The greater diversity of prey species results in a greater likelihood that prey species will have overlapping (not uniquely identifiable) FA signatures, making it more difficult to accurately infer the diet.

The findings in this paper suggest that large-scale taxonomic differences (i.e., the relative contributions of fish, crayfish, frog, or mollusc) in the river otter diet have significantly different FA profiles, and the QFASA estimates similarly have small standard errors. Many finer-scale differences such as individual species may also be discernible with sufficiently large samples, though the PCA analyses indicate that not all species will be discernible from FA signatures alone. The analyses indicated which species are likely to have the most similar signatures, such as the mollusc species, the three ridge, the round pigtoe, and the Wabash pigtoe, none of which had significantly different signatures along any of the first five PC.

The results of this study provide an initial inference of the otter diet. This inference is limited, and a more accurate inference of the diet requires knowledge of how otters metabolize different FA. However, experimental studies with controlled diet have demonstrated that QFASA has a high rate of accurately inferring diet [14], though this accuracy may diminish with the complexity of the diet. Although a more accurate inference may be gained by studying fat metabolism, such studies are costly, and for many species such as otters and many marine species, capturing and restraining the animals in order to perform the studies is often difficult or impossible. Thus, although the inference is limited, these limitations are counterbalanced by the relative ease and lower cost of the methods described herein. Furthermore, it is important to recognize that, molluscs, which have generally been absent from previous otter diet studies, presumably because remains do not show up in fecal samples, were estimated here to constitute as much as 32% of the otter diet in Illinois. In regions where mollusc species are endangered, these findings have important implications for wildlife management.

This study did not analyze all candidate prey taxa. In particular we did not analyze insect, mammal, or avian fatty acid profiles, but instead focused on those species most commonly reported in the river otter diet. The results here must therefore be considered the upper limits of the proportion of each taxon in the Illinois river otter diet. Furthermore, there is likely to be some seasonal variation in the otter diet, and none of our otter specimens were obtained during the summer months (Jun-Aug).

Abbreviations

ANOVA: Analysis of variance; AOCS: American Oil Chemists' Society; FA: Fatty Acid(s); GLC: Gas–liquid Chromatography; PC: Principle Component(s); PCA: Principal Components Analysis; QFASA: Quantitative Fatty Acid Signature Analysis; SS: Sum of Squares; USA: United States of America.

Competing interests

The authors have no financial or non-financial competing interests.

Authors' contributions

DSP: Performed all laboratory analyses, and took the lead in statistical analyses and drafting the manuscript. NMP: Involved with the conception, design, and acquisition of data as well as with revising the manuscript for critical intellectual content. JN: Involved with the conception, design, and acquisition of data as well as with revising the manuscript for critical intellectual content. All authors read and approved the final manuscript.

Acknowledgements

Funding for this project was provided by the Federal Aid in Wildlife Restoration Projects W-167-R and W-146-R with additional support provided by the University of Illinois, Departments of Animal Sciences and Illinois Natural History Survey. Additional thanks to Samantha K. Carpenter, Kevin Cummings, Chris Phillips, David Wahl, and Max Wolter for providing specimens, and to John P. Jerrell, Marcos Tavárez and Hunter Galloway for assistance in the laboratory.

Author details

[1]Illinois Natural History Survey, University of Illinois Urbana-Champaign, 1816 S. Oak Street, Champaign, IL 61820, USA. [2]Department of Animal Sciences, University of Illinois at Urbana-Champaign, 1503 S. Maryland Drive, Urbana, IL 61801, USA.

References

1. Bluett RD, Nielsen CK, Gottfried RW, Miller CA, Woolf A: **Status of the river otter (Lutra canadensis) in Illinois, 1998–2004.** Trans Illinois State Acad Sci 2004, **97:**209–217.
2. Raesly EJ: **Progress and status of river otter reintroduction projects in the United States.** Wildl Soc Bull 2001, **29:**856–862.
3. Greer KR: **Yearly food habits of the river otter in the Thompson lakes region, northwestern Montana, as indicated by scat analyses.** Am Midl Nat 1955, **54:**299–313.
4. Melquist WE, Hornocker MG: **Ecology of river otters in west central Idaho.** Wildl Monogr 1983, **83:**3–60.
5. Kruuk H: Otters: Ecology, Behaviour and Conservation. New York: Oxford University Press; 2006.
6. Carrs DN, Parkinson SG: **Errors associated with Lutra lutra faecal analysis. I. Assessing general diet from spraints.** J Zool 1996, **238:**301–317.
7. Crimmins SM, Roberts NM, Hamilton DA: **Effects of prey size on scat analysis to determine river otter Lontra canadensis diet.** Wildl Biol 2009, **15:**449–453.
8. Iverson SJ, Field C, Bowen WD, Blanchard W: **Quantitative fatty acid signature analysis: a new method of estimating predator diets.** Ecol Monogr 2004, **74:**211–235.
9. Hooker SK, Iverson SJ, Ostrum P, Smith SC: **Diet of northern bottlenose whales inferred from fatty-acid and stable isotope analyses of biopsy samples.** Can J Zool 2001, **79:**1442–1454.
10. Worthy GAJ: An investigation into the possible relationship between killer whale (Orcinus orca) predation and the continuing decline of the Steller sea lion (Eumetopias jubatus) population. Final report to the Pollock Conservation Cooperative Research Center; 2008. PCCRC purchase order no. FP204077. https://www.sfos.uaf.edu/pcc/projects/01/horning/final_report.pdf.
11. Loseto LL, Stern GA, Connelly TL, Deibel D, Gemmill B, Prokopowicz A, Fortier L, Ferguson SH: **Summer diet of beluga whales inferred by fatty acid analysis of the eastern Beaufort Sea food web.** J Exp Mar Biol Ecol 2009, **374:**12–18.
12. Iverson SJ, Arnould JPY, Boyd IL: **Milk fatty acid signatures indicate both major and minor shifts in the diet of lactating Antarctic fur seals.** Can J Zool 1997, **75:**188–197.
13. Iverson SJ, Frost KJ, Lowry LF: **Fatty acid signatures reveal fine scale structure of foraging distribution of harbor seals and their prey in Prince William Sound, Alaska.** Mar Ecol Prog Ser 1997, **151:**255–271.
14. Nordstrom CA, Wilson LJ, Iverson SJ, Tollit DJ: **Evaluating quantitative fatty acid signature analysis (QFASA) using harbour seals Phoca vitulina richardsi in captive feeding studies.** Mar Ecol Prog Ser 2008, **360:**245–263.
15. Carpenter SK, Mateus-Pinilla NE, Singh K, Lehner A, Satterthwaite-Phillips D, Bluett RD, Rivera NA, Novakofski JE: **River otters as biomonitors for

organochlorine pesticides, PCBs, and PBDEs in Illinois. *Ecotoxicol Environ Saf* 2014, **100**:99–104.

16. Käkelä R, Hyvärinen H: **Site-specific fatty acid composition in adipose tissues of several northern aquatic and terrestrial mammals.** *Comp Biochem Physiol* 1996, **115B**:501–514.

17. Leszczynski DE, Pikul J, Easter RA, McKeith FK, McLaren DG, Novakofski J, Jewell DE, Bechtel PJ: **Characterization of lipid in loin and bacon from finishing pigs fed full-fat soybeans or tallow.** *J Anim Sci* 1992, **70**:2175–2181.

18. AOCS: *Official Methods and Recommended Practices of the AOCS,* Am. Oil. 5th edition. Champaign, IL: Chem. Soc., Champaign; 1998.

19. R Development Core Team: *R: A Language and Environment for Statistical Computing.* Vienna, Austria: R Foundation for Statistical Computing; 2012. http://www.R-Project.org/.

20. Hartigan JA, Wong MA: **A K-means clustering algorithm.** *Appl Stat* 1979, **28**:100–108.

21. Husson F, Lê S, Pagès J: *Exploratory Multivariate Analysis by Example Using R.* Boca Raton: CRC Press; 2011.

22. Lawrence T, Fowler V, Novakofski J: *Growth of Farm Animals.* 3rd edition. Cambridge, MA: CABI; 2012.

7

Influence of ruminal degradable intake protein restriction on characteristics of digestion and growth performance of feedlot cattle during the late finishing phase

Dixie May[1], Jose F Calderon[1], Victor M Gonzalez[1], Martin Montano[1], Alejandro Plascencia[1], Jaime Salinas-Chavira[2], Noemi Torrentera[1] and Richard A Zinn[3*]

Abstract

Two trials were conducted to evaluate the influence of supplemental urea withdrawal on characteristics of digestion (Trial 1) and growth performance (Trial 2) of feedlot cattle during the last 40 days on feed. Treatments consisted of a steam-flaked corn-based finishing diet supplemented with urea to provide urea fermentation potential (UFP) of 0, 0.6, and 1.2%. In Trial 1, six Holstein steers (160 ± 10 kg) with cannulas in the rumen and proximal duodenum were used in a replicated 3 × 3 Latin square experiment. Decreasing supplemental urea decreased (linear effect, $P \leq 0.05$) ruminal OM digestion. This effect was mediated by decreases (linear effect, $P \leq 0.05$) in ruminal digestibility of NDF and N. Passage of non-ammonia and microbial N (MN) to the small intestine decreased (linear effect, $P = 0.04$) with decreasing dietary urea level. Total tract digestion of OM (linear effect, $P = 0.06$), NDF (linear effect, $P = 0.07$), N (linear effect, $P = 0.04$) and dietary DE (linear effect, $P = 0.05$) decreased with decreasing urea level. Treatment effects on total tract starch digestion, although numerically small, likewise tended (linear effect, $P = 0.11$) to decrease with decreasing urea level. Decreased fiber digestion accounted for 51% of the variation in OM digestion. Ruminal pH was not affected by treatments averaging 5.82. Decreasing urea level decreased (linear effect, $P \leq 0.05$) ruminal N-NH and blood urea nitrogen. In Trial 2, 90 crossbred steers (468 kg ± 8), were used in a 40 d feeding trial (5 steers/pen, 6 pens/ treatment) to evaluate treatment effects on final-phase growth performance. Decreasing urea level did not affect DMI, but decreased (linear effect, $P \leq 0.03$) ADG, gain efficiency, and dietary NE. It is concluded that in addition to effects on metabolizable amino acid flow to the small intestine, depriving cattle of otherwise ruminally degradable N (RDP) during the late finishing phase may negatively impact site and extent of digestion of OM, depressing ADG, gain efficiency, and dietary NE.

Keywords: Cattle, Degradable protein, Digestion, Growth performance

Background

Because of its low cost per unit of N compared with most sources of natural protein, urea is a primary source of supplemental N in conventional steam-flaked corn-based finishing diets for feedlot cattle [1]. In a review of nutrition consultant recommendations across 11 states in the USA, Vasconcelos and Galyean [2] observed that on average, flaked corn-based finishing diets contained 13.5% CP with

1.2% of supplemental urea (approximately 64% DIP). Although dietary formulation in this manner is expected to meet urea fermentation potential (UFP) for optimal microbial growth, it may exceed protein requirements for cattle growth, particularly during the late finishing phase. Preston [3] proposed the feasibility of restricting protein supplementation during the late finishing phase as a means of minimizing N excess and associated environmental impact [1,4] without detrimentally affecting cattle performance. However, the impact of this practice on digestive function and cattle growth-performance has received limited research attention. The aim of this study was to

* Correspondence: razinn@ucdavis.edu
[3]Department of Animal Science, University of California, Davis E. Holton Rd, El Centro, CA 92242, USA
Full list of author information is available at the end of the article

evaluate the influence of UFP for optimal microbial growth on characteristics of digestion and growth performance of feedlot cattle during the late finishing phase.

Methods

All procedures involving animal care and management were in accordance with and approved by the University of California, Davis, Animal Use and Care Committee.

Trial 1

Six Holstein steers (160 ± 10 kg) with cannulas in the rumen and proximal duodenum [5] were used in 3×3 replicated Latin square experiment. Burroughs *et al.* [6] proposed that amount of degradable intake protein (DIP) necessary to optimize microbial growth was equivalent to the net microbial protein synthesis. Accordingly, the urea fermentation potential of the diet (percentage of additional urea that may be added to the diet in order to optimize microbial growth) would be equivalent to: (0.104TDN- DPI)/ 2.8, where TDN is expressed as a percentage, and DPI is expressed as the percentage of RDP in the basal diet before urea supplementation. Accordingly, treatments consisted of a steam flaked corn-based finishing diet adjusted for restriction of rumen DIP to provide urea fermentation potentials of 0 (UFP-0), 0.6 (UFP-0.6) and 1.2% (UFP-1.2). Composition of experimental diets is shown in Table 1. Chromic oxide (0.40%, DM basis) was included in diets as a digesta marker. Dry matter intake was restricted to 4.0 kg/d (2.2% of BW daily), and feed was offered in equal portions at 0800 and 2000 daily. The three experimental periods consisted of a 10-d diet adjustment period followed by a 4-d collection period. During the collection period duodenal and fecal samples were taken from all steers, twice daily as follows: d 1, 1050 and 1450; d 2, 0900 and 1500; d 3, 0730 and 1330, and d 4, 0600 and 1200. Individual samples consisted of approximately 750 mL of duodenal chyme and 200 g (wet basis) of fecal material. Samples from each steer and within each collection period were composited for analysis. During the final day of each collection period, 4 h after feeding, ruminal and blood samples were collected from each steer via ruminal cannula and caudal venous respectively. Ruminal fluid pH was determined by inserting a pH electrode into the freshly collected samples. The ruminal fluid sample was divided into two parts: 40 mL was measured into a plastic bag, placed in an ice bath, and carried to a laboratory for determination of N-NH in fresh ruminal fluid [7]. The remainder was strained through four layers of cheesecloth. Ten mL of freshly prepared 25% (wt/vol) metaphosphoric acid was added to 40 mL of strained ruminal fluid, 10 mL were then centrifuged ($17,000 \times g$ for 10 min), and supernatant fluid was stored at $-20°C$ for VFA analysis. Upon completion of the trial, ruminal fluid was obtained via the ruminal cannula from all steers and composited for

Table 1 Diet composition of experiment 1 and 2[1]

Item	Urea fermentation potential		
	0	0.6	1.2
Ingredient (g/kg of DM)			
Steam flaked corn	797.5	803.0	809.0
Sudangrass hay	50.0	50.0	50.0
Alfalfa hay	50.0	50.0	50.0
Urea	12.5	7.0	1.0
Cane molasses	50.0	50.0	50.0
Yellow grease	20.0	20.0	20.0
Limestone	14.0	14.0	14.0
Trace mineral salt[2]	4.0	4.0	4.0
Magnesium oxide	2.0	2.0	2.0
Monensin[3]	0.022	0.022	0.022
Nutrient composition (DM basis)[4]			
NE$_m$ (Mcal/kg)	2.23	2.24	2.25
NE$_g$ (Mcal/kg)	1.56	1.56	1.58
DE (Mcal/kg)	3.86	3.86	3.89
CP (g/kg)	130.0	115.0	99.1
RDP (g/kg of CP)	648	600	530
NDF (g/kg)	125.0	125.0	125.0
Calcium (g/kg)	6.6	6.6	6.6
Phosphorus (g/kg)	2.8	2.8	2.8

[1]Chromic oxide (0.40%) was added in substitution of corn grain as a digesta marker in Trial 1. RDIP, rumen degradable intake protein. UFP, estimated urea fermentation potential.
[2]Trace mineral salt contained: CoSO4, 0.068%; CuSO4, 1.04%; FeSO4, 3.57%; ZnO, 1.24%; MnSO4, 1.07%; KI, 0.052%; and NaCl, 92.96%.
[3]Rumensin80 (Elanco Animal Health, Greenfield, IN).
[4]Based on tabular values for individual feed ingredients (NRC, [17]).

microbial isolation via differential centrifugation [8]. The microbial isolates were prepared for analysis by oven drying at 70°C and then grinding with mortar and pestle. Feed, duodenal, and fecal samples were prepared for analysis by oven drying at 70°C and then grinding in a laboratory mill. Samples were then oven dried at 105°C until no further weight loss occurred and stored in tightly sealed glass jars. Samples were subjected to all or part of the following analysis: DM (oven-drying at 105°C until no further weight loss), ash, N-NH, Kjeldahl N [9], NDF-adjusted for insoluble ash [10], purines [11], starch [12] and VFA concentrations of ruminal fluid (gas chromatography; [13]), GE (adiabatic bomb calorimetry), and chromic oxide [14]. Duodenal flow and fecal excretion of DM were calculated based on marker ratio, using chromic oxide. Microbial organic matter (MOM) and microbial N (MN) leaving the abomasum were calculated using purines as a microbial marker [11]. Organic matter fermented in the rumen was considered equal to OM intake minus the difference between the amount of total OM reaching the duodenum and MOM reaching the duodenum. Feed N escape to the small intestine was considered equal to total N leaving the abomasum minus N-NH, microbial, and endogenous N (0.195 g/kg $W^{0.75}$; [15]). Methane production (mol/mol of glucose equivalent fermented) was estimated based on the theoretical fermentation balance for observed molar distribution of VFA [16]. Whole blood samples were centrifuged and the plasma frozen for BUN analysis. The blood samples collected were centrifuged and the plasma analyzed for Blood Urea Nitrogen (BUN) by slide method using Vitros Bun/Urea DT60 II (Ortho Clinical Diagnostics, Inc., Rochester, NY), and ruminal N-NH [7]. The effects of the urea level on characteristics of digestion in cattle were analyzed as a 3 × 3 replicated Latin square design using the MIXED procedure (SAS Inst. Inc., Cary, NC). The fixed effect consisted of treatment, and random effects consisted of steer and period. The statistical model for the trial was as follows:

$$Y_{ijk} = \mu + R_l + S_{i(l)} + P_{j(l)} + T_k + E_{ijk},$$

where: Y_{ijk} is the response variable, μ is the common experimental effect, R_l is the replicated effect, S_i is the steer effect within replicate, P_j is the period effect within replicate, T_k is the treatment effect and E_{ijk} is the residual error. Treatment effects were tested using the following contrasts: 1) linear effect of the urea level, and 2) quadratic effect of the urea level, which were determined according to SAS (SAS Inst., Inc., Cary, NC; Version 9.1).

Trial 2

Ninety crossbred steers with an average initial weight of 468 ± 8 kg were used in a 40 d finishing trial to evaluate the treatment effects on growth performance. Steers had a purchase weight of 214 ± 14 kg and had been on feed 197 d before initiation of the study. Steers had been implanted with Synovex-S (Zoetis, Florham Park, NJ) upon arrival into the feedlot and with Revalor-S (Merck Animal Health, Summit, NJ) on d 98. Ten d prior to initiation of the study steers were weighed, reimplanted with Revalor-S, blocked by weight and randomly allotted within weight groupings to 18 pens (5 steers/pen). Pens were 43 m^2, with 22 m^2 of overhead shade, automatic waterers, and 2.4 m long fence-line feed bunks. Dietary treatments were the same as those used in Experiment 1. All steers received the UFP-0 diet for 10 d prior to initiation of the trial. Diets were prepared at weekly intervals and stored in plywood boxes located in front of each pen. Steers were allowed free access to dietary treatments. Fresh feed was provided twice daily. Individual steers were weighed upon initiation and completion of the trial. In the calculation of steer performance live weights were reduce 4% to adjust for digestive tract fill. Estimates of steer performance were based on pen means. Net energy values for each diet were calculated from estimates of

energy gain (EG, Mcal/d) based on growth-performance; EG = 0.0557 $BW^{0.75}$ ($ADG^{1.097}$), where EG is the daily energy deposited (Mcal/d), BW is the mean shrunk body weight (full weight × 0.96) and maintenance energy expended (EM, Mcal/d); EM = 0.077 $BW^{0.75}$ [18]. Dietary NEg was derived from NE_m by the equation: NE_g = 0.877 NE_m - 0.41 [19]. Dry matter intake is related to energy requirements and dietary NE_m according to the equation: DMI = EG / NE_g), and can be resolved for estimation of dietary NE by means of the quadratic formula: $x = \frac{-b \pm \sqrt{b^2 - 4ac}}{2a}$, where x = NE_m, a = -0.877 DMI, b = 0.877 EM + 0.41 DMI + EG, and c = -0.41 EM [19].

All steers were harvested on the same day. Each carcass was weighed at time of slaughter to determine dressing percentage [20]. Performance (gain, gain efficiency, and dietary energetics) and carcass data were analyzed as a randomized complete block design; the experimental unit was the pen. The MIXED procedure of SAS [21] was used to analyze the variables. The fixed effect consisted of treatment, and pen was the random component. Treatments effects were tested using the following contrasts: 1) linear effect of the urea level, and 2) quadratic effect of the urea level, which were determined according to SAS [21].

Results and discussion

The influence of dietary treatments on ruminal and total tract digestion is shown in Table 2. Decreasing supplemental urea decreased (linear effect, $P \leq 0.05$) ruminal OM digestion. This effect was mediated by decreases in ruminal digestibility of NDF (linear effect, $P = 0.05$), starch (linear effect, $P = 0.09$) and N (linear effect, $P = 0.04$). Likewise, Zinn et al. [22] observed decreased ruminal digestion of OM, NDF and starch in response to decreasing urea supplementation of a steam-flaked corn-based finishing diet fed to feedlot steers [22].

Passage of non-ammonia N to the small intestine decreased (linear effect, $P = 0.04$) with decreasing dietary urea level. This effect was due to decreased (linear effect, $P = 0.04$) MN synthesis. Taking into consideration energy intake alone, predicted flow of MN to the small intestine was 48g/d ([17], Level 1). Accordingly, with decreasing urea level, the observed flow of MN to the small intestine was 85, 73, and 65% of predicted flow for UFP-0, UFP-0.6, and UFP-1.2, respectively. This decline in net synthesis is consistent with [19] who observed that MN flow to the small intestine declines with decreasing DIP below 100 g/kg of total tract digestible OM. For the present study, DIP averaged 95, 81, and 61g/kg total tract digestible OM for UFP-0, UFP-0.6, and UFP-1.2, respectively. Thus, it is apparent that as DIP intake drops below 95 g/kg digestible OM there is not sufficient compensation in ruminal N recycling to maintain microbial growth, and as microbial growth declines, likewise, ruminal OM digestion declines.

Table 2 Influence of dietary treatments on characteristics of digestion

Item	Urea fermentation potential			P - value		SEM
	0	0.6	1.2	Linear	Quadratic	
Steer replications	6	6	6			
Intake (g/d)						
DM	3556	3553	3551			
OM	3343	3359	3378			
NDF	453	455	457			
N	66.2	57.6	48.2			
Starch	1907	1919	1932			
GE (Mcal/d)	15.2	15.2	15.3			
Flow to the duodenum (g/d)						
OM	1655	1749	1894	0.05	0.76	83.0
NDF	338	370	456	0.05	0.55	42.0
Starch	385	441	533	0.08	0.77	61.0
Total N	76.4	68.8	64.4	0.04	0.70	4.0
Microbial N	40.6	34.7	31.4	0.04	0.69	3.0
NH-N	2.30	2.04	1.58	0.06	0.72	0.3
Non-ammonia N	74.1	66.7	62.9	0.04	0.66	3.8
Feed N	24.7	23.2	22.7	0.20	0.71	0.9
Ruminal digestibility, %						
OM	62.6	58.3	53.23	0.04	0.91	0.3
NDF	25.3	18.6	0.30	0.05	0.54	0.9
Starch	79.8	77.0	72.4	0.09	0.78	0.3
Feed N	62.6	59.7	52.9	0.04	0.64	0.5
Microbial efficiency[1]	19.4	17.8	17.5	0.06	0.40	0.7
N efficiency[2]	1.12	1.16	1.30	0.05	0.44	0.07
Fecal excretion (g/d)						
OM	624	705	756	0.06	0.77	48.0
NDF	282	319	348	0.06	0.87	25.0
Starch	35.9	49.0	56.6	0.10	0.78	9.4
Total N	20.9	22.2	21.8	0.48	0.44	1.0
GE (Mcal/d)	3.28	3.66	3.90	0.05	0.75	0.84
Postruminal digestibility (% of flow to duodenum)						
OM	62.2	59.7	60.0	0.37	0.49	1.9
NDF	15.3	11.2	22.8	0.41	0.33	7.3
Starch	90.5	89.1	89.3	0.60	0.69	1.9
Total N	72.6	67.6	66.2	0.08	0.53	2.7
Total tract digestibility (% of intake)						
OM	81.3	79.0	77.6	0.06	0.75	1.4
NDF	37.8	29.8	23.9	0.07	0.85	5.3
Starch	98.1	97.4	97.1	0.11	0.77	0.5
Total N	68.5	61.5	54.9	0.04	0.97	4.5
DE, %	78.5	76.0	74.5	0.05	0.74	1.4
DE, Mcal/kg	3.36	3.26	3.20	0.05	0.73	0.06

[1]Microbial N, g/kg OM fermented.
[2]Nonammonia N flow to the small intestine as a fraction of N intake.

There were no treatment effects ($P = 0.20$) on passage of feed N to the small intestine. Notwithstanding decreased non-ammonia N flow to the small intestine with decreasing urea level, ruminal N efficiency (non-ammonia N flow to the small intestine as a fraction of N intake) increased (linear $P < 0.05$), reflecting increased contribution of recycled N into microbial protein synthesis, consistent with the observation that ruminal N flux increases inversely with dietary N concentration [23]. Observed DIP (Table 2) averaged 103% of expected based on tabular values ([17]; Table 1) for the three dietary treatments.

Total tract digestion of OM (linear effect, $P = 0.06$), NDF (linear effect, $P = 0.07$), N (linear effect, $P = 0.04$) and dietary DE (linear effect, $P = 0.05$) decreased with decreasing urea level. Treatment effects on total tract starch digestion, although numerically small, likewise tended (linear effect, $P = 0.11$) to decrease with decreasing urea level. Decreased fiber digestion accounted for 51% of the variation in OM digestion. In a previous study involving steam-flaked corn-based finishing diets in which urea was the sole source of supplemental N [22], increasing urea level from 1.0 to 1.6% of the steam-flaked corn in the diet (an upper level similar to that of the present study; Table 1) likewise enhanced total tract OM and fiber digestion. In contrast Zinn and Shen [19] observed removal of urea from a steam-flaked corn-based growing-finishing diet markedly depressed ruminal OM digestion and flow of MN to the small intestine but did not affect total tract OM digestion. Treatment effects on apparent N digestion were largely a function of the N content of the diet brought about by changes in dietary urea level [24].

Treatment effects on ruminal pH, VFA molar proportions, and BUN are shown in Table 3. Ruminal pH (measured

4-h postprandium) was not affected ($P = 0.51$) by treatments, averaging 5.82. Upon hydrolysis, dietary urea can have an appreciable alkalizing effect on ruminal pH during the first hour post-feeding [25]. However by 4 h postprandium, the effect of urea supplementation of corn-based diets on ruminal pH has been negligible [22,26,27].

Decreasing urea level decreased (linear effect, $P < 0.01$) ruminal N-NH. The N-NH concentration has been reported to increase immediately after feeding for 2 to 3 h [28,29]. Satter and Roffler [30] observed a close relationship ($R^2 = 0.92$) between the level of dietary CP and ruminal N-NH concentration at given dietary TDN. Likewise, in the present study dietary CP explained 88% of the variation ruminal N-NH concentration. Blood urea nitrogen (BUN) concentration 4 h postprandium also decreased (linear effect, $P < 0.01$) with decreasing urea supplementation. Blood urea nitrogen is also closely associated dietary CP and ruminal N-NH concentrations [31,32]. Consistent with Zinn et al. [22], decreasing urea level increased ruminal acetate:propionate molar ratio (linear effect, $P = 0.05$), and estimated methane production (mol/mol glucose equivalent fermented; linear effect, $P = 0.04$).

Treatment effects on growth performance of feedlot steers are shown in Table 4. Decreasing urea level did not affect DMI ($P = 0.32$), but decreased ADG (linear effect, $P < 0.01$), gain efficiency (linear effect, $P < 0.01$), and dietary NE (linear effect, $P = 0.03$). Few research has evaluated the influence of marked RDP restriction on growth-performance and dietary NE in feedlot cattle fed steam-flaked corn-based finishing diets. As with the present study, Zinn et al. [22] observed a linear increase in urea resulted in a linear increase in ADG, gain efficiency and dietary NE linearly increased. The UFP of the basal

Table 3 Treatment effects on ruminal pH, VFA molar proportions and BUN

| Item | Urea fermentation potential | | | P – value | | |
	0	0.6	1.2	Linear	Quadratic	SEM
Ruminal pH	5.75	5.86	5.84	0.51	0.59	0.10
Ruminal N-NH (mg/dL)	5.37	4.69	3.89	0.05	0.91	0.56
Total VFA (mM)	95.9	105	94.2	0.83	0.21	5.4
Ruminal VFA (mol/100 mol)						
Acetate	46.8	49.2	57.5	0.08	0.54	4.6
Propionate	36.1	30.1	20.3	0.04	0.74	5.7
Isobutyrate	1.19	1.17	0.89	0.30	0.60	0.23
Butyrate	12.0	15.1	17.7	0.17	0.93	3.1
Isovalerate	1.53	1.77	0.83	0.19	0.20	0.41
Valerate	2.36	2.63	2.83	0.42	0.95	0.48
Acetate:propionate	1.34	1.85	2.94	0.05	0.63	0.59
Methane[1]	0.35	0.44	0.60	0.04	0.70	0.09
BUN (mg/dL)	4.43	2.80	1.45	<0.01	0.53	0.25

[1]Methane production (mol/mol of glucose equivalent fermented) was estimated based on the theoretical fermentation balance for observed molar distribution of VFA [16].

Table 4 Treatment effects on growth performance and carcass weight of feedlot steers

Item	Urea fermentation potential			P-value		SEM
	0	0.6	1.2	Linear	Quadratic	
Days on test	40	40	40			
Pen replicates	5	5	5			
Live weight (kg)[1]						
Initial	469	465	470	0.83	0.28	3.22
Final	510	502	501	0.17	0.54	4.40
DMI (kg/d)	7.32	6.99	6.94	0.32	0.67	0.26
ADG (kg/d)	1.04	0.93	0.78	<0.01	0.71	0.06
G:F	0.142	0.134	0.112	<0.01	0.32	0.005
Diet NE (Mcal/kg)						
Maintenance	2.37	2.33	2.18	0.03	0.37	0.21
Gain	1.67	1.64	1.50	0.03	0.37	0.21
Observed/expected NE						
Maintenance	1.07	1.05	0.98	0.03	0.37	0.02
Gain	1.09	1.07	0.98	0.03	0.37	0.03
HCW (kg)	336	331	330	0.16	0.67	3.02
Dressing (%)	65.9	66.0	65.8	0.80	0.67	0.23

[1]Initial and final weights were reduced 4% to adjust for digestive tract fill.

unsupplemented diet in that trial was 1.36%, indicating that cattle performance may be enhanced when level of urea supplementation exceeded that necessary for maximal ruminal microbial protein synthesis. As with steam-flaked corn, urea supplementation of dry rolled corn-based finishing diets to meet the UFP also enhanced ADG and gain efficiency [26,33].

The decrease in dietary NE due to restriction of rumen degradable intake protein observed in the growth performance trial (Table 4) is consistent with the decrease in dietary DE observed in the metabolism trial (Table 2). However, why cattle didn't simply compensate for this difference in NE by increasing energy intake to maintain their growth potential is puzzling. A comparison of requirements and estimated supply of metabolizable protein and the amino acids methionine and lysine for the various dietary treatments is given in Table 5. As per NRC [17], metabolizable protein supply was estimated as 80% of undegraded intake crude protein plus microbial crude protein entering small intestine in Trial 1 (Table 2), adjusted for level of intake of steers in Trial 2 (Table 4). Metabolizable amino acid supply was based on diet composition (Table 1) and corresponding tabular amino acid composition of RUP for individual feed ingredients and average amino acid composition of ruminal bacteria [17]. Metabolizable protein and amino acid requirements were based on average body weight and daily weight gain (Trial 2; NRC, [17], Level 1). As expected, estimated metabolizable protein and amino acid supply decreased with increasing UFP. Across treatments, estimated

metabolizable protein supply exceeded requirements by an average of 11%. Nevertheless, metabolizable protein supply for UFP-0.6 and UFP-1.2 were less (2 and 8%, respectively) than the estimated requirement to achieve daily weight gain observed with UFP-0 treatment. Particularly notable is the very close association between metabolizable methionine and lysine and requirements versus supply, indicative that daily weight gain may have been closely

Table 5 Treatment effects on metabolizable protein and amino acid supply[1] versus requirements[2]

Item	Urea fermentation potential		
	0	0.6	1.2
Metabolizable protein, g/d			
Supply	688	600	565
Requirement	613	574	493
Metabolizable methionine, g/d			
Supply	12.4	10.6	9.9
Requirement	12.3	11.5	9.9
Metabolizable lysine, g/d			
Supply	39.2	32.9	30.1
Requirement	39.3	36.7	31.5

[1]Metabolizable protein supply estimated as 80% undegraded intake crude protein and microbial crude protein entering small intestine (Trial 1), adjusted for level of intake. Metabolizable amino acid supply based on diet composition and corresponding tabular amino acid composition of undegradable intake protein for individual feed ingredients and average amino acid composition of ruminal bacteria (NRC, [17]).
[2]Metabolizable protein and amino acid requirements based on average body weight and daily weight gain (Trial 2; NRC, [17], Level 1).

mediated by supply of these two amino acids. As corn (the major contributor of protein to the basal diet) is a particularly poor source of lysine, and methionine, the diminution of microbial protein synthesis brought about by restriction in RDP, was sufficient to restrict growth.

Conclusion

It is concluded that in addition to effects on net protein flow to the small intestine, depriving cattle of otherwise RDP during the late finishing phase may negatively impact site and extent of OM digestion, depressing ADG, gain efficiency, and dietary NE.

Competing interests
The authors declare that they have no competing interests.

Authors' contribution
DM: PhD student, carried out growth performance and digestion trials, participated on laboratory analyses and manuscript preparation. JFC: Monitoring the growth performance trial. VGV: Carried out the digestion trial, participated on surgery and welfare of cattle. MM: Carried out the digestion trial, participated on samples procedures and manuscript preparation. APJ: Assisted in manuscript preparation. JS: Assisted in manuscript preparation. NT: Carried out the growth performance trial, participated on carcass evaluation and assisted in manuscript preparation. RZ: Experimental design, data analysis, and manuscript preparation. All authors read and approved the final manuscript.

Author details
[1]Instituto de Investigaciones en Ciencias Veterinarias, UABC, Mexicali, Baja California 21100, México. [2]Facultad de Medicina Veterinaria y Zootecnia, UAT, Cd. Victoria, Tamaulipas 87000, México. [3]Department of Animal Science, University of California, Davis E. Holton Rd, El Centro, CA 92242, USA.

References

1. Vasconcelos JT, Cole NA, McBride KW, Gueye A, Galyean ML, Richardson CR, Greene LW: Effects of dietary crude protein and supplemental urea levels on nitrogen and phosphorus utilization by feedlot cattle. *J Anim Sci* 2009, 87:1174–1183.
2. Vasconcelos JT, Galyean ML: Nutritional recommendations of feedlot consulting nutritionists: The 2007 Texas Tech University survey. *J Anim Sci* 2007, 85:2772–2781.
3. Preston RL: Empirical value of the crude protein systems for feedlot cattle. In *Protein Requirements for Cattle: Symposium*. Edited by Owens FN. Stillwater, OK: Oklahoma Experimental Station MP-109, Oklahoma State University; 1982:201–217.
4. Hristov AN, Hanigan M, Cole A, Todd R, McAllister TA, Ndegwaand PM, Rotz A: Review: Ammonia emissions from dairy farms and beef feedlots. *Can J Anim Sci* 2011, 91:1–35.
5. Zinn RA, Plascencia A: Interaction of whole cottonseed and supplemental fat on digestive function in cattle. *J Anim Sci* 1993, 71:11–17.
6. Burroughs W, Nelson DK, Mertens DR: Protein physiology and its application in the lactating cow: The metabolizable protein feeding standard. *J Anim Sci* 1975, 41:933–944.
7. Fawcett JK, Scott JE: A rapid and precise method for the determination of urea. *J Clin Pathol* 1960, 13:156–159.
8. Bergen WG, Purser DB, Cline JH: Effect of ration on the nutritive quality of rumen microbial protein. *J Anim Sci* 1968, 27:1497–1501.
9. Association Official Analytical Chemists (AOAC): *Official methods of analysis*. 17th edition. Gaithersburg, MD: Association Official Analytical Chemists; 2000.
10. Van Soest PJ, Robertson JB, Lewis BA: Methods for dietary fiber, neutral detergent fiber, and nonstarch polysaccharides in relation to animal nutrition. *J Dairy Sci* 1991, 74:3583–3597.
11. Zinn RA, Owens FN: A rapid procedure for purine measurement and its use for estimating net ruminal protein synthesis. *Can J Anim Sci* 1986, 66:157–166.
12. Zinn RA: Influence of steaming time on site digestion of flaked corn in steers. *J Anim Sci* 1990, 68:776–781.
13. Zinn RA: Comparative feeding value of supplemental fat in finishing diets for feedlot steers supplemented with and without monensin. *J Anim Sci* 1988, 66:213–227.
14. Hill FN, Anderson DL: Comparison of metabolizable energy and productive determinations with growing chicks. *J Nutr* 1958, 64:587–603.
15. Ørskov ER, MacLeod NA, Kyle DJ: Flow of nitrogen from the rumen and abomasum in cattle and sheep given protein-free nutrients by intragastric infusion. *Br J Nutr* 1986, 56:241–248.
16. Wolin MJ: A theorical rumen fermentation balance. *J Dairy Sci* 1960, 43:1452–1459.
17. National Research Council (NRC): *Nutrient Requirements of Beef Cattle*. 7th edition. Washington, DC: National Academy of Press; 1996.
18. National Research Council (NRC): *Nutrient Requirements of Beef Cattle*. 6th edition. Washington, DC: National Academy Press; 1984.
19. Zinn RA, Shen Y: An evaluation of ruminally degradable intake protein and metabolizable amino acid requirements of feedlot calves. *J Anim Sci* 1998, 76:1280–1289.
20. United States Department of Agriculture (USDA): *United States Standards for Grading of Carcass Beef*. Washington, DC: Agricultural Marketing Service, United States Department of Agriculture; 1997.
21. Statistical Analysis System (SAS): *SAS/STAT User's Guide: Version 9.1*. Cary, North Caroline: SAS Institute Inc; 2004.
22. Zinn RA, Borquez JL, Plascencia A: Influence of levels of supplemental urea on characteristics of digestion and growth performance of feedlot steers fed a fat-supplemented high-energy diets. *Prof Anim Sci* 1994, 10:5–10.
23. Muscher AS, Schroder B, Breves G, Huber K: Dietary nitrogen reduction enhances urea transport across goat rumen epithelium. *J Anim Sci* 2010, 88:3390–3398.
24. Holter JA, Reid JT: Relationship between the concentrations of crude protein and apparently digestible protein in forages. *J Anim Sci* 1959, 18:1339–1349.
25. Zinn RA, Barrajas R, Montaño M, Ware RA: Influence of dietary urea level on digestive function and growth performance of cattle fed steam-flaked barley- based finishing diets. *J Anim Sci* 2003, 81:2383–2389.
26. Milton CT, Brandt RT Jr, Titgemeyer EC: Urea in dry rolled corn diets: Finishing steers performance, nutrient digestion and microbial protein production. *J Anim Sci* 1997, 75:1415–1424.
27. Brake DW, Titgemeyer EC, Jones ML, Anderson DE: Effect of nitrogen supplementation on urea kinetics and microbial use of recycled urea in steers consuming corn-based diets. *J Anim Sci* 2010, 88:2729–2740.
28. Chumpawadee S, Sommart K, Vongpralub T, Pattarajinda V: Effects of synchronizing the rate of dietary energy and nitrogen release on ruminal fermentation, microbial protein synthesis, blood urea nitrogen and nutrient digestibility in beef cattle. *Asian-Australasian J Anim Sci* 2006, 19:181–188.
29. Seo JK, Yang JY, Kim HJ, Upadhaya SD, Cho WM, Ha JK: Effects of synchronization of carbohydrate and protein supply on ruminal fermentation, nitrogen metabolism and microbial protein synthesis in Holstein steers. *Asian-Aust J Anim Sci* 2010, 23:1455–1461.
30. Satter LD, Roffler RE: Nitrogen requirements and utilization in dairy cattle. *J Dairy Sci* 1975, 58:1219–1237.
31. Hammond AC: Effect of dietary protein level, ruminal protein solubility and time after feeding on plasma urea nitrogen and the relationship of plasma urea nitrogen to other ruminal and plasma parameters. *J Anim Sci* 1983, 57(1):435.
32. Hennessy DW, Nolan JV: Nitrogen kinetics in cattle fed a mature subtropical grass hay with and without protein meal supplementation. *Aust J Agric Res* 1988, 39:1135–1150.
33. Tedeschi LO, Baker MJ, Ketchen DJ, Fox DG: Performance of growing and finishing cattle supplemented with a slow-release urea product and urea. *Can J Anim Sci* 2002, 82:567–573.

8

Effects of varying nursery phase-feeding programs on growth performance of pigs during the nursery and subsequent grow-finish phases

Chai Hyun Lee[1], Dae-Yun Jung[1], Man Jong Park[2] and C Young Lee[1,2]*

Abstract

The present study investigated the effects of varying durations of nursery diets differing in percentages of milk products on growth performance of pigs during the nursery phase (NP) and subsequent grow-finish phase (GFP) to find the feasibility of reducing the use of nursery diets containing costly milk products. A total of 204 21-d-old weanling female and castrated male pigs were subjected to one of three nursery phase feeding programs differing in durations on the NP 1 and 2 and GFP diets containing 20%, 7%, and 0% lacrosse and 35%, 8%, and 0% dried whey, respectively, in 6 pens (experimental units) for 33 d: HIGH (NP 1, 2 and 3 diets for 7, 14, and 12 d), MEDIUM (NP 2 and 3 for 14 and 19 d), and LOW (NP 2 and 3 and GFP 1 for 7, 14, and 12 d). Subsequently, 84 randomly selected pigs [14 pigs (replicates)/pen] were fed the GFP 1, 2 and 3 diets during d 54-96, 96-135, and 135-182 of age, respectively. The final body weight (BW) and average daily gain (ADG) of nursery pigs did not differ among the HIGH, MEDIUM, and LOW groups (14.8, 13.3, and 13.7 kg in BW and 273, 225, and 237 g in ADG, respectively). The average daily feed intake during the nursery phase was greater (p <0.01) in the HIGH group than in the MEDIUM and LOW groups, whereas the gain:feed ratio did not differ across the treatments. The BW on d 182 and ADG during d 54-182 were greater in the HIGH and MEDIUM groups vs. the LOW group (110.0, 107.6, and 99.6 kg in BW, respectively; p <0.01). The backfat thickness and carcass grade at slaughter on d 183 did not differ across the treatments. In conclusion, the MEDIUM program may be inferior to the commonly used HIGH program in supporting nursery pig growth. Nevertheless, the former appears to be more efficient than the latter in production cost per market pig whereas the LOW program is thought to be inefficient because of its negative effect on post-nursery pig growth.

Keywords: Pig, Nursery, Feeding program, Growth, Grow-finish phase

Background

Weaning is a most stressful transition not only socially but also nutritionally in the life of commercial pigs [1,2]. Therefore, weaner pigs are usually provided with palatable diets containing highly digestible milk products such as lactose and dried whey to alleviate the weaning stress [3-5]. However, the milk products are so costly that some researchers compared the relative effects of 'complex' nursery diets containing high percentages of the milk products vs. 'simple' ones containing no or low percentages of the milk products on growth performance of pigs to examine the feasibility of reducing the ratio of the milk products in the nursery diets [6-10]. The rate of weight gain of nursery pigs was greater on the complex than on simple diets in all of these studies. Moreover, the difference in body weight between the two dietary groups of pigs after the nursery phase sustained or narrowed due to varying compensatory growth during the subsequent grow-finish phase. The carcass characteristics of the pigs including the backfat thickness and loin eye area at the market weight were not influenced by the nursery feeding program [8,10]. However, the relative effects of the alternative nursery feeding programs on growth performance as well as the pig production cost under the experimental conditions were rather inconclusive. Further, little is known as to how the complexity of commercial nursery pig diets, which could be mimicked by manipulating the

* Correspondence: cylee@gntech.ac.kr
[1]Department of Animal Resources Technology, Gyeongnam National University of Science and Technology, Jinju 660-758, South Korea
[2]The Regional Animal Industry Center, Gyeongnam National University of Science and Technology, Jinju 660-758, South Korea

feeding duration on each of the nursery phase diets, influences the efficiency of pig production under commercial conditions. The present study was therefore conducted to investigate the effects of varying durations on the nursery diets on growth performance of pigs during the nursery and subsequent grow-finish phases under commercial production conditions and thereby to find the feasibility of reducing the use of the nursery diets containing the costly milk products.

Methods

Nursery pigs

The experimental protocol of the present study conformed to the guidelines of the Institutional Animal Care and Use Committee (IACUC) at Gyeongnam National University of Science and Technology. The animals used in the present study were handled humanely and did not receive any prolonged constraint throughout the experiment. A total of 204 (Yorkshire × Landrace) × Duroc piglets consisting of equal numbers of females and castrated males, which had been provided with 200 g of the nursery phase 2 diet/litter (Table 1) as creep feed during the last 10 d of lactation, were randomly allocated to 6 pens at weaning at 21 d of age, with three pens per sex and 34 animals per pen. One pen of females and another pen of castrated males were placed on one of three phase-feeding programs: HIGH, MEDIUM, and LOW which had varying durations on the nursery phases 1, 2, and 3 diets containing 20%, 7%, and 0% lactose and 35%, 8%, and 0% dried whey, respectively, and a grow-finish phase 1 diet (Tables 1 and 2). The lysine concentrations of the nursery diets were slightly greater than those formerly recommended by NRC [11], but were slightly less than the current NRC [12] recommendations.

The HIGH group of animals received the nursery phase 1 diet to d 6 of the experiment, at which time

Table 1 Declared minimum nutritional values of the commercial nursery and grow-finish diets used in the present study (as-fed basis)

Item	Nursery			Grow-finish			
	Phase 1[1]	Phase 2[2]	Phase 3[3]	High	Medium plane[5]		
				Phase 1[4]	Phase 1	Phase 2	Phase 3
CP[6], %	21.0	21.0	20.5	18.2	17.5	15.5	13.4
Lysine, %	1.40	1.40	1.25	1.10	1.0	0.80	0.70
CF[6], %	7.5	7.5	7.0	6.0	4.0	3.0	2.5
DE[6], Mcal/kg	3.75	3.60	3.60	3.45	3.35	3.30	3.25

[1-3]Contained 20%, 7%, and 0% lactose and 35%, 8%, and 0% dried whey, respectively.
[1-4]Used for nursery pigs between d 21 and 54 of age.
[5]Provided to grow-finish pigs from d 54 through d 182 of age.
[6]CP, crude protein; CF, crude fat; DE, digestible energy.

Table 2 Nursery phase (P)-feeding programs (treatments): experimental design

	7 days		7 days	7 days	12 days	
High[1]	Nursery P 1	T[2]		Nursery P 2	T[3]	Nursery P 3
Medium[1]		Nursery P 2		T[3]		Nursery P 3
Low[1]	Nursery P 2	T[2]		Nursery P 3	T[3]	High-plane grow-finish P 1

[1]Indicated group of nursery pigs received the indicated diets between d 21 and 54 of age. All pigs received the medium-plane grow-finish phases 1, 2, and 3 diets during days 54-96, 96-135, and 135-182 of age, respectively. See Table 1 for nutritional values of the diets.
[2,3]Transition with 20 and 30 kg of mixtures of two diets of the adjoining phases, respectively.

remaining diet was removed from the feeder and three 6.7 kg portions of 2:1, 1:1 and 2:1 mixtures of the phases 1 and 2 diets, respectively, were layered sequentially on the feeder. Thereafter, the animals were provided with the nursery phase 2 diet to d 20 when a total of 30 kg of mixtures of the phases 2 and 3 diets were provided after the weigh-back as had been done at the end of d 6, followed by provision of the phase 3 diet to d 33. As such, the HIGH group animals, which received the transitive mixed diet twice approximately for 2 d each beginning from the ends of d 6 and 20, respectively, were actually on the nursery phases 1, 2, and 3 diets approximately for 7, 14, and 12 d, respectively. The MEDIUM group was provided with the nursery phases 2 and 3 diets for 14 and 19 d, respectively, whereas the LOW group was fed the nursery phases 2 and 3 and grow-finish phase 1 diets for 7, 14, and 12 d, respectively. The transitive mixed diets in the MEDIUM and LOW groups were provided at the ends of d 13 and d 6 and 20, respectively, in the same way as was used in the HIGH group. All animals were weighed by the unit of pen at the beginning and end of the nursery phase. The feed intake in each pen was measured by the weigh-back method at the ends of d 6, 13, 20 and 33.

Growing-finishing pigs

Upon finishing the nursery feeding trial, 84 randomly selected pigs then aged 54 d (14 animals/pen) were numbered using the ear tag, weighed individually, and moved to grower pens with non-tagged pigs altogether in a pen-to-pen manner. These animals were fed the grow-finish phase 1 diet through d 96 and moved again to finisher pens where they received the grow-finish phases 2 and 3 diets successively to d 135 and 182, respectively, as described previously [13,14]. The lysine concentrations of the grower-finisher diets, as in the nursery diets, were in between the previous and current NRC [11,12] recommendations for corresponding stages of pigs. The grower-finisher diets were provided through a mechanical feeder line and hence the intakes of the diets were not measured. The 84 tagged animals were weighed repeatedly on d 96, 135, and 182.

A total of 60 animals (10 animals/pen), estimated carcass weights of which were within or close to the range of those eligible for the grades 1[+] or 1 by the MAFRA [15] standards, were selected upon termination of the feeding trial, transported to a local abattoir on the following morning, given approximately 3 h of lairage, and slaughtered. The carcass weight, backfat thickness, and other carcass characteristics including the structural integrity and color of the muscle and fat, which are used as criteria for carcass grading [15], were measured or evaluated by the grading official.

Statistical analysis

All data were analyzed using SAS (SAS Inst. Inc., Cary, NC, USA). Data obtained from nursery pigs were analyzed as a completely randomized design (CRD) using the MIXED procedure. The statistical model included the nursery feeding program and sex as a fixed effect and a random effect, respectively, with the pen regarded as the experimental unit. Data from growing-finishing animals were analyzed as a CRD with a 2 (sex) × 3 (nursery feeding program) arrangement of treatments using the GLM procedure, with the individual pig taken as the experimental unit. The initial body weights at the beginning of the nursery and grow-finish phases, which are known to be correlated with subsequent growth rates in pigs [13,16,17], were included in the statistical models for the growth performance variables of the nursery and growing-finishing pigs, respectively. Means were separated using the PDIFF option when a main effect or an interaction of the main effects was significant (p <0.05).

Results

Nursery growth performance

The initial and final body weights (BW) as well as the average daily gain (ADG) of the nursery pigs during the 33-day (d) nursery phase did not differ across the dietary treatments (Table 3). The average daily feed intake (ADFI) also did not differ among the three dietary groups during d 1-6, 1-13, or 13-33. However, the ADFI during the entire nursery phase was greater (p <0.01) in the HIGH group than in the MEDIUM and LOW groups by 20% and 17%, respectively, whereas the gain: feed ratio was not influenced by the dietary treatment.

Post-nursery growth

The initial BW on d 54 of age of the animals used for the continuing grow-finish feeding trial was greater (p <0.01) in the HIGH group than in the MEDIUM and LOW groups (15.5, 13.3, and 13.4 kg, respectively) as well as in barrows than in gilts (Table 4). The BW of the animals was not different among the treatments or between the

Table 3 Effects of varying phase-feeding programs on growth performance of nursery pigs

Item	High[1]	Medium[2]	Low[3]	SEM[4]	P-value
Initial BW[4], kg	5.98	5.60	5.93	0.13	0.27
Final BW[4,5], kg	14.83 ± 0.52	13.25 ± 0.68	13.65 ± 0.46		0.42
ADG[4,5], g	273 ± 16	225 ± 20	237 ± 14		0.42
ADFI[4], g					
D[4] 1-6	266	233	184	15	0.12
D 1-13	307	214	257	13	0.07
D 13-33	503	447	433	15	0.14
Overall	426[a]	355[b]	364[b]	4	0.01
Gain: feed	0.62	0.67	0.64	0.04	0.64

[1]Provided with the nursery phases 1, 2 and 3 diets for 7, 14 and 12 days, respectively.
[2]Provided with the nursery phases 2 and 3 diets for 14 and 19 days, respectively.
[3]Provided with the nursery phases 2 and 3 diets and the high-plane grow-finish phase 1 diet for 7, 14 and 12 days, respectively.
[1-3]See Table 1 for nutritional values of the diets used in each feeding program. Data are means or least squares means ± SEM of two pens consisting of 34 females and 34 castrated males, respectively.
[4]SEM, standard error of mean; BW, body weight; ADG, average daily gain; ADFI, average daily feed intake; D, day.
[5]Initial BW was included in the model as a covariate.
[a,b]Means with no common superscripts differ (p <0.05).

two sexes at the end of the early grow-finish stage (d 96), but thereafter it was greater in the HIGH and MEDIUM groups than in the LOW group (70.3, 67.5, and 63.1 kg on d 135 and 110.0, 107.6, and 99.6 kg on d 182 for the HIGH, MEDIUM, and LOW groups, respectively; p <0.01) and also in barrows vs. gilts (p <0.05). Apart from the statistical significance, the numerical difference in BW between the HIGH and MEDIUM groups at the beginning of the grow-finish phase thus persisted through the end of the phase. In contrast, the difference between these two groups and the LOW group widened between d 96 and 182, which was largely attributable to remarkably lower BW of gilts in the LOW group compared with those in the other groups. The ADG during d 54-96, as in d-96 BW, did not differ among the treatments or between the sexes. However, the ADG during d 96-135 and 135-182 as well as overall ADG during d 54-182 were greater in the HIGH and MEDIUM groups vs. the LOW group (752, 733, and 670 g in overall ADG for the HIGH, MEDIUM, and LOW groups, respectively), but there was no difference between the former two groups. Moreover, the ADG during d 96-135 as well as overall ADG was greater in barrows than in gilts. Finally, the dressing percentage, backfat thickness, and quantitated carcass grade of the pigs slaughtered after termination of the feeding trial were not different among the treatments (Table 5).

Discussion

Weanling pigs, in general, are provided with a palatable milk products-based phase 1 nursery diet for a short period

Table 4 Growth performance of the growing-finishing pigs which received the varying nursery phase-feeding programs

Item	High[1]		Medium[2]		Low[3]		P-value		
	B[4]	G[4]	B	G	B	G	FP[4]	S[4]	FP × S
BW[4,5], kg									
D[4] 54	16.3 ± 0.5	14.7 ± 0.5	14.4 ± 0.5	12.1 ± 0.5	13.1 ± 0.5	13.6 ± 0.5	<0.01	<0.01	0.03
D 96	38.8 ± 1.0	36.7 ± 0.8	35.1 ± 0.9	36.9 ± 0.9	37.3 ± 0.8	34.0 ± 0.8	0.07	0.10	0.01
D 135	73.1 ± 1.6	67.5 ± 1.5	65.7 ± 1.5	69.3 ± 1.5	67.9 ± 1.3	58.3 ± 1.4	<0.01	<0.01	<0.01
D 182	111.7 ± 3.1	108.4 ± 2.6	108.4 ± 2.7	106.8 ± 2.7	105.4 ± 2.5	93.7 ± 2.5	<0.01	0.01	0.13
ADG[4,5], g									
D 54-96	592 ± 24	542 ± 20	503 ± 22	546 ± 22	557 ± 20	477 ± 19	0.07	0.10	0.01
D 96-135	878 ± 29	787 ± 25	787 ± 25	835 ± 26	785 ± 23	634 ± 23	<0.01	<0.01	<0.01
D 135-182	843 ± 43	876 ± 37	909 ± 37	808 ± 37	802 ± 34	756 ± 34	0.04	0.21	0.21
Overall	765 ± 24	739 ± 20	739 ± 21	727 ± 21	716 ± 20	624 ± 20	<0.01	0.01	0.13

[1]Provided with the nursery phases 1, 2 and 3 diets for 7, 14 and 12 days, respectively.
[2]Provided with the nursery phases 2 and 3 diets for 14 and 19 days, respectively.
[3]Provided with the nursery phases 2 and 3 diets and the high-plane grow-finish phase 1 diet for 7, 14 and 12 days, respectively.
[1-3]See Table 1 for nutritional values of the diets. Data are least squares means ± standard errors of means of 14 animals in each sex.
[4]B, barrow; G, gilt; FP, feeding program during the nursery phase; S, sex; BW, body weight; D, day; ADG, average daily gain.
[5]BW on d 54 was included in the model as a covariate in the analysis of subsequent BW and ADG.

to maximize their intake and accordingly to minimize the post-weaning growth check resulting primarily from insufficient feed intake [1,2,5]. Most nursery feeding programs of commercial feed manufacturers in Korea, like the HIGH program of the present study, recommend the use of the phase 1 diet for 1 week and subsequently phases 2 and 3 diets containing lower percentages of the milk products for 2 weeks each. If the nursery pig's intake of the milk products is restricted by providing them with diets containing no or lower percentages of milk products, their weight gain is also reduced. The reduced weight gain of nursery pigs due to nutritional insufficiency is usually made up for during subsequent grow-finish phase to varying extents depending on the animals' genetic background and grow-finish feeding program in addition to the time, duration, and severity of the nutritional insufficiency [7,18].

The nursery pigs which were fed the nursery phases 2 and 3 diets without (MEDIUM) or with the grow-finish phase 1 diet (LOW) ate less than those fed the nursery phase 1 diet as well as the phases 2 and 3 (HIGH). This resulted in a lesser weight gain in the former than in the latter although the nursery ADG did not differ statistically among the three groups due to the limited number of replicates (pens). The magnitude of the BW difference between the HIGH and MEDIUM groups at the beginning of the grow-finish phase persisted to the end of this phase, although final BW at termination of the feeding trial did not differ statistically between the two groups. On the other hand, according to unpublished data of Dr. B. C. Chae (Kangwon National University, Korea; personal communication), pigs weaned at 4 weeks of age gained less weight on a medium vs. high nursery feeding program

Table 5 Carcass characteristics of the finishing pigs which received the varying nursery phase-feeding programs

Item	High[1]		Medium[2]		Low[3]		SEM[4]	P-value		
	B[4]	G[4]	B	G	B	G		FP[4]	S[4]	FP × S
Live wt[4], kg	111.5	108.4	109.2	107.8	107.1	94.8	2.8	<0.01	0.02	0.12
Carcass wt, kg	83.3	81.4	82.4	79.1	79.9	71.5	2.1	<0.01	0.01	0.29
Dressing, %	74.8	75.1	75.7	73.4	74.7	75.4	1.0	0.89	0.63	0.27
BFT[4,5], mm	21.4	20.9	23.8	21.9	23.2	21.9	1.3	0.38	0.22	0.86
Carcass grade[6]	2.14 ± 0.25	2.80 ± 0.30	2.00 ± 0.27	1.75 ± 0.33	2.40 ± 0.30	2.00 ± 0.66		0.14	0.99	0.23

[1]Provided with the nursery phases 1, 2 and 3 diets for 7, 14 and 12 days, respectively.
[2]Provided with the nursery phases 2 and 3 diets for 14 and 19 days, respectively.
[3]Provided with the nursery phases 2 and 3 diets and the high-plane grow-finish phase 1 diet for 7, 14 and 12 days, respectively.
[1-3]See Table 1 for nutritional values of the diets used in each feeding program. Data are means of 10 animals in each sex, except for the carcass grade in which data are least squares means ± standard errors of means.
[4]B, barrow; G, gilt; SEM, standard error of mean; FP, feeding program during the nursery phase, S, sex; wt, weight; BFT, backfat thickness.
[5]Average of the measurements between the 11 and 12th ribs and at the last rib adjusted for a 110-kg live weight.
[6]Assigned 3, 2 and 1 of arbitrary point units to carcass grades 1[+], 1 and 2, respectively. Grade 2 carcasses which did not reach the minimum weight (79 kg) eligible for the grade 1 [15] were excluded from the analysis

similar to the present MEDIUM vs. HIGH during the nursery phase, but the pigs previously on the former caught up the growth of the others previously on the latter during the subsequent grow-finish phase. Collectively, these results suggest that whether or not the pigs after the medium vs. high nursery feeding program exhibit a compensatory growth during the grow-finish phase may also depend on the age when the animals were weaned.

The BW differential between the HIGH and LOW groups at the outset of the grow-finish phase widened with the advancing stage of the phase, which was more pronounced in gilts than in barrows. There's no convincing explanation about this difference in post-nursery growth performance between the LOW and MEDIUM groups, because neither the ADG nor ADFI during the nursery phase was different between these two groups. It is thus speculated that the LOW program somehow impaired the growth potential of the animals probably because the planes of nutrition of the nursery phases 3 and grow-finish phase 1 diets used in this program were substantially lower than those recommended by NRC [11,12] for nursery pigs during the stages when the diets were provided.

The MEDIUM program cost approximately 6,000 won (approximately 6 US dollar) less for nursery feed per head than the HIGH grogram. On the other hand, the MEDIUM group required approximately 4.5 kg of additional grow-finish diet for 1.5 kg of additional weight gain and also 3 more days on feed costing 2,250 and 750 won per head, respectively, than the HIGH group to reach a 110-kg market weight, assuming that the feed cost per kilogram weight gain during the entire growth-finish phase was equal in both groups. The total production cost per 110-kg market pig was thus assessed to be saved approximately 3,000 won by substitution of the MEDIUM nursery feeding program for the commonly used HIGH program. Relative production cost per market pig for the LOW group, however, could not be extrapolated, because the growth performance of this group during the grow-finish period was much lower than that of the other groups.

Conclusions

Results of the present study indicate that substitution of the MEDIUM nursery feeding program using only the nursery phases 2 and 3 diets for the common HIGH program using the phase 1 diet in addition to the phases 2 and 3 causes a reduced weight gain of nursery pigs and that the reduced BW due to the substitution persists throughout the grow-finish phase. Nevertheless, the former seems to be more efficient than the latter in terms of production cost per market pig whereas the LOW program using the nursery phases 2 and 3 and grow-finish phase 1 diets is thought not to be worthy of further consideration because of its severely negative effect on post-nursery pig growth. Obviously, however, more studies are necessary to confirm the present assessment.

Competing interests

The authors declare that they have no competing interests.

Authors' contributions

CHL, MJP, and CYL designed the experiment. CHL, D-YJ, and CYL have done the feeding trial and analyzed the results. All authors read and approved the final manuscript.

Acknowledgements

This work was supported by Gyeongnam National University of Science and Technology Grant 2014.

References

1. Weary DM, Jasper JJ, Hotzel MJ: Understanding weaning stress. *Appl Anim Behav Sci* 2008, **110**:24–41.
2. Kim JC, Hansen CF, Mullan BP, Pluske JR: Nutrition and pathology of weaner pigs: nutritional strategies to support barrier function in the gastrointestinal tract. *Anim Feed Sci Technol* 2012, **173**:3–16.
3. Maxwell CV Jr, Carter SD: Feeding the weaned pig. In *Swine Nutrition*. 2nd edition. Edited by Lewis AJ, Southern LL. New York: CRC Press; 2001:691–715.
4. DeRouchey JM, Goodband RD, Tokach MD, Nelssen JL, Dritz SS: Nursery swine nutrient recommendations and feeding management. In *National Swine Nutrition Guide*, Pork Center of Excellence (USPCE). Ames, IA, USA; U.S: Iowa State University; 2010:65–79.
5. Kim SW, Hansen JA: Diet formulation and feeding programs. In *Sustainable Swine Nutrition*. Edited by Chiba LI. Ames, IA, USA: Wiley-Blackwell; 2013:217–227.
6. Tokach MD, Pettigrew JE, Johnston LJ, Overland M, Rust JW, Cornelius SG: Effect of adding fat(or) milk products to the weanling pig diet on performance in the nursery and subsequent growth-finish stages. *J Anim Sci* 1995, **73**:3358–3368.
7. Whang KY, McKeith FK, Kim SW, Easter RA: Effect of starter feeding program on growth performance and gains of body components from weaning to market weight in swine. *J Anim Sci* 2000, **78**:2885–2895.
8. Wolter BF, Ellis M, Corrigan BP, DeDecker JM, Curtis SE, Parr EN, Webel DM: Impact of postweaning growth rate as affected by diet complexity and space allocation on subsequent growth performance of pigs in a wean-to-finish production system. *J Anim Sci* 2003, **81**:353–359.
9. Chae BJ: Strategies for saving the feed cost by improving the feed efficiency in pig production (in Korean). In *Proceedings of "Symposium on Strategies for Reducing Production Cost of Animal Production", organized by Korean Society of Animal Nutrition & Feedstuffs during 2013 Annual Congress of Korean Society of Animal Sciences and Technology*. 2013:9–20.
10. Skinner LD, Levesque CL, Wey D, Rudar M, Zhu J, Hooda S, de Lange CFM: Impact of nursery feeding program on subsequent growth performance, carcass quality, meat quality, and physical and chemical body composition of growing-finishing pigs. *J Anim Sci* 2014, **92**:1044–1054.
11. NRC: *Nutrient Requirements of Swine*. 10th edition. Washington, D.C: National Academy Press; 1998.
12. NRC: *Nutrient Requirements of Swine*. 11th edition. Washington, D.C: National Academy Press; 2012.
13. Ha D-M, Jang K-S, Won H-S, Ha S-H, Park M-J, Kim SW, Lee CY: Effects of creep feed and milk replacer and nursery phase-feeding programs on pre- and post-weaning growth of pigs. *J Anim Sci Technol* 2011, **53**:333–339.
14. Ha D-M, Jung D-Y, Park MJ, Park B-C, Lee CY: Effects of sires with different weight gain potentials and varying planes of nutrition on growth of growing-finishing pigs. *J Anim Sci Technol*. In press.
15. MAFRA: Grading Standards for Livestock Products (in Korean). In *Notification No. 2014-4 of the Ministry of Agriculture, Food and Rural Affairs, Republic of Korea*. 2014.

16. McConnell JC, Eargle JC, Woldorf RC: **Effects of weaning weight,
    co-mingling, group size and room temperature on pig performance.**
    J Anim Sci 1987, **65:**1201–1206.
17. Mahan DC, Lepine AJ: **Effect of pig weaning weight and associated
    nursery feeding programs on subsequent performance to 105 kilograms
    body weight.** *J Anim Sci* 1991, **69:**1370–1378.
18. Lawrence TLJ, Fowler VR, Novakofski JE: *Growth of Farm Animals.* 3rd
    edition. Wallingford, UK: CABI; 2012.

9

Perception of the HACCP system operators on livestock product manufacturers

Jung-Hyun Kim[1], Ki-Chang Nam[2], Cheorun Jo[3] and Dong-Gyun Lim[4*]

Abstract

The purpose of this study was to investigate crucial factors on HACCP system implementation in domestic livestock product plants, and to offer job satisfaction and the career prospect of HACCP system operators. The survey was carried out by selecting 150 HACCP system operators who implemented HACCP system. The respondents claimed that the most important contents in HACCP system operation were to assemble HACCP team (21.8%), and the second was to monitoring (20.0%). Documentation and recording (16.9%) and verification (11.1%) were followed. The respondents answered the major factor in sanitation management was cleaning/washing/disinfection (18.9%) and inspection (18.4%). The results showed that there were significant differences in the prospect of occupation in HACCP system operator by the gender ($p < 0.015$), age, livestock product facilities, service period, and position ($p < 0.001$). The respondents from HACCP system operator were satisfied with their job (73%) and also showed optimistic prospect of occupation (82%).

Keywords: HACCP system operator, Livestock product plants, Survey, Monitoring, Prospect of occupation

Background

The safety of food products has become a major issue of concern. Hazard analysis critical control points (HACCP) is a food safety management system [1], widely acknowledged as the best method of assuring product safety while becoming internationally recognized as a tool for controlling food borne safety hazards (Codex Alimentarius Commission, [2,3]). HACCP is a systematic approach to the identification, evaluation, and control of hazards in those steps in food manufacturing that is critical to food safety (Ropkins et al., [4]).

In several countries, including Korea, hazard analysis critical control point (HACCP) systems have been introduced with regard to product hygiene and safety (Codex Alimentarius Commission, [5]). In Korea, regulatory authorities have introduced HACCP systems on meat processing plant in 1997, slaughter house in 2000, livestock product plant in 2001, milk processing plant, meat sale and distribution in 2004, feed mill in 2005, and animal farm in 2006 [6]. HACCP implementation of the slaughterhouse in livestock product field is only mandatory in

Korea. The HACCP system is being increasingly used in many food industries under regulatory agencies. Most developed countries including USA (1998), EU (1996) and Australia (1997) are implementing HACCP system as an obligation.

It is doubtful if any company can implement HACCP without trained-HACCP team members. This is particularly true for the small-scale company with limited access to information [7]. Competency in HACCP can be effectively gained through training and this must be complemented with the appropriate knowledge of food sanitation and food microbiology. Research has observed that the employment of an experienced, technically qualified person is the single most important factor influencing the implementation of HACCP [8].

The studies on HACCP have recently focused on evaluation of sanitation management performance, benefits of HACCP implementation, and employees' knowledge and performance degree of HACCP in school foodservice sector [9-11]. In the livestock products sector, economic feasibility of HACCP at slaughterhouse [12], and comparative analysis of the prerequisite items for HACCP in livestock product plants [13] have been reported. However, there has not yet been studied about the basic information for job order of priority and career prospect of HACCP

* Correspondence: tousa0994@naver.com
[4]Department of Health Administration and Food Hygiene, Jinju Health College, Jinju 660-757, Korea
Full list of author information is available at the end of the article

system operators (HACCP and microbiological analysis operator) in livestock products sector. Therefore, the objective of this study was not only to investigate crucial factors on HACCP system implementation in domestic livestock product plants, but also to offer job satisfaction and the career prospect of HACCP system operators in livestock product industry.

Methods
Study population
This study was based on data obtained 15 livestock product manufacturer (5 large scale plants, 10 medium scale plants) located in Korea. A survey was conducted with subjects to operate HACCP in livestock product plants. Most of the respondents were HACCP team members (HACCP and microbiological analysis operators) who are in charge of quality control department of the manufacturer. The survey was carried out with 150 respondents who implemented HACCP system. The survey questionnaire was performed by some questions with multiple choices for the answers and consisted of 11 questions including 7 questions of general characteristics, 2 questions of priority of duties in HACCP system operation in plants and 2 questions of a job analysis.

Description of demographic variables
In the study, demographic variable included gender, age, educational and income level, size of livestock product facilities, service period, and position. Age reported at the time of interview was categorized into three groups: less than 29, 30-39, and 40-49. Educational level was divided into college and university. Income level per month was categorized into four groups: less than 200, 200 to 250, 250 to 300, and more than 300 Korean won. Livestock product facilities were divided into two categories: small- or medium and large-scale. The service period was classified into three groups: less than 1, 3-5, and 5-10 years. Position was categorized into four groups: employees, team manager, assistant manager, and president, director. All participants provided written and informed consent to participate in the study. Respondents were individually given adequate time to answer each query in writing.

Statistical analysis
Statistical analysis was performed using the STATA software (version12.0) for Window and Two-sided P-value <0.05 was considered statistically significant. General characteristics of the study participants and priority of duties in HACCP system operation and sanitation management are presented as frequency and percentage. The chi-square tests were performed to investigate the differences among job satisfaction and prospect of occupation in HACCP system operators by age, educational and income level, size of food product facilities, career, and position.

Results and discussion
General characteristics of the study participants are shown in Table 1. Overall, male subjects are 68% and female ones are 32%. Among age variable, age is 20-29 (46.7%), 30-39 (33.3%), and 40-49 (20%) respectively. In educational level, the respondents have acquired a college diploma (58.7%) or university degree (41.3%). Income level per month is under the 3 million won (35.3%), 2 million won (29.3%), 2.5 million won (25.3%), and over 3 million won (10.0%). Size of livestock product facilities is; small and medium-scale facilities were mostly 73.3% and large-scale ones were respectively 26.7%. Career duration is; the highest was less than 1 year (56.7%), whereas 3-5 year was the lowest (8.7%) under a year workers have been working as a current HACCP system operators in manufactures. The position is employees (53.3%), team manager (28%), assistant manager (17.3%) and president (CEO) (1.3%).

Table 2 shows the priority of duties in HACCP system operation in plants. Basically, HACCP system is a science-

Table 1 General characteristics of respondents participated (N = 150)

Characteristics		Frequency (N)	Percent (%)
Gender	Male	102	68
	Female	48	32
Age	20-29	70	46.7
	30-39	50	33.3
	40-49	30	20.0
Educational background	≥College diploma	88	58.7
	University degree	62	41.3
Income per month (million won)	<2	44	29.3
	<2.5	38	25.3
	<3	53	35.3
	≥3	15	10.0
Size of processing facilities	Large-scale	40	26.7
	Small and medium-scale	110	73.3
Service period (yrs.)	<1	85	56.7
	3-5	13	8.7
	5-10	52	34.7
Position	Employees	2	53.3
	Team manager	42	28
	Assistant manager	26	17.3
	President, CEO	80	1.3
Total		150	100

Table 2 Priority of duties in HACCP system operation in plants

HACCP system operation	Frequency (N)	Percent (%)
Assemble HACCP team	98	21.8
Monitoring	90	20.0
Documentation and recording	76	16.9
Verification	50	11.1
Establish a corrective actions	29	6.4
Establish critical limits for each CCP	28	6.2
Describe product and identify intended use	27	6.0
Construct flow diagram and on-site confirmation of it	20	4.4
Conduct a Hazard Analysis	17	3.8
CCP determination	15	3.3
Total	450	100

based system created to identify specific hazards and actions to control them in order to ensure food safety. It is also a systematic process: a sequence of twelve tasks has been described, in which the seven basic HACCP principles are included (Codex Alimentarius Commission, [14]). The respondents claimed that the most important contents in HACCP system operation were to assemble HACCP team (21.8%), and the second was to monitoring (20.0%). Documentation and recording (16.9%) and verification (11.1%) were followed. The reason for choosing first on assembling HACCP team might be due to the size of food companies. While relatively large-scale manufacturer will find it easier to find human resource and technical assistance, small and medium-scale businesses find it more difficult because they lack appropriate human resources, technical knowledge and experience to introduce HACCP into practice. Particularly, small food processors tend to employ the staff they need to carry out production tasks, to think only in terms of productivity rather than safety and to understand the HACCP system as complicated and unnecessary to produce food products. Therefore, the introduction of HACCP into these companies is more difficult than in large ones to use HACCP [15,16]. It is confirmed that small-scale ones were less likely to invest in hygiene and food safety than larger ones [17]. Thus, HACCP operator with the ability to manage if the HACCP system is working correctively is urgently required.

Monitoring in HACCP means checking that the preventative measure at a CCP is under control to prevent hazard [18]. Monitoring was the second selected reason (20.0%) for the implementation of the HACCP system. The third most frequent given reason was a documentation and recording (16.9%). Records related to steps and procedures of HACCP must be fully completed and signed by responsible person, which adds an extra task

to the routine work of food processing. Managers and staff, particularly in small businesses, require a great deal of paper work [19]. Verification was the fourth selected reason (11.1%) for the implementation of the HACCP system. The Codex Alimentarius defines verification as the application of methods, procedures, tests and other evaluations in addition to monitoring to determine compliance with the HACCP plan [19]. As stated in the sixth principles, verification includes all activities (e. g. auditing, food analysis and test), which are focused on determining that all health hazards are controlled [19]. As mentioned in Table 2, 4.4% reported to have a constructing flow diagram and on-site confirmation of it. Food factory layout must be designed to achieve a smooth flow of operations keeping the amount of handling of food materials to the minimum possible.

The World Health Organization has published a definition for prerequisites (WHO, [20]) "practices and conditions needed prior to and during the implementation of HACCP and which are essential for food safety" and again mentions that these are described in Codex Alimentairus Commission's General Principles of Food Hygiene and other Codes of Practice. The concepts of prerequisite program (PRP) and how it will benefit HACCP had been reported by Wallace and Williams [21]. It has been recommended that before HACCP is utilized, a prerequisite program is needed [22].

Table 3 lists the priority of duties in sanitation management related to pre-requisites programs in plants. Sanitation management consisted of 11 tasks, where the importance of the field is hygiene control. Duties of importance are like these; cleaning/washing/disinfection (18.9%), inspection (18.4%), water supply (16.4%), pest control (10.9%), and employee hygiene (7.8%). It can be quite affected an entire sanitation management such as cleaning, washing and sterilizing; cause by neglecting the

Table 3 Priority of duties in sanitation management in plants

Sanitation management	Frequency (N)	Percent (%)
Cleaning/Washing/disinfection	85	18.9
Inspection	83	18.4
Water supply	74	16.4
Pest control	49	10.9
Employee hygiene	35	7.8
Ventilation	34	7.6
Record of receiving raw materials	29	6.4
Facility and equipment	28	6.2
Transportation	16	3.6
Recall	13	2.9
Storage	4	0.9
Total	450	100

management of hygiene. For example, it could be caused germs; dirt hands, a rust knife, cutting board and bacterial pollution. That is, it is thought that they were required to have high level of sanitary duties; keeping a clean knife, sterilizing cutting board, and washing hands for the final product of the process. Equipment should also be designed and constructed so that cleaning, maintenance and inspection are facilitated. Well designed and structured premises with reliable equipment could help in maintaining hygienic conditions, improving cleanliness and cleaning effectiveness and controlling pest infestations [23]. However, food premises with congested and unhygienically designed food preparation rooms are frequently found. Normally, this is the case in small businesses that have been increasing their productivity without the consequent expansion of their facilities and installations, or businesses that are crowded with staff and machinery to satisfy workloads. In those situations, the implementation of HACCP is far more complicated due to the difficulty of controlling basic sanitary standards [19].

Our results suggest that food product plants in Korea were more likely to implement HACCP to improve hygiene ability rather than for other reasons. This result might be related with some reasons. First, Korean consumers showed increased the knowledge about food hygiene result from bovine spongiform encephalopathy (BSE), foot and mouth disease (FMD), and avian Influenza (AI) etc. Secondly, food hygiene might be the most

important factors for livestock product plants employers and employees because SSOP was compulsorily applied for food products processing plant in Korea.

Table 4 represents the job satisfaction in HACCP system operators by gender, age, size of food product facilities, service period, and position. Approximately 83% of respondents indicated that they were satisfied with their jobs. Overall, both male and female were satisfied with their jobs. The proportion of male who thought of themselves as "Agree" in job satisfaction is higher than that of female. Concerning the age, the highest proportion of "Agree" was the age group 40-49 (100%) whereas the highest that of "Disagree" was the age group 20-29 (15.7%); however the job satisfaction increased with linearly age. The proportion of workers at large scale company who thought of themselves as "Agree" in is higher than that of small scale one. Regarding service period, there was a U-shaped association between service period and job satisfaction although there was no significant difference across the service period. The highest proportion of "Agree" by position was team manager and over (100.0%), whereas employees were the lowest (13.8%). The Chi-square test showed that the gender, age, livestock product facilities, and position significantly affected to job satisfaction (p < 0.001). The respondents showed an optimistic attitude on job satisfaction as HACCP system operator.

Table 5 depicts the prospect of occupation in HACCP system operators by gender, age, size of food product

Table 4 Job satisfaction in HACCP system operators by gender, age, size of facilities, service period, and position

		Disagree (%)	Moderate (%)	Agree (%)	Total (N)	x^2	df	p
Gender	Male	11(10.8)	6(5.9)	85(83.3)	102	39.353	2	0.000
	Female	0(0)	23(47.9)	25(52.1)	48			
	Total	11(7.3)	29(19.3)	110(73.3)	150			
Age (yrs)	20-29	11(15.7)	15(21.4)	44(62.9)	70	24.442	4	0.000
	30-39	0(0)	14(28.0)	36(72.0)	50			
	40-49	0(0)	0(0)	30(100.0)	30			
	Total	11(7.3)	29(19.3)	110(73.3)	150			
Size of facilities	Large-scale	0(0)	0(0)	40(100.0)	40	19.835	2	0.000
	Small-scale	11(10.0)	29(26.4)	70(63.6)	110			
	Total	11(7.3)	29(19.3)	110(73.3)	150			
Service period (yrs.)	<1	11(12.9)	15(17.6)	59(69.4)	85	14.039	4	0.007
	3-5	0(0)	0(0)	13(100.0)	13			
	5-10	0(0)	14(26.9)	38(73.1)	52			
	Total	11(7.3)	29(19.3)	110(73.3)	150			
Position	President, Director	0(0)	0(0)	2(100.0)	2	42.727	6	0.000
	Assistant manager	0(0)	0(0)	42(100.0)	42			
	Team manager	0(0)	0(0)	26(100.0)	26			
	Employees	11(13.8)	29(36.3)	40(50.0)	80			
	Total	11(7.3)	29(19.3)	110(73.3)	150			

Table 5 Prospect of occupation in HACCP system operators by gender, age, size of facilities, service period, and position

		Moderate (%)	Agree (%)	Total (N)	x^2	df	p
Gender	Male	13(12.7)	89(87.3)	102	5.963	1	0.015
	Female	14(29.2)	34(70.8)	48			
	Total	27(18.0)	123(82.0)	150			
Age (yrs)	20-29	0(0)	70(100.0)	70	24.442	4	0.000
	30-39	227(54.0)	23(46.0)	50			
	40-49	0(0)	30(100.0)	30			
	Total	27(18.0)	123(82.0)	150			
Size of facilities	Large-scale	0(0)	40(100.0)	40	11.973	1	0.001
	Small-scale	27(24.5)	83(75.5)	110			
	Total	27(18.0)	123(82.0)	150			
Service period (yrs.)	<1	0(0)	85(100.0)	85	62.054	2	0.000
	3-5	0(0)	13(100.0)	13			
	5-10	27(51.9)	25(48.1)	52			
	Total	27(18.0)	123(82.0)	150			
Position	President, Director	0(0)	2(100.0)	2	27.710	3	0.000
	Assistant manager	0(0)	4(100.0)	4(100.0)			
	Team manager	13(50.0)	13(50.0)	26(100.0)			
	Employees	14(17.5)	66(82.5)	80(100.0)			
	Total	27(18.0)	123(82.0)	150			

facilities, service period, and position. The same trends shown in Table 4 were observed except for service period. A total of 82% of respondents showed optimistic prospect of occupation. Subjects aged 20-29 and 40-49 reported that HACCP system operator would be hopeful, whereas those aged 30-39 thought of the same as now. Furthermore, it was higher than those aged 20-29 and 40-49. The proportion of workers at large scale who thought of themselves as "Agree" in is higher than that of small scale. Regarding service period, the lowest proportion of "Agree" by career was 5-10 years group and over (48.1%), whereas less than 1 and 3-5 years group were the highest (100.0%). Especially, the prospect of occupation in HACCP system operator decreased with service period and there was significant difference across service period. The results of Chi-square testing showed that there were significant differences in the prospect of occupation in HACCP system operator by the gender (p < 0.015), age, livestock product facilities, service period, and position (p < 0.001).

The main role of HACCP system operators is to encourage and motivate supervisory staff and food handlers on different aspects of the HACCP concept. Consequently, they will need to attribute responsibilities between personnel involved with the implementation of the system. This must be done in accordance to the difficulty of the operation and the capabilities of the person who is going to be responsible for it [19].

Conclusions

In conclusion, most of respondents answered an optimistic attitude on job satisfaction as HACCP system operator. The respondents from HACCP system operator were also showed optimistic prospect of occupation. Results from this study could be used to better educate HACCP system operators and industry implementers.

Competing interests
The authors declare that they have no competing interests.

Authors' contributions
DGL carried out the HACCP survey including the design of the study and all authors participated in and drafted the manuscript. JHK performed the statistical analysis and all authors read and approved the final manuscript.

Acknowledgement
This study was financially supported by a grant from Jinju Health College Research Foundation.

Author details
[1]Department of Public Health, Graduate School of Public Health, Seoul National University, Seoul 151-742, Korea. [2]Department of Animal Science and Technology, Sunchon National University, Suncheon 540-742, Korea. [3]Department of Agricultural Biotechnology and Research Institute of Agriculture and Life Science, Seoul National University, Seoul 151-921, Korea. [4]Department of Health Administration and Food Hygiene, Jinju Health College, Jinju 660-757, Korea.

References
1. Al-Kandari D, Jukes JD: **Incorporating HACCP into national food control systems analyzing progress in the United Arab Emirates.** *Food Control* 2011, **22**:851–861.
2. Codex Alimentarius Commission (CAC): *Recommended international code of practice: General principles of food hygiene*; 2003. CAC/RPP 1e1969, Revision 4.
3. Wallace C, Powell S, Holyoak L: **Development of methods for standardized HACCP assessment.** *Brit Food J* 2005, **107**(10):723–742.
4. Ropkins K, Beck A: **Evaluation of worldwide approaches to the use of HACCP to control food safety.** *Trends in Food Sci Technol* 2000, **11**:10–21.
5. Codex Alimentarius Commission (CAC): *Food hygiene basic texts. Codex Alimentarius -Joint FAO/WHO food standards programs.* 3rd edition. Lanham, MD: Bernan Association; 2001.
6. QIA: *Animal Plant and Fisheries Quarantine and Inspection Agency.* Republic of Korea; 2012. Available: http://www.qia.go.kr/livestock/clean/livestock_livestock_food.jsp.
7. Taylor E: **HACCP in small companies: benefit or burden.** *Food Control* 2001, **12**:217–222.
8. Holt G: *Researcher investigating barriers to the implementation of GHP in SMEs.* Personal Communication; 1999. Oct. 1999.
9. Moon HK, Ryu K: **Usage status survey on some essential facilities, equipment and documentary records for HACCP implementation in contract foodservice.** *J Kor Soc Food Sci Nutr* 2004, **33**:1162–1168.
10. Kim GM, Lee SY: **A study on the sanitation management status and barriers to HACCP system implementation of school foodservice institutions in Seoul metropolitan area.** *J Comm Nutr* 2008, **13**:405–417.
11. Park JS, Park SI: **A survey of washing and sanitizing methods for the pre-preparation of fruits at a school foodservice in the Seoul and Kyunggi area.** *Kor J Food Ctr* 2009, **24**:39–50.
12. Kwak CK, Kim TK, Park SH, Jang JK: **Economic feasibility of HACCP at slaughter plants.** *Korean J Agr Econ* 2002, **29**:1–18.
13. Hong CH, Cho DH: **Comparative analysis of the prerequisite items applicable to the HACCP in livestock processing plants.** *J Food Hygiene Safety* 2008, **23**:19–25.
14. Codex Alimentarius Commission (CAC): *Recommended International Code of Practice-General Principles of Food Hygiene*; 1997. CAC/RCP 1-1969, Rev.3 and anex.
15. Panisello PJ, Quantick PC, Knowles MJ: **Towards the implementation of HACCP; results of a UK regional survey.** *Food Control* 1999, **10**:87–90.
16. Mortlock MP, Peters AC, Griffith C: **Food hygiene and hazard analysis critical control point in the United Kingdom food industry: practices, perceptions and attitudes.** *J Food Prot* 1999, **62**:786–792.
17. Gormley RT: **RTD needs and opinions of European food SMEs.** *Farm and Food* 1995, **5**:27–30.
18. NACMCF: **Generic HACCP for raw beef. The national advisory committee on microbiological criteria for food.** *Food Microbiol* 1993, **10**:449.
19. Panisello PJ, Quantick PC: **Technical barriers to HACCP.** *Food Control* 2001, **12**:165–173.
20. World Health Organization (WHO): *Training Considerations for the Application of the Hazard Analysis Critical Control Points System to Food Processing and Manufacturing*, WHO Document, WHO/FNU/FOS/93.3, Division of Food and Nutrition. Geneva: WHO; 1993.
21. Wallace C, Williams T: **Pre-requisites: A help or a hindrance to HACCP.** *Food Control* 2001, **12**:235–240.
22. Seward S: **Application of HACCP in food service.** *Irish J Agr Food Res* 2000, **39**:221–227.
23. Forsythe SJ, Hayes PR: *Food hygiene, microbiology and HACCP.* 3rd edition. Maryland: Aspen; 1998.

10

Effects of different physical forms of concentrate on performance, carcass characteristics, and economic analysis in hanwoo steers

Sung Il Kim[1], Bo cheon Seo[2], In Surk Jang[2], Ouk Kim[3], Chang Bon Choi[4] and Keun Ki Jung[5*]

Abstract

This study was performed to investigate the effects of different forms of concentrate fed to Hanwoo steers on performance, carcass characteristics, and economic performance. Forty-two Hanwoo steers (average age of 5.1 ± 0.8 mo. with body weight of 147.05 ± 10.85 kg) were randomly allotted into FC (animals fed flakes for entire experimental period) and GC (animals fed grounded concentrate during growing and fattening phases followed by flaked concentrate during finishing phase) groups for 758 d after reaching an age of 30.0 ± 0.82 mo. There was no difference in body weight (BW) or ADG between the treatments until fattening ($15 \sim 22$ mo.) phase. However, by finishing phase ($23 \sim 30$ mo.), the GC group (739.24 kg BW and 0.67 kg ADG) showed greater ($P < 0.05$) BW and ADG than the FC group (702.93 kg BW and 0.59 kg ADG). Steers in the GC group also showed greater ($P < 0.05$) BW and ADG than the FC group throughout the entire experimental period ($5 \sim 30$ mo.). There was no significant difference in carcass weight or backfat thickness between the treatments. *M. Longissimus dorsi* area of the GC group (91.00 cm^2) was greater ($P < 0.05$) than that of the FC group (83.59 cm^2). Marbling score and percentage of 1^{++} meat quality grade were 14.0 and 48.0% higher in the GC group compared to the FC group. There was no significant difference in physicochemical characteristics, including moisture and crude protein levels, between the treatments. Gross income per head excluding operating expenses was 59.3% greater in the GC group (1,647,512 won) compared to the FC group (1,034,343 won).

Keywords: Hanwoo steers, Flaked concentrate, Grounded concentrate, Economic performance

Background

The current trend in feeding systems for Hanwoo steers in Korea involves administration of animal feeds in the form of pellets or flakes rather than as grounded feeds. Pellets are formed by grounding and compressing raw ingredients into the shape of a pellet while flakes are processed using high heat and pressure. Hanwoo steers are fed grain-oriented compound feeds to increase their intake and efficiency [1]. However, production costs of Hanwoo farms are rapidly increasing due to skyrocketing prices of grains and animal feeds. As such, there is increasing demand to fortify the competitiveness of the Hanwoo industry by cutting down on production costs, and the most effective method may be simplifying the entire animal feed process. Processing of feedstuffs increases

the gelatinization and digestibility of starch, ultimately improving feed efficiency [2]. Gelatinization of starch can be increased by pelleting and flaking by 35% and 50%, repectively [3,4]. The range of this processing effect and feed efficiency depend on the type of grain, processing method, origin of feed ingredients, and breeding period of cattle [5,6]. Therefore, it is necessary to appropriately adjust the processing method of feed with regard to the breeding period of Hanwoo steers. To this end, this study investigated the effects of administration of various physical forms of grain feed during different feeding phases on performance, carcass characteristics, and economic performance in Hanwoo steers, thereby reducing production costs with efficient feed processing methods.

Materials and methods

Experimental animals and design

Forty-two Hanwoo steers (average age of 5.1 ± 0.8 mo. with body weight of 147.05 ± 10.85 kg) were randomly

* Correspondence: kkjung@ynu.ac.kr
[5]Moksan Hanwoo Reserch Institute, 108-7, Bujeok-ri, Apryang-myun, Gyeongsan, Gyeongsangbuk-do 712-821, Korea
Full list of author information is available at the end of the article

allotted into FC (animals fed flakes for entire experimental period; growing, fattening, and finishing phases, 5 ~ 30 mo. of age) and GC (animals fed grounded concentrate during growing, 5 ~ 14 mo. of age, and fattening phases, 15 ~ 22 mo. of age, followed by flakes during finishing phase, 23 ~ 30 mo. of age). Hanwoo steers in the FC (18 steers divided into three pens) and GC (24 steers divided into four pens) groups were administered the assigned diets for 758 d after reaching an age of 30.0 ± 0.82 mo. (Table 1). The animals were sacrificed for meat production and parts of the carcass were used for the analysis with consent from the farmers in this study.

Experimental diets
Experimental diets were formulated by an animal feed manufacturing company located in Kimhae, Korea. Roughages used in this study included timothy, alfalfa, and tall fescue. Chemical compositions of the experimental diets are shown in Table 2 (concentrate) and 3 (roughage). Table 4 shows the physicochemical characteristics of corns. Amounts of concentrate and roughages used in the experimental diets were determined by considering the growth stage and nutrient requirements of the animals (Table 5).

Feeding management
Each treatment group was placed in a 5.0 m × 10.0 m pen (six animals per pen) and administered the assigned diets twice per day. All animals had *ad libitum* access to water. Feed intake was recorded every day, and animals were weighed every month throughout the experiment. Animals were cared and managed according to Korean traditional farm regulations.

Meat quality measurement
At the end of the experimental period, animals were fasted for 24 h, weighed, and slaughtered at a commercial abattoir located in Ansung, Kyunggi province, Korea. Carcass measurement were obtained after chilling for 24 h at 4°C. Carcass yield and quality were graded by meat graders using the criteria provided by Livestock Quality Assessment [7].

Table 1 Feeding regimen of concentrate diet for the entire experiment

Treatment	Phases		
	Growing[1]	Fattening[2]	Finishing[3]
FC	Flaked & pelleted diet	Flaked & pelleted diet	Flaked & pelleted diet
GC	Ground diet	Ground diet	Flaked & pelleted diet

[1]Feeding period: 5.1 to 13.9 months of age.
[2]Feeding period: 13.9 to 22.1 months of age.
[3]Feeding period: 22.1 to 30.0 months of age.

Evaluation of carcass chemical composition
a. Chemical composition
Chemical composition, including moisture, ash, crude protein, and fat contents, were analyzed according to the AOAC methodology [8]. Moisture content (%) of loin muscle samples (2 g) was measured by homogenizing and drying samples at 105°C in an oven and then measuring weight loss after drying. Total lipids were analyzed by the soxhelt extraction method. Crude protein content was measured by the Kjeldahl method. Briefly, 0.5 g of loin samples was digested at 450°C for 5 h, distilled by addition of 50% NaOH, and titrated with HCL, after which the total protein amount was calculated by multiplying% N by 6.25.

b. Meat color
Meat color, including Hunter L (lightness), a (redness), and b (yellowness), was determined by a Chroma Meter (CR-10, Minolta Corporation, LTD, Japan).

c. Melting point
Melting point was measured by the slip-point method. Briefly, lipids were extracted from meat samples by cutting them into small pieces with a Hanil Mini Cooking Cutter (Hanil electric co. HMC-150 T), homogenization with chloroform and methanol (2:1 v/v) solution, filtration, and then evaporation with nitrogen. Capillary tubes (100 mm, open) were filled to a height of 1 cm from one end and then placed in a freezer (−20°C) until lipids were firm (about 24 h). After removal from the freezer, the capillary tube was place on a warm incubator, and the temperature was increased at a rate of 1°C per min. with stirring until the lipids melted.

Economic analysis
We analyzed the economic values of Hanwoo steers used in this experiment by calculating the average carcass prices at four different slaughter points. Profits from by-products were also considered as economic values. Feed costs for both the concentrate and roughage used in this analysis were applied as the actual purchase price of the farm where this experiment was performed. Costs for purchasing the calves, bedding, medicine, utilities (water and heating), and castration were averaged based on the number of animals used in this experiment.

Statistics
Data was analyzed by t-test of SAS [9]. Probability values less than 0.05% were considered significant. Data of feed intake and feed conversion rate from the breeding group were excluded from the significance test.

Table 2 Chemical composition of concentrate diets

Composition	Concentrate					
	Growing		Fattening		Finishing	
	Flaked	Ground	Flaked	Ground	Flaked	Flaked
	———%, as - fed ———					
Moisture	11.78 ± 0.12[1]	11.82 ± 0.05	12.67 ± 0.06	12.90 ± 0.04	12.59 ± 0.09	12.59 ± 0.09
Crude protein	16.22 ± 0.01	16.33 ± 0.02	15.33 ± 0.04	15.18 ± 0.03	13.26 ± 0.03	13.26 ± 0.03
Crude fat	3.06 ± 0.01	2.63 ± 0.04	3.03 ± 0.03	3.19 ± 0.04	3.17 ± 0.06	3.17 ± 0.06
Crude fiber	11.74 ± 0.25	13.13 ± 0.08	10.15 ± 0.06	10.17 ± 0.10	9.24 ± 0.39	9.24 ± 0.39
Crude ash	5.44 ± 0.06	6.03 ± 0.05	4.89 ± 0.02	5.04 ± 0.06	4.57 ± 0.20	4.57 ± 0.20
NFE	51.76 ± 1.61	50.06 ± 1.32	53.93 ± 1.11	53.52 ± 1.46	57.17 ± 0.95	57.17 ± 0.95
Ca	0.77 ± 0.04	0.66 ± 0.01	0.59 ± 0.01	0.63 ± 0.02	0.54 ± 0.06	0.54 ± 0.06
P	0.45 ± 0.00	0.44 ± 0.00	0.40 ± 0.00	0.43 ± 0.00	0.41 ± 0.02	0.41 ± 0.02
NDF[2]	29.04 ± 0.36	33.54 ± 0.21	26.06 ± 0.07	25.33 ± 0.07	27.85 ± 1.37	27.85 ± 1.37
ADF[3]	16.18 ± 0.19	15.63 ± 0.24	14.75 ± 0.10	14.34 ± 0.07	13.58 ± 0.38	13.58 ± 0.38
TDN[4]	68.0	68.0	70.0	70.0	72.0	72.0

[1] Means ± standard error.
[2] Neutral derergent fiber.
[3] Acid detergent fiber.
[4] Calculated.

Results and discussion

Performance

Changes in body weight (BW) and ADG in steers fed the experimental diets are shown in Table 6. There was no significant difference in BW or ADG between the treatments during growing (5 ~ 14 mo.) and fattening (15 ~ 22 mo.) phases. However, by finishing phase (23 ~ 30 mo.), the GC group (739.24 kg BW and 0.67 kg ADG) showed greater ($P < 0.05$) BW and ADG than the FC group (702.93 kg BW and 0.59 kg ADG). Consistent with this result, steers in the GC group also showed greater ($P < 0.05$) BW and ADG than the FC group throughout the entire experimental period (5 ~ 30 mo.). There was no significant difference in feed intake for either concentrate or roughage between the treatments during growing and fattening phases (Table 7). Feed intake for concentrate was 7.4% higher in the GC group compared to the FC group, whereas the feed conversion rate was 6.7% lower in the GC group compared to the FC group during finishing phase. The GC group showed a 3.0% greater feed

Table 3 Chemical composition of roughages

Composition	Roughages			
	Timothy hay	Alfalfa hay	Tall fescue straw	Ryegrass straw
	———%, as-fed basis ———			
Moisture	8.22 ± 0.07[1]	9.64 ± 0.18	9.76 ± 0.44	7.72 ± 0.04
Crude protein	7.87 ± 0.18	17.79 ± 0.16	7.06 ± 0.59	5.39 ± 0.16
Crude fat	1.89 ± 0.02	1.97 ± 0.02	0.77 ± 0.00	1.10 ± 0.05
Crude fiber	32.77 ± 0.28	27.71 ± 0.32	32.57 ± 1.63	32.34 ± 0.19
Crude ash	6.37 ± 0.12	9.20 ± 0.10	5.97 ± 0.76	5.82 ± 0.07
NFE[2]	42.88 ± 0.74	33.69 ± 0.97	43.87 ± 2.54	47.63 ± 0.19
Ca	0.28 ± 0.00	1.48 ± 0.02	0.20 ± 0.01	0.36 ± 0.01
P	0.16 ± 0.00	0.22 ± 0.00	0.08 ± 0.01	0.13 ± 0.00
NDF[3]	59.99 ± 0.30	37.70 ± 0.43	59.42 ± 1.93	60.88 ± 0.10
ADF[4]	34.36 ± 0.23	30.71 ± 0.68	34.20 ± 1.78	34.45 ± 0.15
TDN[5]	54.61	53.55	34.18	52.82

[1] Means ± standard error.
[2] Nitrogen-free extract.
[3] Neutral derergent fiber.
[4] Acid detergent fiber.
[5] Calculated.

Table 4 Physicochemical characteristics and distribution of particle size in flake and grounded corns

Items	Flaked corn	Ground corn
Flake thickness[1], mm	3.29 ± 0.04[2]	—
Starch gelatinization[3],%	31.93 ± 1.69	—
Density, g/ℓ	506.8 ± 2.0	693.5 ± 2.5
Particle size[4],%		
Sieve mesh		
6 ~ 8	—	36.1 ± 2.8
14 ~ 18	—	45.3 ± 4.3
25 ~ 40	—	10.9 ± 1.3
60 ~ 100	—	6.2 ± 0.6
Under 100	—	1.5 ± 0.4

[1]Measured by vernier calipers.
[2]Means ± standard error.
[3]Determined by diastase method.
[4]Percents retained on screen.

intake for concentrate as well as a 3.3% lower feed conversion rate compared to the FC group for the entire experimental period. There was no significant difference in feed intake for roughage between the treatment groups. These results indicate that the physical form of concentrate have no affect on the ADG or feed intake of steers during growing and fattening phases. Zinn and Barajas [10] also reported that steers administered various densities of corn and barley for 86 d showed no difference in body weight gain or ADG. Consistent with these results, administration of various densities of sorghum flakes (412, 360, 309, and 257 g/L) did not affect the ADG of steers during growing phase [11]. However, in our study, steers from the GC group during finishing phase showed higher (P < 0.05) BW and ADG than those from the FC group. This result might be associated increased feed intake in the GC group. Furthermore, in a previous report, steers fed mashed concentrate during growing phase and then

Table 5 Feeding program for Hanwoo steers in the experiment

Fattening phase	Age in mon.	Body weight range (kg)	Daily gain (kg)	Feeding level (body weight,%)	Concentrate fed (kg/hd/d, as-fed basis)			Roughage fed (kg/hd/d, as-fed basis)		
					Growing	Fattening	Finishing	Timothy hay	Alfalfa hay	Straw
Growing	5	147 ~ 162	0.50	0.90	1.4			1.5	0.5	
	6	162 ~ 188	0.85	1.25	2.0			2.0	1.0	
	7	188 ~ 214	0.85	1.40	2.6			3.0	1.0	
	8	214 ~ 241	0.90	1.50	3.2			3.4	1.0	
	9	241 ~ 268	0.90	1.50	3.6			3.5	1.0	
	10	268 ~ 295	0.90	1.53	4.1			4.0	1.0	
	11	295 ~ 322	0.90	1.55	4.6			4.0	1.0	
	12	322 ~ 351	0.95	1.61	5.2			4.5	0.5	
	13	351 ~ 379	0.95	1.70	6.0			4.5	0.5	
Fattening	14	379 ~ 408	0.95	1.84	3.5	3.5		4.5		
	15	408 ~ 436	0.95	1.96		8.0		3.5		0.5
	16	436 ~ 466	1.00	2.06		9.0				3.0
	17	466 ~ 496	1.00	2.08		9.7				3.0
	18	496 ~ 526	1.00	2.02		10.0				2.5
	19	526 ~ 553	0.90	1.90		10.0				2.3
	20	553 ~ 579	0.85	1.81		10.0				2.0
	21	579 ~ 604	0.85	1.73		10.0				1.5
	22	604 ~ 628	0.80	1.66		10.0				1.5
Finishing	23	628 ~ 649	0.70	1.51			9.5			1.3
	24	649 ~ 667	0.60	1.39			9.0			1.2
	25	667 ~ 682	0.50	1.35			9.0			1.2
	26	682 ~ 696	0.45	1.25			8.5			1.2
	27	696 ~ 708	0.40	1.22			8.5			1.2
	28	708 ~ 718	0.35	1.13			8.0			1.2
	29	718 ~ 727	0.30	1.11			8.0			1.2
	30	727 ~ 736	0.30	1.03			7.5			1.2

Table 6 Body weight and daily gain of Hanwoo steers by treatment

Items	FC[1]	GC[2]	T-test[3]
No. of heads	18	24	
Body weight (kg)			
Initial (5 mo)	146.7 ± 1.35	147.4 ± 0.99	0.9253
Growing (14 mo)	370.7 ± 1.81	374.5 ± 1.22	0.6985
Fattening (22 mo)	562.2 ± 2.47	575.1 ± 1.90	0.2297
Finishing (30 mo)	702.9 ± 2.90	739.2 ± 2.67	0.0483
Average daily gain (kg)			
Growing phase	0.84 ± 0.00	0.85 ± 0.00	0.6629
Fattening phase	0.77 ± 0.01	0.80 ± 0.01	0.3220
Finishing phase	0.59 ± 0.00	0.67 ± 0.01	0.0548
Overall period	0.73 ± 0.00	0.78 ± 0.00	0.0414

[1]Growing (flaked & pelleted diet), fattening (flaked & pelleted diet) & finishing (flaked & pelleted diet).
[2]Growing (ground diet), fattening (ground diet) & finishing (flaked & pelleted diet).
[3]Probability of the T test.

Table 7 Feed intake and feed conversion in Hanwoo steers

Items	FC[1]	GC[2]
Growing phase		
Feed intake (kg/head/day)		
Concentrate	3.52	3.53
Timothy hay	2.60	2.78
Alfalfa hay	0.68	0.70
Sub-total	3.28	3.48
Feed conversion, kg/kg	8.14	8.27
Fattening phase		
Feed intake (kg/head/day)		
Concentrate	7.91	7.89
Timothy hay	1.01	0.91
Tall fescue straw	0.80	0.99
Ryegrass straw	0.10	-
Sub-total	1.91	1.89
Feed conversion, kg/kg	12.83	12.19
Finishing phase		
Feed intake (kg/head/day)		
Concentrate	7.80	8.38
Tall fescue straw	-	1.23
Ryegrass straw	1.23	-
Sub-total	1.23	1.23
Feed conversion, kg/kg	15.39	14.38
Overall period		
Feed intake (kg/head/day)		
Concentrate	6.32	6.50
Timothy hay	1.25	1.28
Alfalfa hay	0.24	0.25
Tall fescue straw	0.26	0.72
Ryegrass straw	0.42	-
Sub-total	2.18	2.25
Feed conversion, kg/kg	11.59	11.22

[1]Growing (flaked & pelleted diet), fattening (flaked & pelleted diet) & finishing (flaked & pelleted diet).
[2]Growing (ground diet), fattening (ground diet) & finishing (flaked & pelleted diet).

switched to flaked concentrate during finishing phase showed a greater (by 0.98 kg) ADG compared to those fed flaked concentrate for the entire period [12]. Flaking improves the feed conversion rate by inducing gelatinization of starch [13]. On the other hand, administration of flaked concentrate for the entire feeding period may decrease feed intake by reducing the rumen pH, which is associated with accelerated degradation of starch [14]. Taken together, feeding steers grounded concentrate during growing and fattening phases and then switching to flaked concentrate finishing phase effectively improved ADG and feed intake.

Carcass characteristics

Carcass weight, backfat thickness, M. Longissimus dorsi area, marbling score, and meat color of Hanwoo steers fed the experimental diets are shown in Table 8. Carcass weight of the GC group (4229.57 kg) was numerically, but not statistically, higher than that of the FC group (405.94 kg). There was no significant difference in backfat thickness between the treatments. M. Longissimus dorsi area of the GC group (91.00 cm²) was greater (P < 0.05) than that of the FC group (83.59 cm²). Marbling score, which is a meat quality trait, was 18.4% higher in the GC group (6.96) compared to the FC group (5.88). No difference was found in meat color, fat color, texture, or maturity between the treatments. The GC group showed a higher percentage of 1⁺⁺ grade (43.5%) compared to the FC group (29.4%) by 48%. Percentage of quality grade over 1⁺ grade was also higher in the GC group (87%) than in the FC group (29.4%) by 50%. Brandt et al [15] also reported that supplementation of steam-flaked corn to steers increases the M. Longissimus dorsi area. However, carcass weight and backfat thickness are not affected by densities of flaked corn [16]. Moreover, consistent with our current results, NIAS [12] reported that administration of powdered concentrate during fattening phase followed by flaked concentrate during finishing phase to steers increased the marbling score by 13.3% compared to the flake-fed group for the entire period. In the current study, steers in the GC group showed improved M. Longissimus dorsi area and marbling score.

Physicochemical characteristics of carcass

Effects of various physical forms of feeds on the physicochemical characteristics of Hanwoo steers are shown in

Table 8 Effects of physical forms of concentrate on carcass characteristics in Hanwoo steers

Items	FC[1]	GC[2]	T- test[3]
Yield traits			
Cold carcass, kg	405.94 ± 2.08[4]	429.57 ± 1.67	0.0542
Backfat thickness, mm	16.88 ± 0.26	17.96 ± 0.22	0.4892
Longissmus muscle area, cm²	83.59 ± 0.48	91.00 ± 0.44	0.0178
Yield index	62.30 ± 0.17	61.71 ± 0.14	0.5568
Yield grade,%			
A	0.0[5]	0.0	
B	58.8	47.8	
C	41.2	52.2	
Quality traits			
Marbling score[6]	5.88 ± 0.12	6.96 ± 0.06	0.0650
Meat color[7]	4.88 ± 0.02	4.65 ± 0.02	0.1014
Fat color[8]	2.94 ± 0.01	2.78 ± 0.02	0.1735
Texture[9]	1.24 ± 0.03	1.04 ± 0.01	0.0729
Maturity[10]	2.35 ± 0.03	2.57 ± 0.02	0.1931
Quality grade,%			
1++	29.4	43.5	
1+	29.4	43.5	
1	17.7	8.7	
2	23.5	4.3	

[1]Growing (flaked & pelleted diet), fattening (flaked & pelleted diet) & finishing (flaked & pelleted diet).
[2]Growing (ground diet), fattening (ground diet) & finishing (flaked & pelleted diet).
[3]Probability of the T test. [4]Mean ± Standard error.
[5]Value in parentheses represents percentage of total heads.
[6]9 = the most abundant, 1 = devoid. [7]7 = dark red, 1 = bright.
[8]7 = yellowish, 1 = white. [9]3 = Coarse, 1 = fine. [10]9 = mature, 1 = youthful.

Table 9. Moisture levels of the *M. Longissimus dorsi* muscle from steers in the FC and GC groups were 62.91 and 61.48%, respectively. Contents of crude protein in *M. Longissimus dorsi* muscle from steers in the FC and GC groups were 19.35 and 18.75%, respectively. Crude fat content of the GC group (18.37%) was greater than that of the FC group (15.90%) by 15.5%. The overall range of measured CIE values, including L (lightness), b (yellowness), and h (color), were greater (P < 0.05) in the GC group compared to the FC group. Physicochemical properties of meat are normally affected by moisture and crude fat content [17]. Levels of crude fat and CIE values (L) increase while moisture and crude proten content decrease with an increase in meat quality [18,19]. Physicochemical characteristics of beef might be preferentially related to meat quality grade rather than the physical form or processing method of feed [20]. Melting point of carcass fat was highest in perirenal fat, followed by intramuscular fat and then subcutaneous fat (Table 10). Although not significant, the melting points of subcutaneous fat and intramuscular fat were lower in the GC group

Table 9 Effects of physical forms of concentrate on physicochemical characteristics of *M.longissimus dorsi* muscle in Hanwoo steers

Items	FC[1]	GC[2]	T- test[3]
Moisture,%	62.91 ± 0.19[4]	61.48 ± 0.17	0.2246
Crude fat,%	15.90 ± 0.28	18.37 ± 0.22	0.1240
Crude protein,%	19.35 ± 0.06	18.75 ± 0.05	0.1058
CIE value:[5]			
L	40.43 ± 0.10	43.35 ± 0.07	0.0016
a	23.35 ± 0.06	23.81 ± 0.04	0.3895
b	11.10 ± 0.03	11.77 ± 0.02	0.0242
chroma	25.76 ± 0.06	26.60 ± 0.05	0.1604
hue	25.36 ± 0.04	26.60 ± 0.02	0.0120
Cooking loss,%	29.79 ± 0.05	28.62 ± 0.07	0.2096

[1]Growing (flaked & pelleted diet), fattening (flaked & pelleted diet) & finishing (flaked & pelleted diet).
[2]Growing (ground diet), fattening (ground diet) & finishing (flaked & pelleted diet).
[3]Probability of the T test.
[4]Means ± standard error.
[5]L = lightness, a = redness, b = yellowness.

compared to the FC group by 4.0 and 3.0%, respectively. Carcass fat melting point was the greatest in perireral fat, followed by intramuscular fat and then subcutaneous fat. Melting point is highly correlated with fatty acid composition rather than the type or processing method of feed [21]; a higher fatty acid content is associated with a lower melting point and vice versa [22]. Taken together, the GC group with a high marbling score and meat quality grade showed higher carcass physiocochemical properties, including crude fat content and CIE.

Economic analysis

Effects of various physical forms of concentrate on profitability are shown in Table 11. Carcass sale price for the FC and GC groups were 6,036,373 and 6,667,053 won (KRW), repectively. Carcass and by-product sale prices of the GC group were greater than those of the FC group by 10.0%. Total operating expenses, including calf purchase expenses, feed, slaughter, and other expenses, increased by 0.5% in the GC group (5,360,702 won) compared to the FC group (5,334,848 won). Gross income per head excluding operating expenses was 59.3% greater in the GC group

Table 10 Effects of physical forms of concentrate on melting point of carcass fat in Hanwoo steers

Items	FC[1]	GC[2]	T- test[3]
Perirenal fat	38.82 ± 0.05[4]	39.49 ± 0.03	0.5825
Subcutaneous fat	21.39 ± 0.03	20.54 ± 0.02	0.3142
Intramuscular fat	26.81 ± 0.05	26.02 ± 0.02	0.6595

[1]Growing (flaked & pelleted diet), fattening (flaked & pelleted diet) & finishing (flaked & pelleted diet).
[2]Growing (ground diet), fattening (ground diet) & finishing (flaked & pelleted diet).
[3]Probability of the T test.
[4]Means ± standard error.

Table 11 Effects of physical forms of concentrate on profits in Hanwoo steers

Items	FC[1]	GC[2]
Cold carcass, kg	405.94 ± 2.08[3]	429.57 ± 1.67
1. Gross income(A)		
Carcass sales[4]	6,036,373.19	6,667,053.48
By-product sales[5]	332,818	341,162
Total income	6,369,191.19	7,008,215.48
2. Operating cost(B)		
Calves	2,430,952	2,430,952
Concentrate[6]	1,226,259.2	1,226,740.2
Roughage[7]	576,954.3	593,850.3
Butchery expense[8]	227,044	235,522
Self-help funds	20,000	20,000
Miscellaneous expenses[9]	853,638	853,638
Total cost	5,334,847.5	5,360,702.5
3. Profit(A-B)	1,034,343.69	1,647,512.98

[1]Growing (flaked & pelleted diet), fattening (flaked & pelleted diet) & finishing (flaked & pelleted diet).
[2]Growing (ground diet), fattening (ground diet) & finishing (flaked & pelleted diet).
[3]Means ± standard error.
[4]Carcass price, won/kg: $1^{++}B = 17,413$, $1^{+}B = 15,133$, $1B = 14,290$, $1^{++}C = 16,507$, $1^{+}C$ 14,398, 1 = 13,570, 2 13,330, 2 12,370.
[5]Includes intestines, head, legs, hide, blood, and inedible fat.
[6]Concentrate price, won/kg : Growing(F) = 266.8, Growing(M) = 256.0, Fattening(F) = 264.0, Fattening(M) = 253.2, Finishing(F) = 241.6
[7]Roughage price, won/kg: Timothy hay = 407, Alfalfa hay = 360, Tall fescue straw = 240, Ryegrass straw = 240.
[8]Butchery expense: tax, dissection operation, stamp duty, inspection & grading fee.
[9]Miscellaneous expenses: hired labor, bedding materials, electricity, transport, water service & veterinary & medicine.

(1,647,512 won) compared to the FC group (1,034,343 won). This increase in total revenue in the GC group can be attributed to an elevated carcass weight and percentage of meat grade above 1^{+}. Total operating expenses in the GC group were 7.5% greater than those in the FC group due to increased feed intake following replacement of grounded feed by flaked concentrate during finishing phase. Operating expenses in the GC group were also higher due to elevated slaughter expenses due to increased carcass weight. In conclusion, administration of grounded concentrate during growing and fattening phases followed by flaked concentrate during finishing phase improves the profit and productivity of Hanwoo steers.

Conclusion

This study was performed to investigate the effects of different forms of concentrate fed to Hanwoo steers on performance, carcass characteristics, and economic performance. In conclusion, it is plausible that feeding steers grounded concentrate during growing and fattening phases followed by flaked concentrate during finishing phase can improve ADG and feed intake. This feeding strategy increases the profit and productivity Hanwoo steers.

Competing interests
The authors declare that they have no competing interests.

Authors' contributions
SIK and BCS participated in the design of study and made farm visits. ISC and OK carried out the laboratory work. CBC performed the statistical analysis. KKJ made farm visits. All authors helped to draft the article and approved the final manuscript.

Acknowledgements
The authors sincerely thank Songchon farmers for their support during this study and for time spent during sampling and examination.

Author details
[1]Department of Animal Science, Gyeonbuk Provincial College, Yecheong-eup 757-807, Korea. [2]Gyeongnam National University of Science and Technolony, Jinju 757-803, Korea. [3]Department of Animal Science, Dong-A University, Busan 602-714, Korea. [4]Department of Biotechnolony, Yeungnam University, Gyeongsan 712-749, Korea. [5]Moksan Hanwoo Reserch Institute, 108-7, Bujeok-ri, Apryang-myun, Gyeongsan, Gyeongsangbuk-do 712-821, Korea.

References
1. Jeong J, Jin GL, Shinekhuu J, Ji BJ, Song MK: Effect of method on nutrient availability in ruminants. *Bull Biotechnol* 2009, 2:13–21.
2. Hale WH, Cuitun L, Saba WJ, Taylor B, Theurer B: Effect of steam processing and flaking milo and barley on performance and digestion by steers. *J Anim Sci* 1966, 25:392–396.
3. Han YK, Maeng WJ, Park GK, Paik IK, Ohh SJ, Choi YJ: *Feed Processings*. Seoul, Korea: Sun Jin Mun Hwa Sa; 1993:473–475.
4. Theurer CB, Lozano O, Alio A, Delgado-Elorduy A, Sadik M, Huber JT, Zinn RA: Steam-processed corn and sorghum grain flaked at different densities alter ruminal, small intestinal, and total tract digestibility of starch by steers. *J Anim Sci* 1999, 77:2824–2831.
5. Neock JE, Tamminga S: Site of digestion of starch in the gastrointestinal tract of dairy cows and its effect on milk yield and composition. *J Dairy Sci* 1991, 74:3598–3629.
6. Owens FN, Secrist DS, Hill EJ, Gill DR: The effect of grain source and grain processing on performance of feedlot cattle: a review. *J Anim Sci* 1997, 75:868–879.
7. Livestock Quality Assessment: *Grading of Carcasses of Methods, Standards and Enforcement*; 2007.
8. AOAC: *Official Methods of Analysis*. 15th edition. Arlington, Virginia USA: Association of Official Analytical Chemist; 1990.
9. SAS: *SAS User's Guide:Statistics*. NC: SAS Institute Inc. Cary; 2002.
10. Zinn RA, Barajas R: Influence of flake density on the comparative feeding value of a barley-corn blend for feedlot cattle. *J Anim Sci* 1997, 75:904–909.
11. Swingle RS, Eck TP, Theurer CB, De la Llata N, Poore MG, Moore JA: Flake density of steam-processed sorghum grain alters performance and sites of digestibility by growing-finishing steers. *J Anim Sci* 1999, 77:1055–1065.
12. National Institue of Aniamal Science: Development of technologies for production of high-quality beef from Hanwoo steers. In *Effects of Physical Form of Concentrate on Productivity of Hawoo by Feeding Period*. 1998:545.
13. Theurer CB: Grain processing effect on starch utilization by ruminants. *J Anim Sci* 1986, 63:1649–1662.
14. Ministry of Ariculture, Food and Rural Affairs: Development of technologies to improve competitiveness of Hanwoo. In *Development of feeding program for production of high-quality beef from Hanwoo steers*. 2005:85–92.
15. Brandt RT Jr, Kuhl GL, Campbell RE, Kastner CL, Stroda SL: Effects of steam-flaked sorghum grain or corn and supplemental fat on feedlot performance, carcass traits, longissimus composition, and sensory properties of steers. *J Anim Sci* 1992, 70:343–348.
16. Hales KE, McMeniman JP, Leibovich J, Vasconcelos JT, Quinn MJ, May ML, DiLorenzo N, Smith DR, Galyean ML: Effects of varying bulk densities of steam-flaked corn and dietary roughage concentration on in vitro

fermentation, performance, carcass quality, and acid–base balance
measurements in finishing steers. *J Anim Sci* 2010, **88**(3):1135–1147.

17. Kelly RF, Fontenot JP, Graham PP, Wilkinson WS, Kineaid CM: **Estimates of
carcass composition of beef cattle fed at different plane of nutrition.**
J Anim Sci 1968, **27**:620–627.

18. Savell JW, Cross HR, Smith GC: **Percentage ether extractable fat and
moisture content of beef longissimus muscle as related to USDA
marbling score.** *J Food Sci* 1986, **51**:838–840.

19. Nelson JL, Dolezal HG, Ray FK, Morgan JB: **Characterization of certified
angus beef steaks from the round, loin, and chuck.** *J Anim Sci* 2004,
82:1437–1444.

20. Cameron PJ, Zembayashi M, Lunt DK, Mitsuhashi T, Mitsumoto M, Dzawa S,
Smith SB: **Relationship between Japanese Beef Marbling Standard and
intramuscular lipid in the m. longissimus thoracis of Japanese Black and
American Wagyu cattle.** *Meat Sci* 1994, **38**:361–364.

21. Wood JD: **Fat deposition and the quality of fat tissue in meat animals.** In
Fats in Animal Nutrition. Edited by Wiseman J. 1984:407–435.

22. Enser M, Wood JD: *Effect of Time of Year on Fatty Acid Composition and
Melting Point of UK Lamb*, Proceeding of the 39th International congress of
meat Science and Technology. 2nd edition. 1993:74.

11

Influence of 1α, 25-dihydroxyvitamin D$_3$ [1, 25(OH)$_2$D$_3$] on the expression of Sox 9 and the transient receptor potential vanilloid 5/6 ion channels in equine articular chondrocytes

Ismail M Hdud[1,3] and Paul T Loughna[1,2]*

Abstract

Background: Sox 9 is a major marker of chondrocyte differentiation. When chondrocytes are cultured *in vitro* they progressively de-differentiate and this is associated with a decline in Sox 9 expression. The active form of vitamin D, 1, 25 (OH)$_2$D$_3$ has been shown to be protective of cartilage in both humans and animals. In this study equine articular chondrocytes were grown in culture and the effects of 1, 25 (OH)$_2$D$_3$ upon Sox 9 expression examined. The expression of the transient receptor potential vanilloid (TRPV) ion channels 5 and 6 in equine chondrocytes *in vitro*, we have previously shown, is inversely correlated with de-differentiation. The expression of these channels in response to 1, 25 (OH)$_2$D$_3$ administration was therefore also examined.

Results: The active form of vitamin D (1, 25 (OH)$_2$D$_3$) when administered to cultured equine chondrocytes at two different concentrations significantly increased the expression of Sox 9 at both. In contrast 1, 25 (OH)$_2$D$_3$ had no significant effect upon the expression of either TRPV 5 or 6 at either the protein or the mRNA level.

Conclusions: The increased expression of Sox 9, in equine articular chondrocytes *in vitro*, in response to the active form of vitamin D suggests that this compound could be utilized to inhibit the progressive de-differentiation that is normally observed in these cells. It is also supportive of previous studies indicating that 1α, 25-dihydroxyvitamin D$_3$ can have a protective effect upon cartilage in animals *in vivo*. The previously observed correlation between the degree of differentiation and the expression levels of TRPV 5/6 had suggested that these ion channels may have a direct involvement in, or be modulated by, the differentiation process *in vitro*. The data in the present study do not support this.

Keywords: Cartilage, Vitamin D, Sox 9, TRPV, Chondrocyte

Background

The active metabolite of vitamin D, 1, 25 (OH)$_2$D$_3$ has been reported to be involved in the regulation of cellular functions such as differentiation, proliferation, and the immune system [1]. Similar to other steroids, 1, 25 (OH)$_2$D$_3$ exerts physiological actions at two levels by genomic and non-genomic mechanisms. The genomic mechanism exerts its effect on the cells by binding to specific receptors known as vitamin D receptors, which in turn bind to retinoid X receptors to form heterodimeric complexes [2]. These heterodimeric complexes in turn bind to VDREs (vitamin D responsive elements) of the upstream region of the responsive gene leading to activation/suppression of the target gene. In contrast, the non-genomic effect of 1, 25 (OH)$_2$D$_3$ is mediated though plasma membrane receptors of the target cells which are involved in ion channel activity regulation and also activate signal transduction pathways [3,4]. Exogenous 1, 25 (OH)$_2$D$_3$ in mice increased chondrocyte proliferation as well as enhancing

* Correspondence: paul.loughna@nottingham.ac.uk
[1]School of Veterinary Medicine and Science, Faculty of Medicine and Health Sciences, University of Nottingham, Sutton Bonington Campus, Leicestershire LE12 5RD, UK
[2]Medical Research Council-Arthritis Research UK Centre for Musculoskeletal Ageing Research, University of Nottingham, Leicestershire, UK
Full list of author information is available at the end of the article

cartilage matrix mineralization [5]. The profound effects of vitamin D on calcium metabolism are well established and it has been reported that it can protect against cartilage loss in osteoarthritis in humans and rats [6].

Sox 9 (Sry-type high mobility group domain) transcription factor is expressed in articular chondrocytes, central nervous and urogenital systems [7,8]. It is expressed in all primordial cartilage tissues during embryogenesis [8,9] and functions to induce differentiation of stem cells to chondrocytes [10], regulate cartilage development and phenotypic maintenance during embryonic development [11]. Studies indicate that Sox 9 is involved in chondrocyte differentiation by regulating the expression of cartilage-specific genes such as collagen type $II\alpha 1$ [8,12], $XI\alpha 2$ [13], aggrecan [14], collagen link protein [15], cartilage oligomeric matrix protein and Cd-rap [16]. It has been shown that Sox 9 expression is maintained in articular chondrocytes and reduced in osteoarthritic chondrocytes [17].

Haploinsufficiency of Sox 9 results in campomelic dysplasia, a severe syndrome resulting in inadequate cartilage formation during development leading to severe dwarfism [18]. In chimaeric mice, Sox $9^{-/-}$ the prechondrogenic mesenchyme cells were prevented from differentiating into chondrocytes and lost their ability to express chondrocyte specific genes [19,20]. Sox 9 expression is also significantly reduced in osteoarthritic cartilage compared to normal healthy cartilage [17,21].

In vitro propagation of articular chondrocytes results in de-differentiation that is characterized by gradual loss of chondrocytic phenotype and acquisition of a fibroblastic phenotype [22] which is associated with a rapid decline in Sox 9 expression [23]. We have previously shown in equine chondrocytes that there is an association between the degree of de-differentiation and the expression and the transient receptor potential vanilloid (TRPV) channels 5 and 6 [24].

The transient receptor potential (TRP) superfamily is a non-selective cation ion channels with relative calcium selectivity. The TRPV sub-family is divided into two groups. TRPV1-4 channels are non-selective ion channels with modest permeation to calcium. This group can be activated by different stimuli such as heat/cold, chemical/mechanical stresses and binding to second messengers [25,26]. The other group comprises of TRPV5 and TRPV6 channels that are highly calcium selective and tightly regulated by cytosolic Ca^{2+} concentration [27,28]. The TRPV5 channel is implicated in Ca^{2+} reabsorption from the kidney, whereas TRPV6 channel is involved in Ca^{2+} absorption in the intestine [29]. Expression of both channels in human articular chondrocytes at mRNA level has been reported [30]. We also demonstrated their expression at the protein levels in equine articular chondrocytes (EAC) [24]. A correlation between the expression of TRPV5/6 channels and administered 1, 25 $(OH)_2D_3$ concentration has

also reported in intestinal endothelial cells [31,32], renal cells [33], osteoblasts [34,35]. Calbindin-D9K is a cytosolic Ca^{2+} binding protein, a member of cellular proteins found in the cells with high affinity for Ca^{2+} ions. Calbindin-D9k knockout mice models demonstrated that 1, 25 $(OH)_2D_3$ intake increased the expression of both channels [36]. The aim of the present study was to examine the effects of 1, 25 $(OH)_2D_3$ on the expression of Sox 9 and TRPV 5 and 6 in cultured equine chondrocytes.

Methods
Chondrocyte isolation
Articular chondrocytes were isolated from equine articular cartilage removed from load bearing synovial joints (metacarpophalangeal joints) of skeletally mature animals obtained on the day of slaughter from a local abattoir; these animals were euthanized for purposes other than research. All experiments were performed with local institutional ethical approval (University of Nottingham, School of Veterinary Medicine and Science Ethical Committee), in strict accordance with national guidelines. Articular cartilage slices were collected in serum free DMEM medium supplemented with 2% antibiotics (50 U/ml penicillin and 50 µg/ml streptomycin) (Invitrogen, UK). Cartilage slices were washed in phosphate buffer saline (PBS) supplemented with 10% antibiotics (50 U/ml penicillin and 50 µg/ml) for 30 min with agitation, followed by enzymatic digestion in freshly prepared 0.1% (v/w) collagenase type I from *clostridium histolyticum* enzyme dissolved in serum free medium supplemented with 2% antibiotics at relative humidity of 95%, 5% CO_2, 37C for 18 h. Undigested cartilage debris were removed from the cell/medium suspension by filtering the mixture through a nylon filter strainer of 70 µm pore size (BD Bioscience, Europe). Supernatant was spun to isolate the chondrocytes, followed by three washes using PBS containing 10% antibiotics. Finally cells were suspended in DMEM contains 2% antibiotics and 10% FCS and cultured at 37C, 95% humidity and 5% CO_2 until confluent. Cells viability was assessed by the trypan blue exclusion test (Sigma Aldrich, UK).

Vitamin D treatment
To explore the influence of the active form of vitamin D (1, 25, α-dihydroxy vit D) on equine articular chondrocytes, cells were cultivated at 2×10^5 cells/well until sub-confluence. Chondrocytes were treated with different concentrations of 1, 25 $(OH)_2D_3$ (1×10^{-9} and 1×10^{-12}) [37] in serum free-medium for 24 h. The concentration of DMSO was below 0.1%. At the end of 1, 25 $(OH)_2D_3$ treatment, cells were washed three times sterile PBS and whole cell lysate was collected for protein expression using western blotting.

Western blotting

Total cellular protein lysate was isolated using radio-immunopreceptation assay (RIPA) buffer (150 mM NaCl, 50 mM Tris HCl, pH 7.5, 5 mM ethylene glycol tetraacetic acid (EGTA), 1% Triton, 0.5% sodium deoxycholate and 0.1% sodium dodecyl sulphate) supplemented with phosphatase and protease inhibitors cocktail (Roche Diagnostic, Mannheim, Germany) on ice. Protein concentration was quantified by the Bradford method. 25 μg of total protein lysate was mixed with sample buffer (0.5 M Tris HCl, pH 6.8, 100% glycerol, 20% SDS, 0.5% bromophenol blue and 5% β-mercaptoethanol), separated on a 4/1-% polyacrylamide gel, then electrically transferred to PVDF membrane (Invitrogen, UK) by semi-dry apparatus (Bio-Rad, UK). Transferred proteins were blocked for non-specific binding in 5% (w/v) non-fat milk diluted in Tris base buffer saline with 0.1% tween 20 (TBS-T), followed by incubation with designated antibodies overnight. Membranes were washed five times in TBST, followed by 1 h incubation with goat anti-rabbit IgG conjugated with horseradish peroxidase (HRP) (Dako, UK) secondary antibody at room temperature. Finally, five washes were carried out for 5 min each followed by developing the membrane with the Amersham ECL western blot enhanced chemiluminescence kit (GE Healthcare, UK) and visualized by exposing to X-ray film.

Quantitative PCR

The quantitative real time PCR was carried out using a LightCycler 480 PCR System (Roche Diagnostics) using SYBR green DNA-binding fluorescent dye. 20 μl reactions were made in an optical 96-well reaction plate in triplicate and contained: a mixture of template cDNA (5 μl), sense and anti-sense gene specific primers (0.8 μl) (20 pmol), SYBR Green detection reagent and 3.4 μl RNA-free water. Reaction plate was sealed with ABI-prism optical adhesive cover, and spun at 2000 rpm for 2 min. The expression of target genes were normalised against GAPDH (Glyceraldehyde 3-phosphate dehydrogenase) and HPRT (Hypoxanthine-guanine phosphoribosyltransferase) using comparative cycle of threshold (Ct) value method.

Statistical analysis

Data values are presented as the mean SEM. Each experiment was performed in triplicate. The relative expression on the graphs represents the mean of a combination of three experiments. The differences between animals were analyzed utilizing Students t-test followed by the Bonferroni correction. P values less than or equal to 0.05 were considered statically significant.

Results and discussion

The active form of vitamin D (1, 25 $(OH)_2D_3$) has been shown to play an important role in Ca^{2+} homeostasis.

Intracellular Ca^{2+} concentration is involved in several chondrocyte functions including ECM biosynthesis. This study was designed to explore the possible influence of the active form of vitamin D (1, 25 $(OH)_2D_3$) treatment on the expression of Sox 9 and members of the TRP subfamily that are known as epithelial Ca^{2+} channels (TRPV5 and TRPV6 channels).

The influence of two doses (1 10^{-9} and 1 10^{-12}) of 1, 25 $(OH)_2D_3$ treatment for 24 h on the expression of the transcription factor (Sox 9) at the protein level in equine chondrocytes was examined. Incubation of chondrocytes for 24 h with 1 10^{-12} of 1, 25 $(OH)_2D_3$ significantly increased the expression of Sox 9 by nearly 3 fold, whereas treatment with 1 10^{-9} of 1, 25 $(OH)_2D_3$ induced more than 2 fold increase (P < 0.001) compared to the non-treated cells (Figure 1).

In chondrocytes, the mechanisms by which Sox 9 expression is regulated is of great interest due to the critical role played by this transcription factor in controlling the chondrocyte phenotype. The findings of the present study shows an elevation of Sox 9 protein in response to treatment with 1, 25 $(OH)_2D_3$. Isolation of chondrocytes results in their de-differentiation and loss of chondrocyte phenotype to fibroblast phenotype [22]. Recent studies have reported changes in the pattern of expression of some proteins during the course of chondrocytes de-differentiation. Cultivation of de-differentiated chondrocytes in alginate gel results in restoring the expression of transcription factor Sox 9 and chondrocyte differentiation markers such as collagen type II and aggrecan [38]. The current study indicated that treatment of cultured chondrocytes with 1, 25 $(OH)_2D_3$ results in restoring of the transcription factor Sox 9, which could indicate chondrocyte re-differentiation and restoration of the chondrocyte phenotype.

The physiological activities of Sox 9 protein were demonstrated by mouse genetic studies which indicated the importance of Sox 9 expression level in determining chondrocyte phenotype during development [39,40]. These studies demonstrated that chondrocyte differentiation and cartilage development were severely affected by knock-out or knock-in of a single allele of Sox 9. Moreover, the campomelic dysplasia is a genetic disorder characterized by multiple developmental abnormalities including cartilage induced by haplo-insufficiency of Sox 9 [18].

In human normal articular chondrocytes, Sox 9 expression was progressively reduced by passage in monolayer cell culture [41]. Moreover its expression is reduced in osteoarthritic chondrocytes compared to the normal articular chondrocytes [23] and has therefore been suggested to contribute in osteoarthritis disease processes by altering the ECM gene expression [42]. Thus, improving the expression of Sox 9 by 1, 25 $(OH)_2D_3$ treatment could

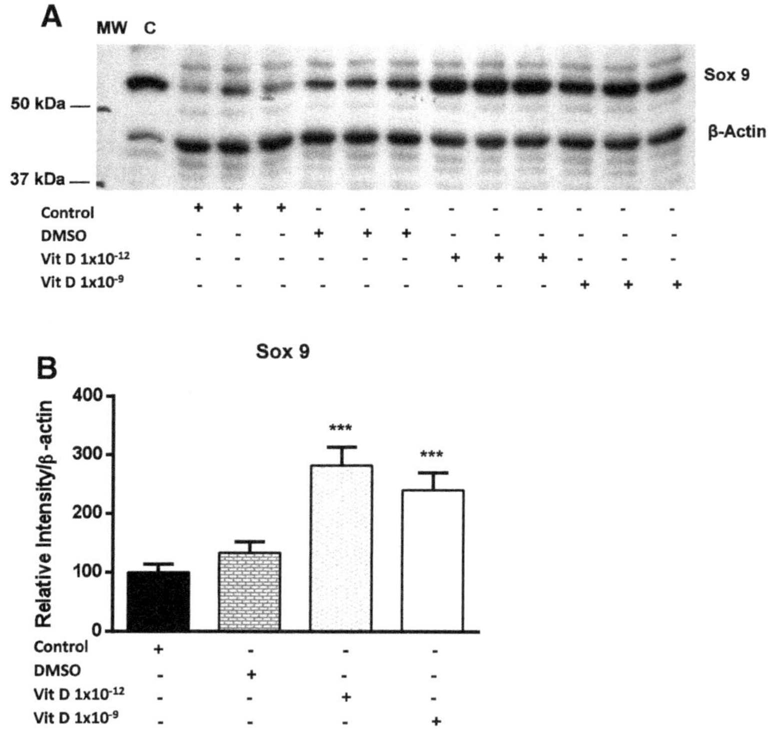

Figure 1 The effect of 1, 25 (OH)$_2$D$_3$ on protein expression of Sox 9 in equine articular chondrocytes following 24 h of incubation. (A) Western blot analysis of protein expression levels using an antibody raised against the Sox 9 protein. Freshly isolated equine whole chondrocyte lysate was used as a positive control C. β-actin was used as an internal load control. **(B)** the level of expression of the protein following 24 h of incubation with different doses (1 ?10 $^{-9}$ and 1 ?10 $^{-12}$) of 1, 25 (OH)$_2$D$_3$ treatment in addition to the solvent (DMSO). Values are presented as means ?S.E. ***Indicates statistically significant at $P < 0.001$.

provide a new insight into the prevention and/or treatment of osteoarthritis.

In fetal cartilage as well as in mature cartilage, the main role of Sox 9 is to maintain the chondrocyte phenotype in addition to inhibition of hypertrophic chondrocyte differentiation [43]. Therefore, down regulation of Sox 9 expression was suggested to be a precondition for hypertrophic alteration occurring in degenerative cartilage [42,44]. As Sox 9 protein is augmented by 1, 25 (OH)$_2$D$_3$ treatment, this study suggests that 1, 25 (OH)$_2$D$_3$ treatment could be utilized to improve the ECM proteins biosynthesis by enhancing the anabolic activities of articular chondrocytes. After further passages, chondrocytes have been shown to exhibit a more pronounced de-differentiated phenotype and lower levels of Sox 9 [41]. Further studies on the effects of 1, 25 (OH)$_2$D$_3$ upon these cells would be of great interest.

We have previously shown that the expresson of both TRPV5 and 6 channels, at the protein level, are inversely related to chondrocyte de-differentiation [24]. In this study we therefore examined the influence of 1, 25 (OH)$_2$D$_3$ treatment on the expression of TRPV5/6 channels on primary equine articular chondrocytes isolated at passage two. The findings of the current study indicated that no

statistically significant changes were observed on the expression of TRPV5 and TRPV6 channels at the protein level following treatment with different doses of 1, 25 (OH)$_2$D$_3$ (Figure 2). To ensure that 1, 25 (OH)$_2$D$_3$ had no effect on the expression of these channels we also examined its effect upon their transcript levels but again no significant changes were observed (Figure 3). Previous studies conducted on mouse investigated the effect of 1, 25 (OH)$_2$D$_3$ on epithelial Ca^{2+} transport including TRPV5 and 6 channels following 1, 25 (OH)$_2$D$_3$ treatment [45,46]. These studies demonstrated that the expression of TRPV5 and TRPV6 channels in kidney and duodenum were stimulated by binding of VDR (vitamin D receptor) at transcriptional level. TRPV6 was reported to have several classes of VDREs (vitamin D response elements) in humans and the mouse [45,47,48].

In the vitamin D receptor knockout mice model, the level of gene expression pattern of the TRPV6 and 5 channels in duodenum is dramatically down-regulated, in contrast no changes are observed in gene expression of other Ca^{2+} transporters [49,50]. The role of vitamin D in the stimulation of Ca^{2+} transporters has emerged recently. Therefore, the correlation between 1, 25 (OH)$_2$D$_3$ treatment and Ca^{2+} transport proteins were investigated

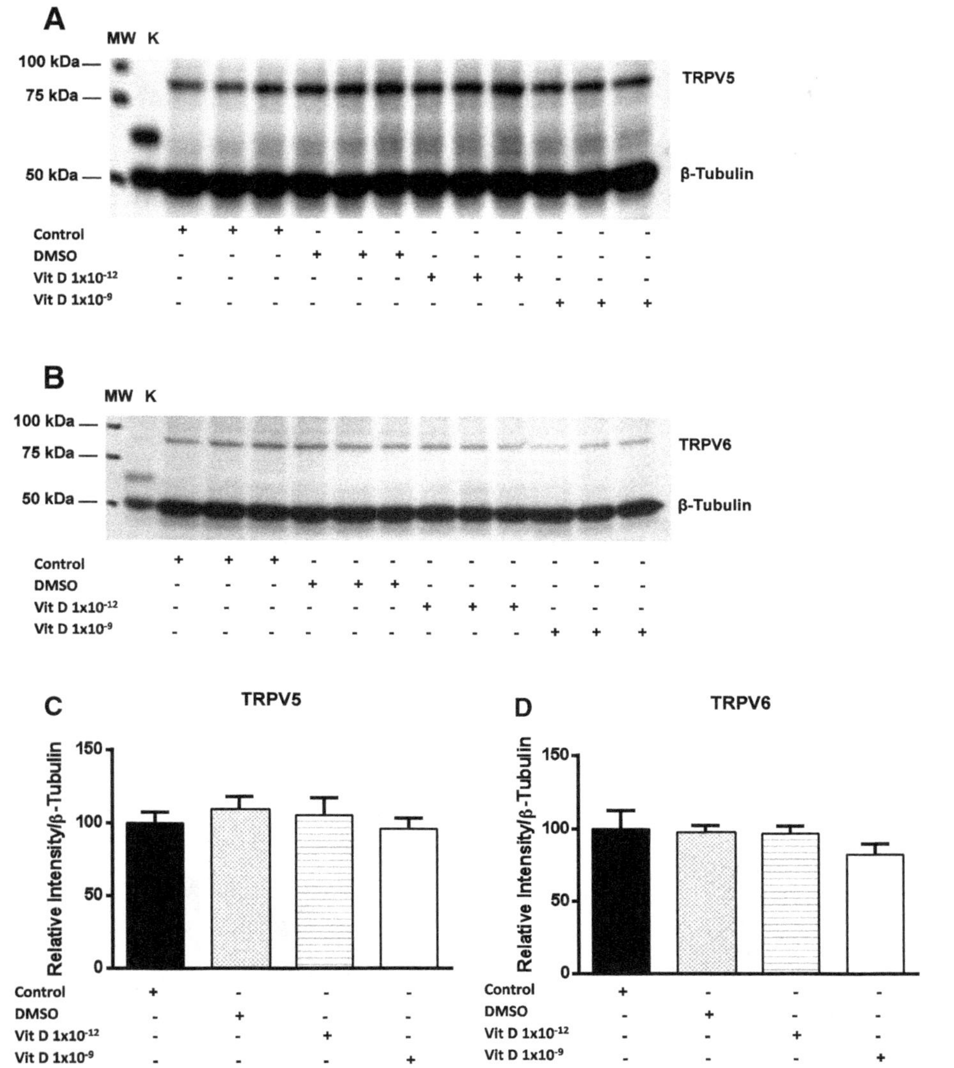

Figure 2 Expression of TRPV5 and TRPV6 channels in equine articular chondrocytes at the protein level. The changes in expression of TRPV5 **(A)** and TRPV6 **(B)** channels following treatment with 1, 25 vit D were assessed by densitometry analysis of western blots and normalized to the expression of housekeeping protein (β-Tubulin). Equine kidney (K) lysate was used as a positive control. Levels of expression at the protein level of the TRPV5 **(C)** and TRPV6 **(D)** channels following 24 h of incubation with different doses of 1, 25 $(OH)_2D_3$ in addition to the solvent (DMSO). Data presented as a mean ?S.E.

in several cell types. 1, 25 $(OH)_2D_3$ was demonstrated to increase the expression of TRPV6 channel, calbinding-D9k and PMCA1b genes in Caco-2 cell lines [36,51]. This finding was not consistent with the current study, where no changes were observed on the expression of either TRPV5 or 6 channels at mRNA level or protein levels. As stated above we have previously observed an inverse correlation between TRPV 5 and 6 levels and differentiation state. The current study suggests that the changes in levels may not be directly linked to the de-differentiation process. This however, needs further study as does the mechanism by which Sox-9 is regulated by 1,

25 $(OH)_2D_3$. Ca^{2+} ions and Ca^{2+} channels have been implicated in chondrocyte metabolism [52,53]. In vitro studies show that augmented extracellular Ca^{2+} promotes differentiation toward a hypertrophic phenotype, in contrast Ca^{2+} reduction improves ECM protein synthesis including collagen type II and aggrecan and delays hypertrophy [54]. Therefore, investigating the effect of the 1, 25 $(OH)_2D_3$ on the expression of Ca^{2+} channels could be considered as a candidate for treatment of joint diseases.

The results of the current study do not, however, suggest an involvement of TRPV5/6 ion channels in the regulation of vitamin D mediated up-regulation of sox-9.

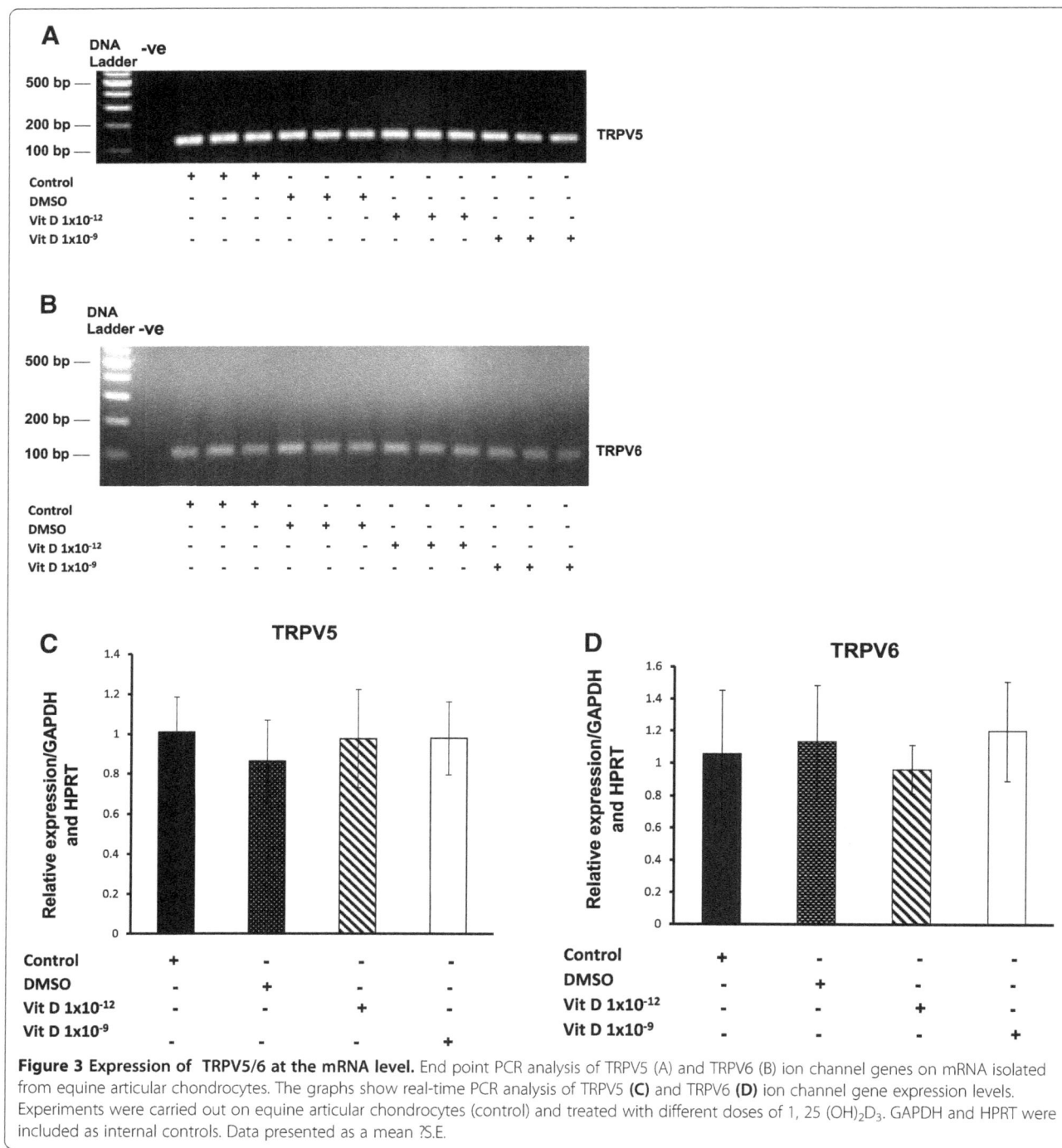

Figure 3 Expression of TRPV5/6 at the mRNA level. End point PCR analysis of TRPV5 (A) and TRPV6 (B) ion channel genes on mRNA isolated from equine articular chondrocytes. The graphs show real-time PCR analysis of TRPV5 **(C)** and TRPV6 **(D)** ion channel gene expression levels. Experiments were carried out on equine articular chondrocytes (control) and treated with different doses of 1, 25 (OH)$_2$D$_3$. GAPDH and HPRT were included as internal controls. Data presented as a mean ?S.E.

Conclusions

This study examined the effects of the active form of vitamin D (1, 25 (OH)$_2$D$_3$) upon the expression of transcription factor Sox 9 and the calcium sensitive TRPV 5 and 6 channels, in equine articular chondrocytes, *in vitro*. An increased expression of Sox 9, in these chondrocytes, in response to the active form of vitamin D suggests that this compound acts to inhibit the progressive de-differentiation that is normally observed in these cells. It is also supportive of previous studies indicating that 1α, 25-dihydroxyvitamin D$_3$ can have a protective effect upon cartilage in animals *in vivo*. There was no effect of vitamin D (1, 25 (OH)$_2$D$_3$) upon the expression of either TRPV channel at either the protein or mRNA level. We have previously observed a correlation between the degree of differentiation and the expression levels of TRPV 5/6 and suggested that these ion channels may have a direct interaction with the differentiation process *in vitro*. The data in the present study do not support this.

Competing interests

The authors declare that they have no competing interests.

Authors contributions

ISM and PTL designed the experiment and analyzed the results. ISM carried out cell culture work. Both authors wrote the manuscript and made subsequent changes.

Author details

[1]School of Veterinary Medicine and Science, Faculty of Medicine and Health Sciences, University of Nottingham, Sutton Bonington Campus, Leicestershire LE12 5RD, UK. [2]Medical Research Council-Arthritis Research UK Centre for Musculoskeletal Ageing Research, University of Nottingham, Leicestershire, UK. [3]School of Veterinary Medicine and Science, Tripoli University, Tripoli, Libya.

References

1. Boyan BD, Sylvia VL, Dean DD, Del Toro F, Schwartz Z: **Differential regulation of growth plate chondrocytes by 1alpha,25-(OH)2D3 and 24R,25-(OH)2D3 involves cell-maturation-specific membrane-receptor-activated phospholipid metabolism.** *Crit Rev Oral Biol Med* 2002, **13**(2):143 154.
2. Tetlow LC, Smith SJ, Mawer EB, Woolley DE: **Vitamin D receptors in the rheumatoid lesion: expression by chondrocytes, macrophages, and synoviocytes.** *Ann Rheum Dis* 1999, **58**(2):118 121.
3. Lehmann B, Meurer M: **Vitamin D metabolism.** *Dermatol Ther* 2010, **23**(1):2 12.
4. Fleet JC: **Genomic and proteomic approaches for probing the role of vitamin D in health.** *Am J Clin Nutr* 2004, **80**(6 Suppl):1730S 1734S.
5. Xue Y, Karaplis AC, Hendy GN, Goltzman D, Miao D: **Exogenous 1,25-dihydroxyvitamin D3 exerts a skeletal anabolic effect and improves mineral ion homeostasis in mice that are homozygous for both the 1alpha-hydroxylase and parathyroid hormone null alleles.** *Endocrinology* 2006, **147**(10):4801 4810.
6. Castillo EC, Hernandez-Cueto MA, Vega-Lopez MA, Lavalle C, Kouri JB, Ortiz-Navarrete V: **Effects of vitamin D supplementation during the induction and progression of osteoarthritis in a rat model.** *Evid Based Complement Altern Med* 2012, **2012**:156563.
7. Morais Da Silva S, Hacker A, Harley V, Goodfellow P, Swain A, Lovell-Badge R: **Sox9 expression during gonadal development implies a conserved role for the gene in testis differentiation in mammals and birds.** *Nat Genet* 1996, **14**(1):62 68.
8. Zhao Q, Eberspaecher H, Lefebvre V, De Crombrugghe B: **Parallel expression of Sox9 and Col2a1 in cells undergoing chondrogenesis.** *Dev Dyn* 1997, **209**(4):377 386.
9. Wright E, Hargrave MR, Christiansen J, Cooper L, Kun J, Evans T, Gangadharan U, Greenfield A, Koopman P: **The Sry-related gene Sox 9 is expressed during chondrogenesis in mouse embryos.** *Nat Genet* 1995, **9**(1):15 20.
10. Shintani N, Hunziker EB: **Chondrogenic differentiation of bovine synovium: bone morphogenetic proteins 2 and 7 and transforming growth factor beta1 induce the formation of different types of cartilaginous tissue.** *Arthritis Rheum* 2007, **56**(6):1869 1879.
11. Sive JI, Baird P, Jeziorsk M, Watkins A, Hoyland JA, Freemont AJ: **Expression of chondrocyte markers by cells of normal and degenerate intervertebral discs.** *Mol Pathol* 2002, **55**(2):91 97.
12. Lefebvre V, Huang W, Harley VR, Goodfellow PN, de Crombrugghe B: **SOX 9 is a potent activator of the chondrocyte-specific enhancer of the pro alpha1(II) collagen gene.** *Mol Cell Biol* 1997, **17**(4):2336 2346.
13. Bridgewater LC, Lefebvre V, de Crombrugghe B: **Chondrocyte-specific enhancer elements in the Col11a2 gene resemble the Col2a1 tissue-specific enhancer.** *J Biol Chem* 1998, **273**(24):14998 15006.
14. Sekiya I, Tsuji K, Koopman P, Watanabe H, Yamada Y, Shinomiya K, Nifuji A, Noda M: **SOX9 enhances aggrecan gene promoter/enhancer activity and is up-regulated by retinoic acid in a cartilage-derived cell line, TC6.** *J Biol Chem* 2000, **275**(15):10738 10744.
15. Kou I, Ikegawa S: **SOX9-dependent and -independent transcriptional regulation of human cartilage link protein.** *J Biol Chem* 2004, **279**(49):50942 50948.
16. Xie WF, Zhang X, Sakano S, Lefebvre V, Sandell LJ: **Trans-activation of the mouse cartilage-derived retinoic acid-sensitive protein gene by Sox 9.** *J Bone Miner Res* 1999, **14**(5):757 763.
17. Brew CJ, Clegg PD, Boot-Handford RP, Andrew JG, Hardingham T: **Gene expression in human chondrocytes in late osteoarthritis is changed in both fibrillated and intact cartilage without evidence of generalised chondrocyte hypertrophy.** *Ann Rheum Dis* 2010, **69**(1):234 240.
18. Wagner T, Wirth J, Meyer J, Zabel B, Held M, Zimmer J, Pasantes J, Bricarelli FD, Keutel J, Hustert E, Wolf U, Tommerup N, Schempp W, Scherer G: **Autosomal sex reversal and campomelic dysplasia are caused by mutations in and around the SRY-related gene SOX9.** *Cell* 1994, **79**(6):1111 1120.
19. Bi W, Deng JM, Zhang Z, Behringer RR, de Crombrugghe B: **Sox 9 is required for cartilage formation.** *Nat Genet* 1999, **22**(1):85 89.
20. Hargus G, Kist R, Kramer J, Gerstel D, Neitz A, Scherer G, Rohwedel J: **Loss of Sox 9 function results in defective chondrocyte differentiation of mouse embryonic stem cells in vitro.** *Int J Dev Biol* 2008, **52**(4):323 332.
21. Cucchiarini M, Thurn T, Weimer A, Kohn D, Terwilliger EF, Madry H: **Restoration of the extracellular matrix in human osteoarthritic articular cartilage by overexpression of the transcription factor SOX9.** *Arthritis Rheum* 2007, **56**(1):158 167.
22. Benya PD, Padilla SR, Nimni ME: **Independent regulation of collagen types by chondrocytes during the loss of differentiated function in culture.** *Cell* 1978, **15**(4):1313 1321.
23. Tew SR, Clegg PD, Brew CJ, Redmond CM, Hardingham TE: **SOX9 transduction of a human chondrocytic cell line identifies novel genes regulated in primary human chondrocytes and in osteoarthritis.** *Arthritis Res Ther* 2007, **9**(5):R107.
24. Hdud IM, El-Shafei AA, Loughna P, Barrett-Jolley R, Mobasheri A: **Expression of transient receptor potential vanilloid (TRPV) channels in different passages of articular chondrocytes.** *Int J Mol Sci* 2012, **13**(4):4433 4445.
25. Hdud IM, Mobasheri A, Loughna PT: **Effect of osmotic stress on the expression of TRPV4 and BKCa channels and possible interaction with ERK1/2 and p38 in cultured equine chondrocytes.** *Am J Physiol Cell Physiol* 2014, **306**(11):C1050 C1057.
26. Hdud IM, Mobasheri A, Loughna PT: **Effects of cyclic equibiaxial mechanical stretch on alpha-BK and TRPV4 expression in equine chondrocytes.** *SpringerPlus* 2014, **3**:59.
27. Vennekens R, Hoenderop JG, Prenen J, Stuiver M, Willems PH, Droogmans G, Nilius B, Bindels RJ: **Permeation and gating properties of the novel epithelial Ca(2+) channel.** *J Biol Chem* 2000, **275**(6):3963 3969.
28. Nilius B, Vennekens R, Prenen J, Hoenderop JG, Bindels RJ, Droogmans G: **Whole-cell and single channel monovalent cation currents through the novel rabbit epithelial Ca2+ channel ECaC.** *J Physiol* 2000, **527**(Pt 2):239 248.
29. Hoenderop JG, Nilius B, Bindels RJ: **Calcium absorption across epithelia.** *Physiol Rev* 2005, **85**(1):373 422.
30. Gavenis K, Schumacher C, Schneider U, Eisfeld J, Mollenhauer J, Schmidt-Rohlfing B: **Expression of ion channels of the TRP family in articular chondrocytes from osteoarthritic patients: changes between native and in vitro propagated chondrocytes.** *Mol Cell Biochem* 2009, **321**(1 2):135 143.
31. Taparia S, Fleet JC, Peng JB, Wang XD, Wood RJ: **1,25-Dihydroxyvitamin D and 25-hydroxyvitamin D mediated regulation of TRPV6 (a putative epithelial calcium channel) mRNA expression in Caco-2 cells.** *Eur J Nutr* 2006, **45**(4):196 204.
32. Teerapornpuntakit J, Dorkkam N, Wongdee K, Krishnamra N, Charoenphandhu N: **Endurance swimming stimulates transepithelial calcium transport and alters the expression of genes related to calcium absorption in the intestine of rats.** *Am J Physiol Endocrinol Metab* 2009, **296**(4):E775 E786.

33. Embark HM, Setiawan I, Poppendieck S, van de Graaf SF, Boehmer C, Palmada M, Wieder T, Gerstberger R, Cohen P, Yun CC, Bindels RJ, Lang F: Regulation of the epithelial Ca2+ channel TRPV5 by the NHE regulating factor NHERF2 and the serum and glucocorticoid inducible kinase isoforms SGK1 and SGK3 expressed in Xenopus oocytes. Cell Physiol Biochem 2004, 14(4 6):203 212.

34. Zanello LP, Norman AW: Multiple molecular mechanisms of 1 alpha, 25(OH)2-vitamin D3 rapid modulation of three ion channel activities in osteoblasts. Bone 2003, 33(1):71 79.

35. Zhang M, Wang JJ, Chen YJ: Effects of mechanical pressure on intracellular calcium release channel and cytoskeletal structure in rabbit mandibular condylar chondrocytes. Life Sci 2006, 78(21):2480 2487.

36. Lee GS, Jung EM, Choi KC, Oh GT, Jeung EB: Compensatory induction of the TRPV6 channel in a calbindin-D9k knockout mouse: Its regulation by 1,25-hydroxyvitamin D3. J Cell Biochem 2009, 108(5):1175 1183.

37. Fernandez-Cancio M, Audi L, Carrascosa A, Toran N, Andaluz P, Esteban C, Granada ML: Vitamin D and growth hormone regulate growth hormone/insulin-like growth factor (GH-IGF) axis gene expression in human fetal epiphyseal chondrocytes. Growth Horm IGF Res 2009, 19(3):232 237.

38. Kumar D, Lassar AB: The transcriptional activity of Sox9 in chondrocytes is regulated by RhoA signaling and actin polymerization. Mol Cell Biol 2009, 29(15):4262 4273.

39. Akiyama H, Lyons JP, Mori-Akiyama Y, Yang X, Zhang R, Zhang Z, Deng JM, Taketo MM, Nakamura T, Behringer RR, McCrea PD, de Crombrugghe B: Interactions between Sox9 and beta-catenin control chondrocyte differentiation. Genes Dev 2004, 18(9):1072 1087.

40. Kawakami Y, Rodriguez-Leon J, Izpisua Belmonte JC: The role of TGFbetas and Sox9 during limb chondrogenesis. Curr Opin Cell Biol 2006, 18(6):723 729.

41. Tew SR, Hardingham TE: Regulation of SOX9 mRNA in human articular chondrocytes involving p38 MAPK activation and mRNA stabilization. J Biol Chem 2006, 281(51):39471 39479.

42. Aigner T, Gebhard PM, Schmid E, Bau B, Harley V, Poschl E: SOX9 expression does not correlate with type II collagen expression in adult articular chondrocytes. Matrix Biol 2003, 22(4):363 372.

43. Bi W, Huang W, Whitworth DJ, Deng JM, Zhang Z, Behringer RR, de Crombrugghe B: Haploinsufficiency of Sox9 results in defective cartilage primordia and premature skeletal mineralization. Proc Natl Acad Sci U S A 2001, 98(12):6698 6703.

44. Gebhard PM, Gehrsitz A, Bau B, Soder S, Eger W, Aigner T: Quantification of expression levels of cellular differentiation markers does not support a general shift in the cellular phenotype of osteoarthritic chondrocytes. J Orthop Res 2003, 21(1):96 101.

45. Nijenhuis T, Hoenderop JG, van der Kemp AW, Bindels RJ: Localization and regulation of the epithelial Ca2+ channel TRPV6 in the kidney. J Am Soc Nephrol 2003, 14(11):2731 2740.

46. Song Y, Peng X, Porta A, Takanaga H, Peng JB, Hediger MA, Fleet JC, Christakos S: Calcium transporter 1 and epithelial calcium channel messenger ribonucleic acid are differentially regulated by 1, 25 dihydroxyvitamin D3 in the intestine and kidney of mice. Endocrinology 2003, 144(9):3885 3894.

47. Haussler MR, Whitfield GK, Kaneko I, Haussler CA, Hsieh D, Hsieh JC, Jurutka PW: Molecular mechanisms of vitamin D action. Calcif Tissue Int 2013, 92(2):77 98.

48. Hoenderop JG, Muller D, Van Der Kemp AW, Hartog A, Suzuki M, Ishibashi K, Imai M, Sweep F, Willems PH, Van Os CH, Bindels RJ: Calcitriol controls the epithelial calcium channel in kidney. J Am Soc Nephrol 2001, 12(7):1342 1349.

49. Bouillon R, Van Cromphaut S, Carmeliet G: Intestinal calcium absorption: molecular vitamin D mediated mechanisms. J Cell Biochem 2003, 88(2):332 339.

50. Van Cromphaut SJ, Dewerchin M, Hoenderop JG, Stockmans I, Van Herck E, Kato S, Bindels RJ, Collen D, Carmeliet P, Bouillon R, Carmeliet G: Duodenal calcium absorption in vitamin D receptor-knockout mice: functional and molecular aspects. Proc Natl Acad Sci U S A 2001, 98(23):13324 13329.

51. Wood RJ, Tchack L, Taparia S: 1,25-Dihydroxyvitamin D3 increases the expression of the CaT1 epithelial calcium channel in the Caco-2 human intestinal cell line. BMC Physiol 2001, 1:11.

52. Kirsch T, Swoboda B, von der Mark K: Ascorbate independent differentiation of human chondrocytes in vitro: simultaneous expression of types I and X collagen and matrix mineralization. Differentiation 1992, 52(1):89 100.

53. Bonen DK, Schmid TM: Elevated extracellular calcium concentrations induce type X collagen synthesis in chondrocyte cultures. J Cell Biol 1991, 115(4):1171 1178.

54. Koyano Y, Hejna M, Flechtenmacher J, Schmid TM, Thonar EJ, Mollenhauer J: Collagen and proteoglycan production by bovine fetal and adult chondrocytes under low levels of calcium and zinc ions. Connect Tissue Res 1996, 34(3):213 225.

12

Novel SNP in the coding region of the FTO gene is associated with marbling score in Hanwoo (Korean cattle)

Eui-Ryong Chung

Abstract

The *fat mass and obesity associated* (FTO) gene plays an important role in the regulation of energy homeostasis, fat deposition and obesity. For this reason, the FTO gene is a physiological and functional candidate gene for carcass and meat quality traits in beef cattle. The objectives of this study were to identify SNPs in the exonic regions of FTO gene and to evaluate the association of these SNPs with carcass traits in Hanwoo (Korean cattle). In this study, we newly identified two exonic SNPs in Hanwoo population. The g.125550A > T SNP was located in exon 3 and the g.175675C > T SNP was located in exon 6. Genotyping of the two SNP markers was carried out using PCR-RFLP analysis in Hanwoo steers to evaluate their association with carcass traits. As a result, g.125550A > T SNP genotype was significantly associated with effects on marbling score. Animals with the AA and TT homozygous genotypes had a significantly higher marbling score ($p < 0.001$) than those with AT heterozygous genotype, and this was significant after *Bonferroni* correction of the significance threshold ($p = 0.003$). Dominance effect was also observed for the marbling score ($P < 0.05$) with higher marbling score of homozygous animals. However, no significant associations with meat quality traits were observed for the g.175675C > T SNP. Our results suggest that the exonic SNP g.125550A > T in the FTO gene may be used as a DNA marker for the selection of Hanwoo with higher marbling.

Keywords: FTO, Marbling, SNP, Meat quality, Hanwoo

Background

Marbling is the most economically important trait in beef cattle industry of Korea. Generally, marbling means the amount and distribution of intramuscular fat in a cross section of musculus longissimus muscle [1]. In particular, eating quality traits such as taste, juiciness and tenderness of meat are influenced by the amount of intramuscular fat [2]. High levels of marbling improve the palatability and acceptability of beef by affecting the taste and tenderness of the meat [3,4]. Therefore, the challenge to the beef cattle industry in Korea is the production of meat with higher marbling score. Knowledge on the genetic background of fat tissue accumulation and better understanding of the molecular mechanism of marbling are very important in beef production. To identify an informative gene or DNA marker for meat quality traits,

candidate genes can be selected on the basis of the function of the encoded protein in physiological processes controlling energy homeostasis.

Recently, the *fat mass and obesity associated* (FTO) gene have been shown to have a relatively large effect on body mass index (BMI) and obesity-related traits in various human populations, suggesting that FTO associated with the development of fat tissue and adiposity [5-9]. Some studies also suggested that the FTO may play a key role in the regulation of energy homeostasis and associated with increased lipolytic activity in adipose tissue [9,10]. In livestock species, polymorphic variations in the FTO gene are associated with fatness-related traits such as intramuscular fat deposition and backfat thickness [11-15]. In addition, the bovine FTO gene is located near the QTL region affecting meat quality traits on BTA18. Therefore, the FTO gene is considered as a positional and functional candidate gene for meat quality in beef cattle. However, most studies have so far been published for the pig. Moreover, the association analysis between

Correspondence: erchung@sangji.ac.kr
Division of Animal Science and Resources, College of Life Science and Natural Resources, Sangji University, 660 Usandong, Wonju, Gangwondo 220-702, South Korea

single nucleotide polymorphisms (SNPs) within exons of the FTO gene and meat quality traits has not been reported in Hanwoo (Korean cattle). The objectives of our study were to identify new SNPs in the exonic regions of FTO gene and to evaluate the association of these SNPs with meat quality traits in Hanwoo.

Methods

Animals and genomic DNA extraction

A total of 300 steers, which were registered in a national database of cattle and guided with standardized breeding programs provided by the *Hoengseong* Hanwoo population in Korea, were used to genotype and collect carcass traits. To confirm genetic inheritance of the identified SNPs, 51 Hanwoo animals, which were used for the evaluation of performance ability for proven sires in the national progeny testing programs, were genotyped. The carcass data analyzed in the current study included marbling score (MS), meat color (MC), fat color (FC), meat texture (MT), meat maturity (MA), backfat thickness (BF), *Longissimus dorsi* muscle area (LMA), and carcass weight (CW). The means carcass traits values were 5.98 ± 1.73 for MS, 4.91 ± 0.37 for MC, 3.00 ± 0.10 for FC, 1.09 ± 0.28 for MT, 2.03 ± 0.19 for MA, 11.99 ± 4.01 mm for BF, 90.05 ± 9.57 cm^2 for LMA and 427.11 ± 46.96 kg for CW.

Meat samples were collected from 13th thoracic rib to the first lumbar vertebrae of the steers within 24 hr of slaughter and evaluated according to the Animal Product Grading System of Korea. Genomic DNA was extracted

from tail root-hair by using a NaCl precipitation protocol [16] with a slight modification. The DNA sample was suspended in TE buffer (10 mM Tris–HCl, pH 7.4; 1 mM EDTA) and stored at –20°C until analysis.

Resequencing and SNP discovery

The bovine FTO gene is mapped to chromosome 18, which includes nine exons coding 505 amino acids (Figure 1). To identify polymorphisms within the coding region of the Hanwoo FTO gene, the nine pairs of primers (Table 1) were designed to amplify the all exon regions based on the genomic sequence of the bovine FTO gene from NCBI GenBank reference sequences (Genomic sequence: AC_000175.1 chromosome 18 reference *Bos taurus* UMD 3.1 primary assembly) using a web-based software Primer 3.0 program (http://frodo.wi.mit.edu/primer3). To determine SNP identification, pooled DNA samples from the sixty unrelated animals were amplified by PCR using the each primer pair. The PCR amplification was performed in a DNA thermal cycler (Perkin Elmer Cetus, Norwalk, CT). The PCR reaction was performed in a 20 µL reaction mixture containing 0.1 µM of each primer, 1.5 mM MgCl₂, 250 µM of each dNTP and 1 unit of *Taq* DNA polymerase, 10 X reaction buffer and 50 ng of pooled DNA as template. The PCR conditions were at 94°C for 5 min, followed by 35 cycles of 94°C for 30 s, annealing at 55°C to 57°C for each primer for 20 s and 72°C for 1 min, with a final extension at 72°C for 5 min. The PCR products were verified by 2% agarose gel electrophoresis

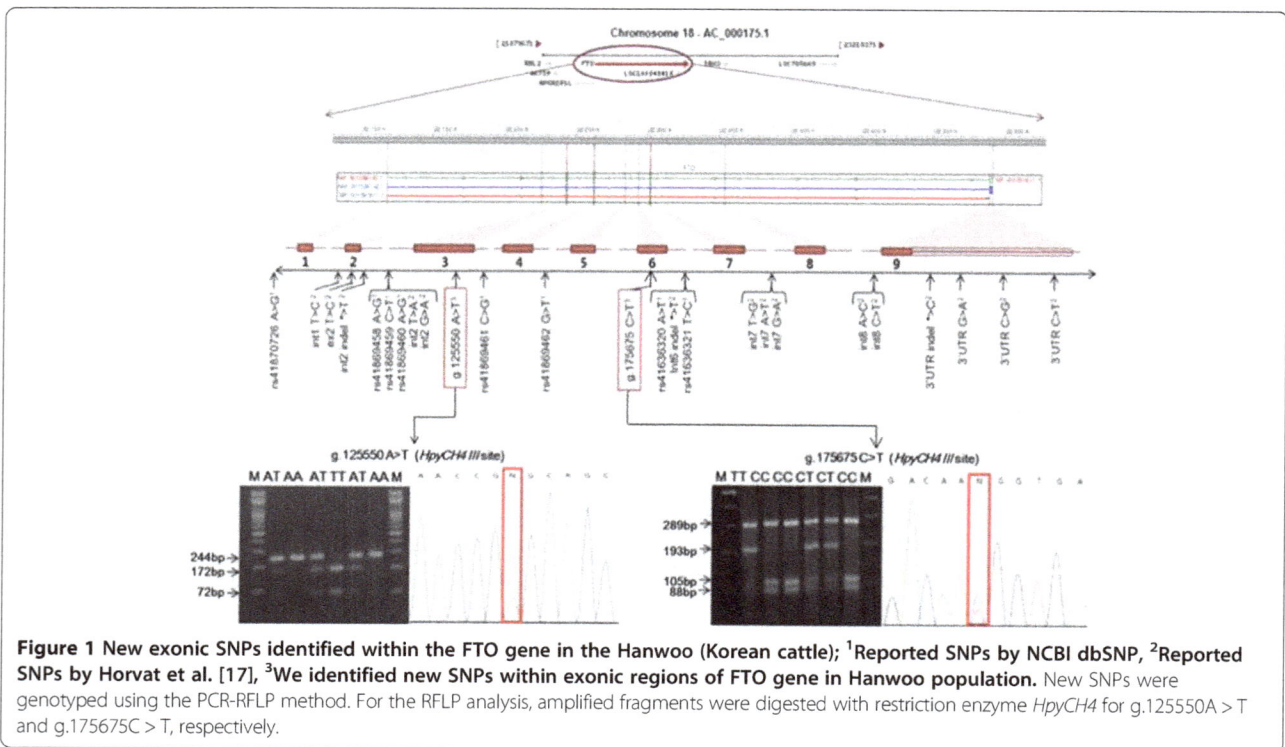

Figure 1 New exonic SNPs identified within the FTO gene in the Hanwoo (Korean cattle); [1]Reported SNPs by NCBI dbSNP, [2]Reported SNPs by Horvat et al. [17], [3]We identified new SNPs within exonic regions of FTO gene in Hanwoo population. New SNPs were genotyped using the PCR-RFLP method. For the RFLP analysis, amplified fragments were digested with restriction enzyme *HpyCH4* for g.125550A > T and g.175675C > T, respectively.

Table 1 The primer sequence used to amplify the exonic sequence variants in the bovine FTO gene

Primer	Primer sequences (5'-3')	Amplified region	Fragment size (bp)	Location	GenBank accession no.
FTO-1	F-GGCTATAACGGCAGCATGAA R-CCAGGTGGTGAAGATGGAAA	exon 1	844	4 ~ 847	
FTO-2	F-AAGGCATTGTCAGCCTGGATG R-ATATGGTACAAGGAGCACCAGG	exon 2	714	107821 ~ 108536	
FTO-3	F-TGTCTGTAATGGAGCTTGGACC R-CAATCATATCTCCAGAGCCTGC	exon 3	906	124883 ~ 125788	
FTO-4	F-CCTTTGCAACTCTCTAGTCCTGTG R-CTCTCCAAACTCCCCAAACTTC	exon 4	1064	144220 ~ 145283	
FTO-5	F-TTTCCAGGCAAGAGTACTGGAG R-CTTGACCTTGACCTTGACCTTG	exon 5	675	166054 ~ 166728	AC_000175.1
FTO-6	F-AACGTGGGGTTCACTTACTGTG R-GCAAGAATACTGGAGTGGGTTG	exon 6	651	175553 ~ 176203	
FTO-7	F-CTGTCATGTGCTTTAGGATCTC R-TGTCTGTGAGTCAACTTCTGTT	exon 7	617	183602 ~ 184218	
FTO-8	F-TAAGGCAGTGGAGCTTTGGA R-TGTGCACAGAAATGAGGCTG	exon 8	821	235720 ~ 236540	
FTO-9	F-TCACAGTGGATTGCCAAGGT R-TGTGGCTCATGCAACTGAAG	exon 9	824	420877 ~ 421700	

and purified with Wizard Prep PCR purification kit (SolGent, Korea). The purified PCR amplicons were directly sequenced in both directions using BigDye™ Terminator V3.1 Cycle Sequencing Kit in an ABI PRISM 3730 Genetic Analyzer (Applied Biosystems, Foster City, CA, USA) according to the manufacturer's instructions. GENESCAN 3.7 ANALYSIS software (Applied Biosystems, USA) was used to assemble the sequences and to identify polymorphisms.

SNP genotyping using PCR-RFLP

Two SNPs, g.125550A > T and g.175675C > T, newly identified in the exons 3 and 6 of the Hanwoo FTO gene respectively, were genotyped using PCR-RFLP method. PCR primers used for PCR-RFLP analysis were 5'-TTCCTCAA GCTCAACAGCTACC-3' and 5'-ACGGTTCCTCTTTC AGGTATGG-3' for the g.125550A > T SNP, and 5'-AGA GTCAGTTCTAGGTGCTGTGGT-3' and 5'-CCACAGT TCTCAGAAGCCCTTA-3' for the g.175675C > T SNP. PCR amplifications were carried out as for the SNP detection protocol. For the RFLP analysis, amplified fragments were digested with restriction enzyme *HpyCH4III* for g.125550A > T and g.175675C > T at 37°C for 3 h, respectively. The digested DNA fragments were separated on 2% agarose gel by electrophoresis with 1 X TBE buffer. The gels were stained with ethidium bromide (EtBr) and the fragments were visualized using a UV transilluminator (Ultra Rum Inc, USA). To define each genotype according to band patterns, the PCR products of different RFLP type

corresponding to each genotype were sequenced and analyzed for nucleotide changes.

Statistical analyses

Allele and genotype frequencies of SNPs and Hardy-Weinberg equilibrium were estimated and tested using PROC ALLELE (SAS Inst. Inc., Cary, NC, USA). Associations between SNP genotypes and phenotypes of carcass traits were analyzed with the following liner mixed model using the MIXED procedure of SAS (SAS Inst. Inc., Cary, NC, USA). The data were analyzed according to the following model: $Y_{ijk} = \mu + G_i + S_j + P_k + \beta A_l + e_{ijkl}$, where $Y_{ijk} =$ the value of carcass traits; $\mu =$ the overall mean for each trait; G = fixed effect of single SNP marker genotype; S = random effect of sire, P = fixed effect of parity; A = fixed effect of age as a covariate; and $e_{ijkl} =$ random residual effect. The *Bonferroni* correction for multiple testing was performed to remove any false positives. The additive and dominance genetic effects were also estimated using REG procedure of SAS according to Falconer and Mackay [18]. This research was followed by internationally recongnized guidelines (Institutional Animal Care and Use Committee, IACUC) for animal experiment.

Results

In this study, we sequenced all exon regions of the bovine FTO gene and newly identified two exonic SNPs in Hanwoo steers (Figure 1). An A to T transition (g.125550A > T SNP) was located in exon 3 and a C to T substitution (g.175675C > T SNP) was located in exon 6

(GenBank accession no. AC_000175). Genotyping of the two novel SNPs in the coding region of the bovine FTO gene was performed by a PCR-RFLP method in Hanwoo population. The allele and genotype frequencies in the two novel SNPs of the FTO gene are shown in Table 2. The genotypic frequencies were as follows: 47.0% AA, 44.3% AT and 8.7% TT for the g.125550A > T SNP; 52.0% CC, 37.3% CT and 10.7% TT for the g.175675C > T SNP. The observed genotype distributions were in good agreement with those expected according to the Hardy-Weinberg equilibrium in this population. Overall average of heterozygosity (He) and polymorphic information contents (PIC) for two SNP markers were calculated to 0.576 and 0.486, respectively. The results of the association analysis for the FTO gene SNP markers with various carcass traits are presented in Table 3. The g.125550A > T SNP genotype was significantly associated with effects on MS. Animals with the AA or TT homozygous genotype had a significantly higher MS ($p < 0.001$) than the animals with AT heterozygous genotype and this was significant after *Bonferroni* correction of the significance threshold ($p = 0.003$). Dominance effect was also observed for the marbling score (P < 0.05) with higher marbling score of homozygous animals. However, no significant association was detected between the g.175675C > T SNP genotype and carcass traits measured in this study.

Discussion

The development of functional genomics has revealed a large number of genes related with fat accumulation and lipid metabolism. Several QTLs for carcass traits and a large number of potential candidate genes based on a known relationship with physiological or biochemical processes and carcass traits have been reported in beef cattle [19-21]. However, a limited number of genetic markers have been recognized for carcass and meat quality traits in beef cattle and these markers explain a relatively small proportion of the genetic variation for a limited number of traits [22]. Fatness related traits such as marbling are classified as a quantitative trait with a high contribution of genetic variation [23]. Thus, identification of DNA polymorphism associated with (or responsible for) fatness traits may be useful for marker assisted selection.

Many regulatory factor genes are involved in the formation of marbling [24], and increased knowledge about the relationship between these genes and the fat accumulation and lipid metabolism is of utmost importance for the improvement of meat quality in beef cattle. The FTO gene is a new candidate gene related to the development of fat tissue and obesity. This gene encodes 2-oxoglutarate-dependent oxygenases, which are involved in various processes, including DNA repair, fatty acid metabolism and posttranslational modifications and is highly expressed in the hypothalamic pituitary adrenal axis, which plays a key role in the control of energy balance [25]. Therefore, the polymorphic variation in this gene may play a causal role in the regulation of energy homeostasis or in the development of fat tissue [5,26]. In addition, the FTO is a transcriptional co-activator, which facilitates transcription from unmethylated and methylation-inhibited gene promoters and enhances C/EBPs binding to DNA, and that it may play a role in the regulation of adiposity [15,27]. These findings suggest close link between FTO and fat deposition and lipid metabolism. In human, genetic variation of this gene was extensively studied and had confirmed that a strong and highly significant association with fatness-related traits such as fat mass, obesity and body mass index in different populations [5-7]. This gene has also become one of the candidate genes for detecting polymorphism associated with the fat deposition in livestock species. Several studies on pig have shown the association between the polymorphism of the FTO gene and intramuscular fat (IMF) [11,13,28]. A significant association between g.276 T > G polymorphism of the 3'-untranslated region and intramuscular fat deposition was reported by Fontanesi et al. [28] in the Italian White Duroc breed. Fan et al. [12] also suggested significant association of the g.-1191A > G SNP in 5′ regulatory region of the porcine FTO gene and intramuscular fat content. These results provided that the porcine FTO gene might play an important function in intramuscular fat and not directly in subcutaneous or abdominal fat deposition [13]. Recently, some previous studies have reported the bovine FTO gene associated with carcass traits in beef cattle breeds. Wei et al. [14] have reported that g.1071C > T SNP in exon 5 within the bovine FTO gene was associated with backfat thinckness and *longissmus*

Table 2 The genotypes and allele frequencies for SNP markers of the FTO gene in Hanwoo

SNP marker	Frequency (%)					He	PIC	HWE	
	Genotype (No. of head)			Allele				χ^2	p-value
g.125550A > T	AA (141)	AT (133)	TT (26)	C	G	0.573	0.478	0.758	0.684
	47.0	44.3	8.7	69.3	30.7				
g.175675C > T	CC (156)	CT (112)	TT (32)	C	T	0.579	0.494	2.955	0.228
	52	37.3	10.7	70.7	29.3				

He: Heterozygosity, PIC: Polymorphic Information Content, HWE: Hardy-Weinberg Equilibrium.

Table 3 The least square means and standard errors for carcass traits with genetic effects according to the FTO genotypes in Hanwoo population

SNP	Traits[1]	Genotype (mean ± SE)			P-value		Genetic effects	
					P_{raw}	$P_{corrected}$	Additive	Dominance
		AA	AT	TT				
	MS/1 ~ 7	6.082 ± 0.194[a]	5.173 ± 0.200[b]	6.000 ± 0.461[a]	<0.001	0.003	0.082 ± 0.501	1.734 ± 0.641[***]
	MC/4 ~ 6	4.931 ± 0.033	4.956 ± 0.034	4.923 ± 0.079	0.846	1.246	0.008 ± 0.086	−0.058 ± 0.110
	FC/2 ~ 4	3.000 ± 0.009	2.958 ± 0.009	3.000 ± 0.022	0.539	0.922	−0.000 ± 0.024	0.028 ± 0.031
g.125550A>T	MT/1 ~ 2	1.068 ± 0.036	1.159 ± 0.037	1.076 ± 0.086	0.209	0.415	−0.008 ± 0.094	−0.173 ± 0.120
	MA/2 ~ 3	2.054 ± 0.034	2.130 ± 0.035	2.153 ± 0.082	0.244	0.478	−0.099 ± 0.089	−0.052 ± 0.0114
	BF/mm	14.301 ± 0.614	14.797 ± 0.631	13.692 ± 1.455	0.730	1.013	0.609 ± 1.579	−1.600 ± 2.022
	LMA/cm^2	92.671 ± 1.132	89.739 ± 1.165	91.230 ± 2.684	0.199	0.348	1.440 ± 2.913	4.423 ± 3.731
	CW/kg	431.045 ± 6.360	432.972 ± 6.541	430.923 ± 15.071	0.976	1.462	0.131 ± 16.358	−3.964 ± 20.947
		CC	CT	TT				
	MS/1 ~ 7	5.932 ± 0.164	5.854 ± 0.258	5.571 ± 0.390	0.695	0.695	0.360 ± 0.424	−0.204 ± 0.668
	MC/4 ~ 6	2.991 ± 0.006	3.000 ± 0.010	3.000 ± 0.016	0.748	0.748	−0.008 ± 0.017	−0.008 ± 0.027
	FC/2 ~ 4	1.101 ± 0.027	1.062 ± 0.043	1.190 ± 0.065	0.272	0.272	−0.088 ± 0.071	0.167 ± 0.112
g.175675C > T	MT/1 ~ 2	2.135 ± 0.027	2.062 ± 0.043	2.000 ± 0.065	0.097	0.097	0.135 ± 0.071	0.010 ± 0.112
	MA/2 ~ 3	30.694 ± 0.086	3.666 ± 0.135	3.476 ± 0.205	0.618	0.618	0.218 ± 0.222	−0.162 ± 0.351
	BF/mm	13.652 ± 0.442	14.125 ± 0.694	14.248 ± 1.049	0.782	0.782	−0.585 ± 1.139	−0.359 ± 1.796
	LMA/cm^2	91.872 ± 0.874	91.666 ± 1.371	93.190 ± 2.074	0.815	0.815	−1.317 ± 2.251	1.730 ± 3.548
	CW/kg	431.525 ± 4.755	441.416 ± 7.456	431.428 ± 11.272	0.520	0.520	0.096 ± 12.235	−19.875 ± 19.289

[1] MS, marbling score; MC, meat color; FC, fat color; MT, meat texture; MA, meat maturity; BF, backfat thickness; LMA, M. *Longissimus dorsi* muscle area; CW, carcass weight.
[***]P <0.001.
[a,b]Within a row, means with different superscripted letters are different (P <0.01).

muscle area traits in five Chinese native cattle breeds. In this study, however, the effect of genotypes of the two novel SNPs was not statistically significant for backfat thickness and *longissmus* muscle area. This might be supported by the fact that Hanwoo breed exhibits very low or negative genetic correlation between marbling and backfat thickness [29,30]. Also, Jevšinek Skok et al. [31] have reported that the T > C SNP in exon 2 of the FTO gene showed a significant effect on growth and carcass traits such as live weight at slaughter, carcass weight and lean weight in Slovenian Simmental population. However, the associations between the SNPs of the FTO gene and intramuscular fat have not been reported in beef cattle. To the knowledge of the authors, the present study is the first report on association between exonic SNP marker within coding regions of the FTO gene and fatness-related traits such as marbling score in beef cattle, suggesting a possible effect of the gene on intramuscular fat deposition. The two exonic SNPs identified in this study by resequencing have not been reported previously and are therefore novel. The bovine FTO gene was near to the QTL region for carcass traits [32,33]. As the A to C substitution in exonic SNP g.125550A > T does not result in an amino acid change due to synonymous mutation, this is unlikely to be the causal mutation for marbling and instead is likely to

be genetically linked to nearby QTL or causative mutations that affect marbling score in beef cattle. The synonymous SNPs could affect gene function because they are under evolutionary pressure, alter the structure, function and expression level of proteins, and can affect mRNA splicing, stability and protein structure as well as folding [34].

Conclusion

In conclusion, we identified two novel exonic SNPs of the FTO gene in Hanwoo population and the g.125550A > T SNP genotype showed significant effect on marbling score. These findings suggest that the FTO gene-specific SNP identified in this study may be useful molecular marker for selection to increase the levels of marbling in Hanwoo. This study also will contribute to a better understanding of the molecular mechanisms of marbling in beef cattle. Further studies are will be needed to confirm the associated effect on other population and breeds.

Competing interests
The author declares that he has no competing interests.

Acknowledgments
This work was supported by Sangji University Research Fund, 2012. The author thanks Dr. Shin and Junsung Kim for their experimental assistance and collecting the data.

References

1. Cameron PJ, Zembayashi M, Lunt DK, Mitsuhashi T, Mitsumoto T, Ozawa S, Smith SB: **Relationship between Japanesebeef marbling standard and intramuscular lipid in the M. longissimusthoracis of Japanese Black and American Wagyu cattle.** *Meat Sci* 1994, **38**:361–364.
2. Platter WJ, Tatum JD, Belk KE, Koontz SR, Chapman PL, Smith GC: **Effects of marbling and shear force onconsumers' willingness to pay for beef strip loin steaks.** *J Anim Sci* 2005, **83**:890–899.
3. Boylston TD, Morgan SA, Johnson KA, Busboom JR, Wright RW Jr, Reeves JJ: **Lipid content and composition of Wagyu and domestic breeds of beef.** *J Agric Food Chem* 1995, **43**:1202–1207.
4. Matsuishi M, Fujimori M, Okitanim A: **Wagyu beefaroma in Wagyu (Japanese Black cattle) beef preferred bythe Japanese over imported beef.** *Anim Sci J* 2001, **72**:498–504.
5. Dina C, Meyre D, Gallina S, Durand E, Korner A, Jacobson P, Carlsson LM, Kiess W, Vatin V, Lecoeur C, Delplanque J, Vaillant E, Pattou F, Ruiz J, Weill J, Levy-Marchal C, Horber F, Potoczna N, Hercberg S, Le Stunff C, Bougneres P, Kovacs P, Marre M, Balkau B, Cauchi S, Chevre JC, Froguel P: **Variation in FTO contributes to childhood obesity and severe adult obesity.** *Nat Genet* 2007, **39**:724–726.
6. Frayling TM, Timpson NJ, Weedon MN, Zeggini E, Freathy RM, Lindgren CM, Perry JR, Elliott KS, Lango H, Rayner NW, Shields B, Harries LW, Barrett JC, Ellard S, Groves CJ, Knight B, Patch AM, Ness AR, Ebrahim S, Lawlor DA, Ring SM, Ben-Shlomo Y, Jarvelin MR, Sovio U, Bennett AJ, Melzer D, Ferrucci L, Loos RJ, Barroso I, Wareham NJ, et al: **A common variant in the FTO gene is associated with body mass index and predisposes to childhood and adult obesity.** *Science* 2007, **316**:889–894.
7. Scuteri A, Sanna S, Chen WM, Uda M, Albai G, Strait J: **Genome-wide association scan shows genetic variant in the FTO gene are associated with obesity-related traits.** *PLoS Genet* 2007, **3**:e115.
8. Andreasen CH, Stender-Petersen KL, Mogensen MS, Torekov SS, Wegner L, Andersen G: **Low physical activity accentuates the effect of the FTO rs9939609 polymorphism on body fat cccumaulation.** *Diabetes* 2008, **57**:95–101.
9. Wahlen K, Sjolin E, Hoffstedt J: **The common rs9939609 gene variant of the fat mass- and obesity-associated gene FTO is related to fat cell lipolysis.** *J Lipid Res* 2008, **49**:607–611.
10. Zhang B, Zhang Y, Zhang L, Wang J, Li Z, Chen H: **Allelic polymorphism detected in the bovine FTO gene.** *Mol Biotechnol* 2011, **49**:257–262.
11. Fontanesi L, Scotti E, Buttazzoni L, Dall'Olio S, Bagnato A, Fiego DPL, Davoli R, Russo V: **Confirmed association between a single nucleotide polymorphism in the FTO gene and obesity-related traits in heavy pigs.** *Mol Biol Rep* 2010, **37**:461–466.
12. Fan B, Du ZQ, Rothschild MF: **The fat mass and obesity associated (FTO) gene is associated with intramuscular fat content and growth rate in the pig.** *Anim Biotechnol* 2009, **20**:58–70.
13. Zhang LF, Miao XT, Hua XC, Jiang XL, Lu YP, Xu NY: **Polymorphism in 5' regulatory region of the porcine fat mass and obesity associated (FTO) gene is associated with intramuscular fat content in a Jinhua x Pietrain F2 reference population.** *J Anim Vet Adv* 2009, **8**:2329–2334.
14. Wei S, Zan L, Ujan JA, Wang H, Yang Y, Adoligbe C: **Novel polymorphism of the bovine fat mass and obesity-associated (FTO) gene are related to backfat thickness and longissimus muscle area in five Chinese native cattle breeds.** *Afr J Biotechnol* 2011, **10**:2820–2824.
15. Dvořáková V, Bartenschlager H, Stratil A, Horák P, Stupka R, Cítek J, Sprysl M, Hrdlicová A, Geldermann H: **Association between polymorphism in the FTO gene and growth and carcass traits in pig crosses.** *Genet Sel Evol* 2012, **44**:13.
16. Miller SA, Dykes DD, Polesky HF: **A simple salting out procedure for extracting DNA from human nucleated cell.** *Nucleic Acids Res* 1988, **16**:1215.
17. Horvat S, Renčelj A, Kunej T, Razpet A, Jevšinek-Skok D, Dovč T, Zgur S, Petrič N, Planinc M, Malovrh S, Kovač M: *Association Analysis of the Fat Mass and Obesity-associated (FTO) Gene SNP Markers with Growth / Carcass Traits in Paternal Half-Sib Families of Slovenian Simmental Cattle. 9th World Congress on Genetics Applied to Livestock Production.* Leipzig, German: 2010.
18. Falconer DS, Mackay TFC: **Introduction to quantitative genetics.** In *Longman scientific and 10 technical.* 4th edition. New York, NY, USA: 1996.
19. Casas E, Shackelford SD, Keele JW, Stone RT, Kappes SM, Koohmaraie M: **Quantitative trait loci affection growth and carcass composition of cattle segregating alternate forms of myostatin.** *J Anim Sci* 2000, **78**:560–569.
20. Stone RT, Deele JW, Shackelford SD, Kappes SM, Koohmaraie M: **A primary screen of the bovine genome for quantitative trait loci affecting carcass and growth traits.** *J Anim Sci* 1999, **77**:1379–1384.
21. Shin SC, Chung ER: **Novel SNPs in the bovine ADIPOQ and PPARGC1A genes are associated with carcass traits in Hanwoo (Korean cattle).** *Mol Biol Rep* 2013, **40**:4651–4660.
22. Dekkers JC: **Commercial application of marker- and gene assisted selection in livestock: strategies and lessons.** *J Anim Sci* 2004, **82**(E-Suppl):E313-328.
23. Switonski M, Stachowiak M, Cieslak J, Bartz M, Grzes M: **Genetics of fat tissue accumulation in pigs: a comparative approach.** *J Appl Genet* 2010, **51**:153–168.
24. Rosen ED, Spiegelman BM: **Molecular regulation of adipogenesis.** *Annu Rev Cell Dev Biol* 2000, **16**:145–171.
25. Gerken T, Girard CA, Tung YC, Webby CJ, Saudek V, Hewitson KS, Yeo GS, McDonough MA, Cunliffe S, McNeill LA, Galvanovskis J, Rorsman P, Robins P, Prieur X, Coll AP, Ma M, Jovanovic Z, Farooqi IS, Sedgwick B, Barroso I, Lindahl T, Ponting CP, Ashcroft FM, O'Rahilly S, Schofield CJ: **The obesity-associated FTO gene encodes a 2-oxoglutarate-dependent nucleic acid demethylase.** *Science* 2007, **318**:1469–1472.
26. Fredriksson R, Hägglund M, Olszewski PK, Stephansson O, Jacobsson JA, Olszewska AM, Levine AS, Lindblom J, Schiöth HB: **The obesity gene, FTO, is of ancient origin, up-regulated during food deprivation and expressed in neurons of feeding-related nuclei of the brain.** *Endocrinology* 2008, **149**:2062–2071.
27. Wu Q, Saunders RA, Szkudlarek-Mikho M, Serna Ide L, Chin KV: **The obesity-associated FTO gene is a transcriptional coactivator.** *Biochem Biophys Res Commun* 2010, **22**(401):390–395.
28. Fontanesi L, Scotti E, Buttazzoni L, Davoli R, Russo V: **The porcine fat mass and obesity associated (FTO) gene is associated with fat deposition in Italian Duroc pigs.** *Anim Genet* 2009, **40**:90–93.
29. Choi TJ, Kim SD, Salces AJ, Baik DH: **Genetic parameter estimation on the growth and carcass traits in Hanwoo (Korean cattle).** *J Anim Sci Technol (Kor)* 2006, **48**:759–766.
30. Won JI, Kim JB, Lee JK: **Evaluation of genetic ability for meat quality in Hanwoo cow.** *J Anim Sci Technol (Kor)* 2010, **54**:259–264.
31. Jevšinek Skok D, Kunej T, Renčelj A, Razpet A, Dovč P, Žgur S, Petrič N, Planinc M, Malovrh Š, Kovač M, Horvat S: **Use of SNP markers within the fat mass and obesity-associated (FTO) gene to Verify pedigrees and determine haplotypes in paternal half-sib families of slovenian simmental cattle.** *Agric Conspec Sci* 2011, **76**:333–336.
32. Casas E, Shackelford SD, Keele JW, Koohmaraie M: **Detection of quantitative trait loci for growth and carcass composition in cattle.** *J Anim Sci* 2003, **81**:2976–2983.
33. Gutierrez-Gil B, Wiener P, Nute GR, Burton D, Gill JL, Wood JD, Williams JL: **Detection of quantitative trait loci for meat quality traits in cattle.** *Anim Genet* 2008, **39**:51–61.
34. Hunt R, Sauna ZE, Ambudkar SV, Gottesman MM, Kimchi-Sarfaty C: **Silent (synonymous) SNPs: should we care about them.** *Methods Mol Biol* 2009, **578**:23–39.

13

Association of genotype of *POU1F1* intron 1 with carcass characteristics in crossbred pigs

Gye-Woong Kim[1], Jae-Young Yoo[2] and Hack-Youn Kim[1*]

Abstract

This study was carried out to investigate the association of *POU1F1* (POU domain, class 1, transcription factor 1, Pit1, renamed as *POU1F1*) gene with backfat thickness (mm), carcass weight (kg), pH, and color values (L^*, a^*, b^*) in crossbred pigs (Landrace x Yorkshire x Duroc). Frequency of the AA genotype indel was at the highest level (66.67%). Frequency of A allele (0.81) was higher than that of b allele (0.19). This population followed Hardy-Weinberg equilibrium. Carcass weights and a^* values of the three genotypes were all significantly different ($p < 0.05$), respectively. However, backfat thickness, L^*, b^*, visual color, and pH of the three genotypes were not significantly different ($p > 0.05$). Visual color was negatively correlated with L^* ($r = -0.521$) and b^* ($r = -0.390$) values, L^* value was correlated with b^* ($r = 0.419$) value, and a^* value was positively correlated with b^* ($r = 0.612$) value. These results indicate that the *POU1F1* gene affected carcass weight and meat redness.

Keywords: *POU1F1*, Carcass characteristics, Genotypes, Gene, Intron 1

Background

As of 2012, the annual average per capita meat consumption in Korea reached 40.6 kg, with pork accounting for the highest average consumption at 19.0 kg [1]. The Food and Agriculture Organization (FAO) revealed that global consumption of pork underwent an annual increase of 2.1% from 1999 to 2010 [2]. Due to an increase in average income, consumers have developed a preference for high quality pork, and research has been conducted to select genetically superior traits in pork [3]. In Korea, a crossbreed of three pig breeds, Landrace, Yorkshire, and Duroc (LY x D), has the largest market share based on its higher number of offspring, faster growth, and higher proportion of meat [4,5] compared to other crossbreeds.

The *POU1F1* (POU domain, class 1, transcription factor 1, *Pit1*) gene is expressed in the pituitary gland of pigs. It is associated with growth regulation as well as secretion of growth hormones, prolactin, and thyroprotein β-subunit [6,7]. *POU1F1* is located on chromosome 13 and consists of six exons and five introns, and it has been shown to influence quantitative traits related to pig growth [market weight, growth rate, and average daily gain (ADG)] as well as carcass traits (carcass weight, backfat thickness, amount of meat) [8-12]. Song et al. [13] reported that genetic change in intron 1 (insertion or deletion of 313 bp) of the *POU1F1* gene is related to pig growth, and the genotype frequencies of intron 1 vary among breeds. Their analysis of 15 breeds, including 11 breeds of Chinese traditional varieties and four traditional breeds, confirmed 100% frequency of the BB genotype in Chinese traditional varieties of Tibetan, Lingao, Rongchang, Songliao Black, and Min. Frequencies of the BB genotype were high in the Meishan, Erhualian, Fenjing, and Leping Spotted breeds, although the highest genotypes were found to be AA type, AB type, and BB type in the Landrace, Pietrain, and Duroc breeds, respectively. Further, A allele was shown to be present at higher frequencies than B allele in all four breeds [13]. Previous studies have reported the relationship between *POU1F1* gene intron region 3 and meat quality characteristics of the carcass [14] as well as that between daily gain and backfat thickness for the Landrace breed according to the *POU1F1* genotype [15]. However, further accurate research is required on *POU1F1* gene intron region 1 in crossbreeds of the three varieties.

Therefore, this study estimated *POU1F1* genotype frequency in 168 crossbred pigs that were raised domestically

* Correspondence: kimhy@kongju.ac.kr
[1]Department of Animal Resources, Kong-Ju National University, # 54 Daehakro, Yesan, Chungnam 340-702, Korea
Full list of author information is available at the end of the article

in order to obtain basic data on meat quality assessment for the establishment of a high-quality pork production system by analyzing meat color and carcass characteristics.

Methods

Test animals

Standard laboratory animals used in this study included 168 crossbred pigs, with a live weight of approximately 110 ± 5 kg. The breeds were Landrace, Yorkshire, and Duroc (LYD or YLD) and were raised in Gyeonggi-do, Chungcheong-do, and Jeolla-do on Nong-Hyup animal feed, which meets the specifications of the National Research Council (NRC).

Sampling and genomic DNA extraction

To determine carcass grades and extract DNA, we transported animals to a slaughterhouse in Chungnam. After securing loin portions as laboratory samples, carcass grades were surveyed, and the meat was frozen and transported to the laboratory. The muscle tissues were then thawed at room temperature ($-25°C$), and an appropriate amount was purified using a QIAamp Mini kit (QIAGEN®, USA) following the manufacturer's protocol. Purified genomic DNA was dissolved in distilled water or TE buffer (Tris-EDTA, pH 8.0) and collected by centrifugation. The amount of genomic DNA was confirmed by performing electrophoresis, followed by storage at $-25°C$ until polymerase chain reaction (PCR) analysis.

Genetic analysis

The primers used for amplification of the *POU1F1* gene are shown in Table 1. The primers used in the study were synthesized by the laboratory that amplified the intron 1 region of the *POU1F1* gene. For the normal genotype (AA type), an amplified product of approximately 1,091 bp in size was obtained, whereas BB type amplified product was 778 bp. The synthesized primers were diluted to a final concentration of 10 pmol.

For the PCR mixture, prime Taq DNA polymerase from GeNet Biosystem (GeNet Bio. Co., Korea) was used. Each PCR mixture was 20 μL and consisted of 1× buffer (0.01 M Tris–HCl, 0.05 M KCl, 0.08% nonidet), 1.5 mM $MgCl_2$, 1 mM of each dNTP, 10 pM of primers, and 2.5 U

of Taq. PCR conditions were as follows: pre-denaturation at 94°C for 5 min, followed by 35 cycles of denaturation for 30 s at 94°C, annealing for 1 min at 55°C, and extension for 1 min at 72°C. PCR was completed with a final extension for 5 min at 72°C. The gene was amplified and analyzed using 1.5% agarose gel electrophoresis with TAE (40 mM Tris-acetate, 1.0 mM EDTA) buffer. The quality of DNA was checked by adding 0.1 mg of ethidium bromide to 1 μL of electrophoresis buffer in the gel and the electrophoresis medium, followed by electrophoresis for approximately 20 min at 100 V. A 100-bp ladder plus (MBI Fermentas Inc., USA) size marker was used as reference for the electrophoretic analysis, and images were obtained under UV illumination.

Carcass trait according to genotype

Backfat thickness, carcass weight, meat color, and pH were surveyed as carcass characteristics. For backfat thickness, grading data from the slaughterhouse were used, whereas weight after slaughtering was measured and used as the carcass weight of each pork sample (live weight, 110 ± 5 kg). For meat color, Hunter L^* (lightness), a^* (redness), and b^* (yellowness) values on the cut surface of the fillet were measured using a color meter (Model NF333; Nippon Denshoku Co., Japan), and the standard meat color board at the time was Y = 92.40, X = 0.3136, and y = 0.3196. The visual color based on the meat color reference standard suggested by the National Pork Producers Council (NPPC) was directly compared for assessment based on seven steps, ranging from a light color score of 1 to a dark color score of 7. The pH of muscle was measured using a glass electrode pH meter (340; Mettler Toledo, Switzerland) after mixing 5 g of sample with 20 mL of distilled water, followed by homogenization for 1 min at 8,000 rpm using a homogenizer (Nissei, Model AM-7; Japan).

Statistical analysis

The gene and genotypes of *POU1F1* intron 1 were analyzed using the SAS package (Statistical Analysis System, version 9.3). For intergroup genetic equilibrium, significance was tested using the χ^2-test. After pigs were slaughtered, backfat thickness (mm), carcass weight (kg), meat color, and pH were determined to calculate the mean and standard deviations of each genotype, and correlation among carcass characteristics was calculated. For comparison of differences among groups, significance was tested at a 5% level by using Duncan's multiple-range tests.

Results and discussion

POU1F1 Intron 1 genetic analysis

The results of PCR analysis of *POU1F1* intron 1 are shown in Figure 1. The PCR-amplified products

Table 1 Nucleotide sequences of primer pairs for PCR amplification of *POU1F1* DNA fragment

Name of primer	Sequence	Size of products
POU1F1 gene in intron 1 region		1091 or 778 bp
	Forward 5'-CAT TCC CAT TCT GCC ATT TG-3' (20 mer)	
	Reverse 5'-CCT GTT GCT GTG TTT CCC AG-3' (22 mer)	

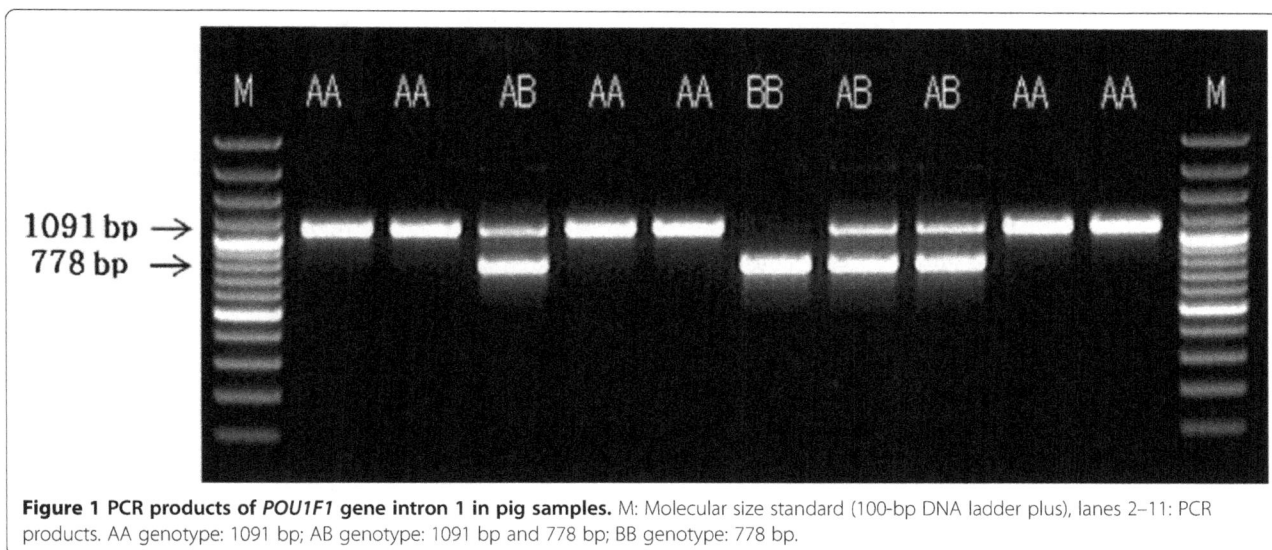

Figure 1 PCR products of *POU1F1* gene intron 1 in pig samples. M: Molecular size standard (100-bp DNA ladder plus), lanes 2–11: PCR products. AA genotype: 1091 bp; AB genotype: 1091 bp and 778 bp; BB genotype: 778 bp.

consisted of three band types, namely AA and BB types as homozygotes and AB type as a heterozygote. Among these, AA type appeared as a single band of approximately 1091 bp, AB type appeared as two bands of 1091 and 778 bp, and BB type was a single band of 778 bp. These results are identical to those reported by Song et al. [13], who confirmed the presence of AA, AB, and BB types in intron 1. On the other hand, a previous study identified AA, AG, and GG types in the *FOU1F1* genotype of Korean beef [16].

The results of the analysis of the *POU1F1* intron 1 SNP genotype and allele frequencies are shown in Table 2. The genotypes were categorized as homozygotes, namely AA and BB types, as well as heterozygotes, which is AB type, along with the existence of alleles A and B. Of the 168 carcasses investigated, 112 were AA type, accounting for the highest percentage (66.67%); 48 (28.57%) and 8 (4.76%) were AA and BB types, respectively. These results are similar to those of Song et al. [13], who showed that the Landrace breed is AA type (0.76). On the other hand, they contradict the genotype frequencies of the Yorkshire breed, which was shown to be AB type (0.58), and Duroc breed, which was shown to be AA type (0.37). The discrepancies in these findings might be attributed to variations in gene distribution between crossbreeds and pure breeds, as crossbreeding involves changes in genotype frequencies. In contrast, the results of this study are similar to those of Franco et al. [15] and Yu et al. [17], who reported that the frequency of AB type is higher than that of AA type. Estimation of genotype frequencies revealed that A allele had a very high expression frequency (0.81), whereas B allele had a very low frequency (0.19). Meanwhile, Franco et al. [15] reported expression frequencies of 0.88 and 0.12 for A and B types in the Landrace breed, respectively. On the other hand, Yu et al. [17] and Song et al. [13] demonstrated high expression frequencies for A allele in the Yorkshire, Duroc, Landrace, and Hampshire breeds, whereas the frequency of B allele was high in the Chinese Meishan breed. These results might be due to differences among each breed. The results of conformity determined on the basis of Hardy–Weinberg law were not

Table 2 Genotypic frequencies of *POU1F1* gene and statistical test for Hardy-Weinberg equilibrium

Genotype	No. of pigs	Percentage (%)	Gene frequency		Statistical test
			A	B	
AA	112 (110)	66.67	0.810	0.190	χ^2-value : 0.908[NS]df = 1
AB	48 (52)	28.57			
BB	8 (6)	4.76			
Total	168 (168)	100	1		-

[NS]: Not-significant (p > 0.05).
( ): No. of expected.

Table 3 Association analysis among genotype of POU1F1 gene and carcass characteristics

Trait	Genotype				Significant F-value
	AA	AB	BB	Mean ± SD	
Back fat Thickness (mm)	13.09 ± 0.92	12.14 ± 1.92	13.33 ± 1.86	12.59 ± 0.77	0.107 NS
Carcass weight (kg)	78.76 ± 1.98[b]	92.00 ± 7.61[a]	92.33 ± 14.68[a]	81.86 ± 2.26	3.474[*]
L^*	39.54 ± 0.91	37.94 ± 1.78	38.11 ± 3.56	39.17 ± 0.79	0.343 NS
a^*	17.42 ± 0.53[ab]	18.36 ± 1.73[a]	14.04 ± 3.55[b]	17.30 ± 0.52	3.258[*]
b^*	10.87 ± 0.40	10.90 ± 0.91	9.78 ± 1.92	10.79 ± 0.36	0.327 NS
pH	5.70 ± 0.62	5.67 ± 0.13	5.59 ± 0.05	5.69 ± 0.05	0.195 NS
Visual color[1]	4.05 ± 0.17	4.63 ± 0.42	5.00 ± 0.07	4.22 ± 0.16	2.102 NS

[1]: 1 = pale, 7 = dark purple red.
[a, b]: Means with different superscripts in the same column differ significantly ($p < 0.05$).
[*]: $p < 0.05$, NS: Not-significant.

significant ($p > 0.05$), and this pork group was shown to maintain genetic equilibrium.

Carcass characteristics according to POU1F1 Intron 1 genotype

The results of the analysis of backfat thickness, carcass weight, Hunter L^*, a^*, and b^* values, meat color (visual color), and pH, which are the main carcass characteristics associated with the POU1F1 SNP genotypes AA, AB, and BB, are shown in Table 3. The AA, AB, and BB genotype groups did not show intergroup significance, with an average backfat thickness of 12.14–13.33 mm ($p > 0.05$). These results are similar to those of Yu et al. [17], who confirmed lack of significant differences among genetic groups for backfat thickness. For carcass weight, the AA genotype group showed a significantly lower weight (78.76 kg) compared to the AB (92.00 kg) and BB (92.33 kg) types ($p < 0.05$). These results are similar to those of Yu et al. [17], who reported that carcass weight is greater in BB type than in AA type and suggested that carcass weight can be influenced by B allele.

No significant differences were observed in terms of meat color based on Hunter L^* (lightness) and b^* (yellowness)

values, which were within the ranges of 37.94–39.54 and 9.78–10.90, respectively ($p > 0.05$). However, the POU1F1 intron 1 gene was found to affect color expression since Hunter a^* value (redness) was the highest in AB type pork ($p > 0.05$). In the meat color (visual color) assessment using the standard meat color board, the AA, AB, and BB genotypes showed values of 4.05, 4.63, and 5.00, respectively, with BB type showing a higher level of redness than AA type. In terms of pH, average frequency was 5.69 for the AB and BB genotypes, although no significant differences were observed between them. This result is similar to the findings of Yu et al. [17], who reported no differences among the three genotypes in terms of meat color.

In terms of average carcass characteristics, Oh et al. [18] obtained different results, with an average backfat thickness of 27 mm. Similarly, Choi et al. [19] reported relatively low Hunter L^* value and relatively high a^* and b^* values in the crossbreeds. Results for pH values were within the range of 5.75–5.49, which is almost the same as that reported by Kim et al. [20] and Oh et al. [18], whereas Choi et al. [19] reported slightly higher pH values. These differences might be attributed to different pig raising methods such as feeding.

Table 4 Correlation coefficients among carcass characteristics in pigs

Items	x_1	x_2	x_3	x_4	x_5	x_6	x_7
Back fat Thickness (x_1)	-	0.069	0.095	0.125	−0.174	0.229	0.066
Carcass weight (x_2)		-	0.101	0.099	−0.216	−0.090	−0.0281
pH (x_3)			-	0.135	−0.109	0.110	−0.097
Visual color (x_4)				-	−0.521[***]	0.003	−0.390[**]
L^* (x_5)					-	−0.021	0.419[*]
a^* (x_6)						-	0.612[***]
b^* (x_7)							-

[*]: $p < 0.05$, [**]: $p < 0.01$, [***]: $p < 0.001$.

Correlation between carcass characteristics

The results of the correlation analysis of carcass characteristics are shown in Table 4. Meat color (visual color) showed a significant negative correlation with L^* ($r = -0.52$, $p < 0.001$) and b^* ($r = -0.39$, $p < 0.01$) values, whereas a low positive correlation was observed for backfat thickness, meat color ($r = 0.12$), and redness ($r = 0.23$). The L^* and b^* ($r = 0.42$, $p < 0.05$) values showed a low positive (+) correlation, whereas a^* and b^* ($r = 0.61$, $p < 0.001$) values showed a very high positive correlation.

These results are similar to those of Kim et al. [21], who reported that color and yellowness as well as redness and yellowness show high correlations, whereas color and redness show a negative correlation. The results on correlations among meat color (visual color), color (L^* value), redness (a^* value), and yellow (b^*) values were very similar to those reported previously [22].

The results of this study suggest that changes in the *POU1F1* intron region 1 are related to carcass weight, meat color, as well as growth. Further extensive study on the relationship between pig growth and changes in *POU1F1* gene expression in relation to carcass characteristics is necessary in order to identify quantitative traits that might possibly be related to pig growth or meat quality of pork.

Conclusion

The results of the analyses of carcass characteristics according to *POU1F1* genotypes of 168 domestically bred pig carcasses are as follows. Analysis of *POU1F1* intron 1 genotypes showed that the AA genotype has the highest frequency (66.7%), whereas frequencies of the AB and BB genotypes were 28.6% and 4.76%, respectively. The A allele showed a very high expression frequency (0.81), whereas B allele was estimated to have a low frequency (0.19). A χ^2-test demonstrated that this allelic frequency was not in Hardy-Weinberg equilibrium. Backfat thickness and pH were not significantly different according to the genotype, but carcass weight was significantly higher in AB (92.00 kg) and BB (92.33 kg) types than in AA type (78.76 kg; $p < 0.05$). Further, Hunter a^* value was higher in AB type than in BB type ($p < 0.05$). In addition, meat color (visual color) showed a negative correlation with L^* (lightness) and b^* (yellowness) values. The results of this study showed that changes in the *POU1F1* gene influence carcass weight, backfat thickness, and meat color. Therefore, extensive investigation of the relationship between carcass characteristics and changes in the *POU1F1* gene is necessary to provide a reliable index for meat quality improvement.

Competing interests

The authors declare that they have no competing interests.

Authors' contributions

All authors participated in the design and laboratory work of the study. All authors helped to draft the manuscript, and all authors read and approved the final manuscript.

Acknowledgments

The authors would like to thank the farmers of Gyeonggi-do, Chungcheong-do, and Jeolla-do for their support during this study and for time spent during sampling and examination.

Author details

[1]Department of Animal Resources, Kong-Ju National University, # 54 Daehakro, Yesan, Chungnam 340-702, Korea. [2]Department of Obstetrics and Gynecology, Ewha Woman's University, Seoul 158-710, Korea.

References

1. Ministry of Agriculture, Food and Rural Affairs (MAFRA): *Information and Data of Agricultural Statistics of Korea*. 2013. Available from http://www.mafra.go.kr/search/ totalSearch. jsp. Accessed May. 14.
2. FAO: *Global Meat Mardets: Challenges. FAO Commodity and Trade Division*. Brussels: FAO; 2003.
3. Kim GW: **Analysis of carcass quality grades according to gender, backfat thickness and carcass weight in pigs.** *J Anim Sci Technol* 2012, **54:**29–33.
4. Hong KC, Kin BC, Son YS, Kom BK: **Effects of the inating system on fattening performance and meat quality in commercial pigs.** *J Anim Sci Technol* 2001, **43:**139–148.
5. Jin SK, Kim IS, Hur SJ, Kim SJ, Jeong KJ: **The influence of pig breeds on qualities of loin.** *J Anim Sci Technol* 2006, **48:**747–758.
6. Cohen LE, Wondisford FE, Radovick S: **Role of pit-1 in the gene expression of growth hormone, prolactin and thyrotropin.** *Endocrinol Metab Clin North Am* 1996, **25:**523–540.
7. Hendriks-Stegeman BI, Augustijn KD, Bakker B, Holthuizen P, Vliet PC, Jansen M: **Combined pituitary hormone deficiency caused by compound heterozygosity for two novel mutations in the POU domain of the Pit-1/POU1F1 gene.** *J Clin Endocrinol Metab* 2001, **86:**1545–1550.
8. Brunsch C, Sternstein I, Reinecke P, Bieniek J: **Analysis of associations of PIT1 genotypes with growth, meat quality and carcass composition traits in pigs.** *J Appl Genet* 2002, **43:**85–91.
9. Moody DE, Pomp D, Newman S, MacNeil M: **Characterization of DNA polymorphisms in three populations of Hereford cattle and their associations with growth and maternal EPD in line 1 Herefords.** *J Anim Sci* 1996, **74:**1784–1793.
10. Renaville R, Gengler N, Vrech E, Prandi A, Massart S, Corradini C, Bertozzi C, Mortiaux F, Burny A, Portetelle D: **PIT1 gene polymorphism, milk yield and conformation traits for Italian Holstein-Friesian bulls.** *J Dairy Sci* 1997, **80:**3431–3438.
11. Stancekova K, Vasicek D, Peskovicova D, Bulla J, Kubed A: **Effect of genetic variability of the porcine pituitary-specific transcription factor (PIT-1) on carcass traits in pigs.** *Anim Genet* 1999, **30:**313–315.
12. Yu TP, Wang L, Tuggle CK, Rothschild MF: **Mapping genes for fatness and growth on pig chromosome 13: a search in the region close to the pig PIT1 gene.** *J Anim Breed Genet* 1999, **116:**269–280.
13. Song CY, Gao B, Teng S, Wang XY, Xie F, Chen G, Wang Z, Jing R, Mao J: **Polymorphisms in intron 1 of the porcine POU1F1 gene.** *J Appl Genet* 2007, **48(4):**371–374.
14. Kim GW, Yoo JY: **Analysis of carcass characteristics in the 3rd intron of pig POU1F1 gene.** *J Anim Sci Technol* 2009, **51(4):**283–288.
15. Franco MM, Antunes RC, Silva HD, Goular LR: **Association of P1T1, GH and GHRH polymorphisms with performance and carcass traits in landrace pigs.** *J Appl Genet* 2005, **46:**195–200.
16. Choi JR, Oh JD, Cho KJ, Lee JH, Kong HS, Lee HK: **Identification and analysis of PIT1 polymorphisms and its association with growth and carcass traits in korea cattles (Hanwoo).** *J Emb Trans* 2007, **22(3):**167–172.
17. Yu TP, Tuggle CK, Schmitz CB, Rothschild MF: **Association of PIT1 polymorphisms with growth and carcass traits in pigs.** *J Anim Sci* 1995, **73:**1282–1288.

18. Oh HS, Kim HY, Yang HS, Lee JI, Joo YK, Kim CU: **Comparison of meat quality characteristics between crossbreeds.** *Korean J Food Sci Ani Resour* 2008, **28**(2):171–180.

19. Choi YS, Park BY, Lee JM, Lee SK: **Comparison of carcass and meat quality characteristics between Korean native black pigs and commercial crossbred pigs.** *Korean J Food Sci Ani Resour* 2005, **25**(3):322–327.

20. Kim IS, Jin SK, Song YM, Park KH, Kang SM, Ha JH, Kim IJ, Park YS, Kim JH: **Quality characteristics of pork by sex on crossbred pigs.** *Korean J Intl Agri* 2006, **18**(1):34–39.

21. Kim GD, Jeong JY, Hur SJ, Yang HS, Jeon JT, Joo ST: **The relationship between meat color(CIE L* and a*), myoglobin content, and their influence on muscle fiber characteristics and pork quality.** *Korean J Food Sci Ani Resour* 2010, **30**(4):626–633.

22. Brewer MS, Zhu LG, Bidner BD, Meisinger J, McKeith FK: **Measuring pork color: Effects of bloom time, muscle, pH and relationship to instrumental parameters.** *Meat Sci* 2001, **57**:169–176.

14

Population genetic structure analysis and effect of inbreeding on body weights at different ages in Iranian Mehraban sheep

Roya Yavarifard, Navid Ghavi Hossein-Zadeh* and Abdol Ahad Shadparvar

Abstract

The objective of this study was to describe the population structure and inbreeding, and to quantify their effects on weights at different ages of Mehraban sheep in Iran. The analysis was based on the pedigree information of 26990 animals and 10278 body weight records from birth to yearling age. Data and pedigree information were collected during 1994 to 2011 by the breeding station of Mehraban sheep. The population structure was analyzed using the CFC program. Inbreeding of all animals was calculated by INBUPGF90 program. All animals were grouped into three classes according to their inbreeding coefficients: the first class included non-inbred animals (F=0); and the second and third classes included inbred animals ($0<F<0.05$ and $F\geq0.05$, respectively). The average inbreeding in Mehraban sheep was 1.69%. Founder equivalent (f_e) values were estimated to be 4244, 3116 and 2965 during 1994-1999, 2000-2005 and 2006-2011, respectively. The effective population sizes (N_e) were 363, 5080 and 5740 during 1994-1999, 2000-2005 and 2006-2011, respectively. Generation interval was 2.15 years for this breed of sheep. Regression coefficients of birth weight, 3-month weight, 6-month weight and yearling weight on lamb inbreeding were estimated to be -6.34±0.69, -14.68±3.33, 48.00±9.43 and 98.65±15.65, respectively. Both positive and negative inbreeding effects were found in the current study. The utilization of a program for designed mating system, in the present flock, could be a suitable approach to keep the level of inbreeding under control.

Keywords: Fat-tailed sheep, Growth traits, Inbreeding depression, Mating program, Pedigree analysis

Background

During the past 50 years, the indigenous genetic reserves are in a critical stage all over the world due to large changes in production systems, change of market demand and intercourse of domestic animals with other breeds. Along with increase in genetic progress, maintaining genetic diversity in the population is very important to adapt with the economic and environmental changes in the future and ensure long-term response to selection for traits that are very important [1]. Effective population size (N_e) is a criterion for determining similarity between alleles of the loci so that there was a common ancestor and determines the level of inbreeding and reduced rate of genetic variation due to gene random drift [2]. Intensive use of a few breeding animals, where the selection intensity is high, could result in greater rates of inbreeding in the population. Therefore, a small number of seedstock, with a strong family relationship, is responsible for the maintenance of almost the whole genetic pool in the population. This is an aspect of great influence in the genealogical analysis of a population structure, because of its effect on the probability of genes lost between generations and the consequent reduction in genetic variability [3].

Discrepancies in ancestral origins and migration events are important causative factors explaining genetic differences between current populations [4-6]. Hence, the optimal management of population is essential in order to prevent from decrease in diversity. Estimation of parameters such as effective population size, inbreeding and coancestry are depended significantly on the genealogy information. Measurement of the effect of inbreeding on economic traits is important in order to estimate the magnitude of change associated with increases in inbreeding. It is apparent that different breeds and populations, as well as different traits vary in their response to inbreeding.

* Correspondence: nhosseinzadeh@guilan.ac.ir
Department of Animal Science, Faculty of Agricultural Sciences, University of Guilan, P. O. Box: 41635 1314, Rasht, Iran

Some populations may show a very pronounced effect of increased inbreeding for a trait, whereas others may not display much of an effect [7,8]. The rate of inbreeding needs to be limited to maintain diversity at an acceptable level, so that genetic variation will ensure that future animals can respond to changes in the environment and to selection. Without genetic variation, animals cannot adapt to these changes [9]. Commonly, negative inbreeding effects, or inbreeding depression, are thought to most frequently occur because of an increase in frequencies of recessive alleles that adversely affect the traits of interest [10]. The increased frequency of recessive alleles leads to a larger number of individuals that are homozygous for the recessive alleles, whereas in non-inbred populations, the recessive allele would more frequently be masked by an advantageous dominant allele [11].

One of the most important breeds of Iranian sheep is Mehraban sheep which is reared in Hamedan province. This breed is adapted to harsh climate and rocky environments in the western regions of Iran. The Mehraban is a fat-tailed carpet wool sheep with light brown, cream or grey color, dark face and neck and primarily used for meat production [12]. The objective of this study was to describe the population structure and inbreeding, and to quantify their effects on weights at different ages of Mehraban sheep in Iran from 1994 to 2011.

Methods

Experimental design and animals

Data set and pedigree information used in this research were collected from the breeding station of Mehraban sheep (Hamedan, Iran) during 1994 2011. The traits included were: Birth weight (BW, n= 10287), 3-month weight (WW, n= 6735), 6-month weight (6 MW, n= 4778), 9-month weight (9 MW, n= 3139) and yearling weight (YW, n= 1985). Ewes were randomly exposed to the rams at about 18 months of age. Matings were controlled and single-sire pens were used allocating 10 15 ewes per ram. Ewes were kept in the flock up to 7 years of age. Ewes usually give births to lambs three times every two years. All lambs were weighed and ear tagged within 12 hours after the birth. Lambs are weaned at approximately 90 days of age. Flocks were grazed during the daytime and housed at night. The lambs were kept indoors and fed manually during the winter. Summary of pedigree information used in this research is presented in Table 1. Also, descriptive statistics of data used in the analysis are shown in Table 2.

Statistical and genetic analyses

The CFC program [13] was used to calculate pedigree statistics and genetic structure analysis of the population. To characterize the population structure, variation changes in inbreeding (ΔF), average coancestry (AC), effective population size (N_e), generation interval (L)

Table 1 Summary of pedigree information for Mehraban sheep

Item	N	%
Individuals in total	26990	100%
Sires	405	15%
Dams	8114	30%
Individuals with known sire	304	11%
Individuals with known dam	11339	42%
Individuals with known sire and dam	3683	14%
Individuals with progeny	18472	68%
Inbred individuals	18872	70%

and parameters derived from the method of analysis of gene origin probability were calculated. The parameters related to the method of analysis of gene origin probability were: founder equivalent (f_e), founder genome equivalent (f_g), effective number of non-founder (N_{enf}), average number of discrete generation equivalents (G_e), maximum number of discrete generation equivalents ($MaxG_e$) and minimum number of discrete generation equivalents ($MinG_e$). Animals were grouped based on their birth years into three classes (1994 1999, 2000 2005, and 2006 2011). This classification was necessary for the CFC program to compute the genetic structure parameters of the population.

To account for unequal founder representation, Lacy [14] estimated the effective number of founders (f_e) as:

$$f_e \quad \left[\sum_{i\ 1}^{f} p_i^2 \right]^{-1}$$

Where p_i is the expected proportional genetic contribution of founder i, calculated by the average relationship of the founder to each animal in the current population, and f is the total number of founders. The parameter f_e indicates the number of equally contributing founders that would produce the same level of genetic diversity as that observed in the current population [15].

Bottlenecks, genetic drift and unequal founder contributions which have a greater impact in small populations

Table 2 Characteristics of data set for Mehraban sheep

	BW	WW	6 MW	9 MW	YW
Number of records	10278	6735	4778	3139	1985
Mean (kg)	3.69	22.16	36.13	45.48	52.70
Standard deviation (kg)	0.76	4.26	6.09	6.43	6.74
Coefficient of variation (%)	20.60	19.22	16.85	14.14	12.79
Minimum (kg)	1.35	8.92	15.06	25.85	28.46
Maximum (kg)	6.03	35.74	58.50	64.91	75.48

BW: birth weight, *WW*: 3-month weight, *6 MW*: 6-month weight, *9 MW*: 9-month weight, *YW*: yearling weight.

can be quantified using the founder genome equivalent (f_g) as follows:

$$ f_g = \left[\sum_{i=1}^{f} \frac{p_i^2}{r_i} \right]^{-1} $$

Where r_i is the expected proportion of founder i s alleles that remain in the current population and can take on a value of 0.5 if one allele is present or 1.0 if two alleles are present, p_i is the expected proportional genetic contribution of founder i, and f is the number of contributing founders [14].

Average number of discrete generation equivalents was determined for total pedigree of each flock according to the following equation:

$$ G_e = \frac{1}{N} \sum_{j=1}^{N} \sum_{i=1}^{n_j} \frac{1}{2^{g_{ij}}} $$

In this equation, n_j is the number of known ancestors for animal j and g_{ij} is the number of generations between animal i (ancestor) and animal j. The depth of the pedigree in each reference population was examined by computing G_e which is the expected number of generations from the base population, to the reference population if generation proceeded discretely [16].

The effective number of non-founders explains the amount of genetic diversity reduced by random genetic drift accumulated in non-founders generations and is calculated using the following equation [15]:

$$ N_{enf} = \left[\frac{1}{f_g} - \frac{1}{f_e} \right]^{-1} $$

Where N_{enf} is the effective number of non-founders. Also, f_g and f_e are founder genome equivalents and founder equivalents, respectively.

The Inbupgf 90 program [17] was used for calculating regular inbreeding coefficients for individuals in the pedigree. Falconer and Mackay [18] established that the average inbreeding coefficient at a given generation t could be estimated using the following equation:

$$ F_t = 1 - (1 - \Delta F)^t $$

Where ΔF is the change in inbreeding from one generation to the next one or new inbreeding. Gonzlez-Recio et al. [19] proposed to operate the above equation, to set the inbreeding coefficient for each individual, as represented below:

$$ \Delta F_i = 1 - \sqrt[t]{1 - F_i} $$

Where F_i is the individual coefficient of inbreeding and t is the equivalent complete generations [20].

The estimate of the effective number (N_e) [17] can be calculated from ΔF, which can be easily computed by averaging the ΔF_i of n individuals included in a given reference subpopulation; therefore, effective number is obtained as:

$$ N_e = \frac{1}{2 \Delta F} $$

This way of computing effective population number is not dependent on the whole reference population mating policy, but on the matings carried out throughout the pedigree of each individual [21].

Generation interval (L) was calculated as the average age of the parents at the birth of their lambs. All the animals were grouped into three classes according to the inbreeding coefficients obtained by their pedigrees: the first class included non-inbred animals ($F = 0$); and the second and third classes included inbred animals ($0 < F < 0.05$ and $F \geq 0.05$, respectively). Moreover, the birth type (single, twin) and lamb sex (male, female) was considered for each of the lambs. Due to the low frequency of triple births, triple lambs were not included in this study.

Trend of inbreeding over time was estimated using the linear regression of individual inbreeding on the birth year using the Reg procedure of SAS [13]. The GLM procedure of SAS was used for determining the fixed factors which had significant effect on the traits investigated. After data verification, defective and doubtful records were deleted (e.g., lambs without weight records or with incomplete records of parentage or with registration numbers lower than the numbers of their parents were left out). The least-squares means were estimated for each trait using the Average Information Restricted Maximum Likelihood (AIREML) algorithm of the Wombat program [20] by fitting six single trait animal models which ignore or include additive direct and maternal genetic and permanent environmental effects. The statistical models included herd-year-season of lambing, lamb sex in 2 classes (male and female), age of dam at lambing in 6 classes (2 7 years old), birth type in 2 classes (single, twin), inbreeding in 3 groups ($F = 0$, $0 < F < 0.05$, $F \geq 0.05$) and interaction between them. The most appropriate model for BW, 9 MW and YW included direct additive genetic and maternal permanent environmental effects and for WW included direct additive genetic effects as well as maternal additive genetic effects and for 6 MW included maternal and direct additive genetic effects as well as covariance between direct additive and maternal additive genetic effects.

Results

Analysis of pedigree

The analysis of pedigree revealed that inbreeding coefficient ranged from 0 to 27% with an average of 1.69%. Table 3 shows the summary statistics for body weight traits in

Table 3 Distribution of records for body weight traits in different inbreeding classes of animals born between 1994 and 2011

Inbreeding class	Traits (kg)									
	BW		WW		6 MW		9 MW		YW	
	N	Mean SE	N	Mean SE	N	Mean SE	N	Mean SE	N	Mean SE
F= 0	1536	3.88 ⁊0.01 [a]	802	22.46 ⁊0.06 [a]	348	34.62 ⁊0.06 [b]	33	39.06 ⁊0.93 [c]	12	50.20 ⁊0.01 [a]
0< F<0.05	8700	3.65 ⁊0.01 [b]	5903	22.12 ⁊0.06 [ab]	4503	36.23 ⁊0.06 [ab]	3095	45.51 ⁊0.11 [b]	1998	52.68 ⁊0.15 [a]
F ≥ 0.05	19	3.32 ⁊0.23 [c]	10	21.01 ⁊0.06 [b]	11	37.48 ⁊0.06 [a]	11	47.66 ⁊0.88 [a]	9	55.35 ⁊1.32 [a]

[a,b,c]Means with similar letters in each sub class within a column do not differ significantly at $P<0.05$. BW: birth weight, WW: 3-month weight, 6 MW: 6-month weight, 9 MW: 9-month weight, YW: yearling weight. F: inbreeding coefficient. SE: standard error.

different inbreeding classes of animals. The analysis of pedigree revealed that 60 animals out of 26990 (0.22%) had a high inbreeding coefficient ($F \geq 0.05$) with a mean value of 22.10% while 24199 out of 26990 animals (89.66%) had medium inbreeding coefficient ($0< F< 0.05$). The remaining lambs (10.12%) were non-inbred. There were significant differences between three classes of inbreeding on BW and animals within first class of inbreeding had greater mean of the trait than two other groups ($P< 0.05$). The WW of animals within first class of inbreeding was higher than those of the lambs belonging to the second and third classes, but only differences were significant between first and third classes ($P< 0.05$). On the other hand, the 6 MW of animals within first class of inbreeding was significantly ($P< 0.05$) lower than those of the lambs in the second and third classes. Also, there were significant differences between three classes of inbreeding on 9 MW and animals within third class of inbreeding had greater mean of the trait than two other groups ($P< 0.05$). In addition, there were no significant differences between three classes of inbreeding on YW.

Table 4 shows the results of the pedigree analysis for the reference population in year groups. The f_e values were 4244, 3116 and 2965 during 1994 1999, 2000 2005 and 2006 2011, respectively. The f_g values were 4211, 2328 and 2118 during 1994 1999, 2000 2005 and 2006 2011, respectively. N_{enf} values were 10057, 9205 and 7422

during 1994 1999, 2000 2005 and 2006 2011, respectively. Therefore, this parameter was decreased over the years. The G_e values were 0.3571, 0.4545 and 0.5359 during 1994 1999, 2000 2005 and 2006 2011, respectively. $MaxG_e$ were 1.625, 1.9375 and 2.77734 during 1994 1999, 2000 2005 and 2006 2011, respectively. The generation interval (L) was 2.15 years in Mehraban sheep. The average coancestries (AC) were 0.000118742, 0.000214769 and 0.000238527 during 1994 1999, 2000 2005 and 2006 2011, respectively. The effective population sizes (N_e) were 363, 5080 and 5740 during 1994 1999, 2000 2005 and 2006 2011, respectively. Changes in inbreeding (ΔF) were 0.00137741, 0.00009843 and 0.00008711 during 1994 1999, 2000 2005 and 2006 2011, respectively. Figure 1 shows the trend of inbreeding coefficients over the years. The inbreeding trend was significantly positive over the years ($P< 0.01$) and its estimate was 0.002 0.00003. Figure 2 shows the pedigree completeness up to 3 generations back. The first ancestor generation of all animals, included in the total data set used, was 15% sire and 57% dam complete.

Inbreeding effects

Table 5 shows the effects of inbreeding on body weight traits of the lambs according to the birth type. Single-born lambs showed no significant difference in all body weights except for 9-month weight which inbred animals were

Table 4 The results of the pedigree analysis for the reference population of Mehraban sheep in year groups

Item/ year	1994-1999	2000-2005	2006-2011	Total years
Number of animals	7745	9340	9905	26990
Founder equivalent (f_e)	7244	3116	2965	-
Founder genome equivalent (f_g)	4211	2328	2118	-
Effective number of non-founders (N_{enf})	10057	9205	7422	-
Effective population size (N_e)	363	5080	5740	11963
Average number of discrete generation equivalents (G_e)	0.3571	0.4545	0.5359	0.4188
Maximum number of discrete generation equivalents ($MaxG_e$)	1.625	1.9375	2.77734	2.777734
Minimum number of discrete generation equivalents ($MinG_e$)	0	0	0	0
Generation interval (L), years	2	2.18	2.27	2.15
Average coancestry (AC)	0.000118742	0.000214769	0.000238527	-
Changes in inbreeding (ΔF)	0.00137741	0.00009843	0.00008711	0.00004179

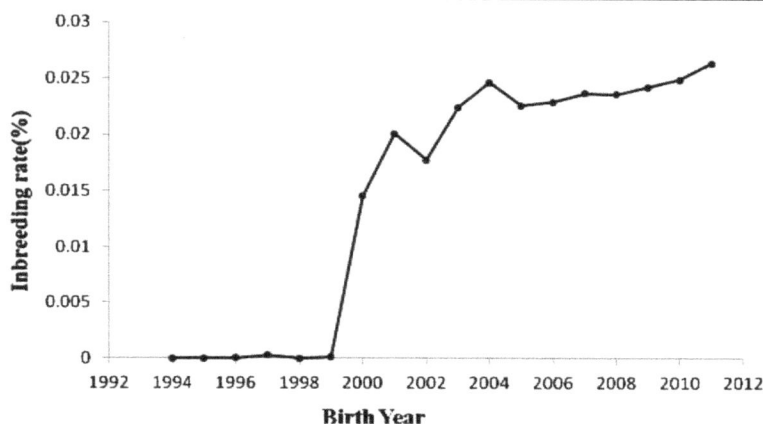

Figure 1 Inbreeding trend over the years.

significantly different from non-inbreds (*P*< 0.05), while twin-born lambs showed a significant difference in all body weights. The BW of twin-born lambs was significantly different in three classes of inbreeding (*P*< 0.05) and animals within first class of inbreeding had greater mean of the trait than two other classes. Twin-born lambs in the first and second classes of inbreeding had greater WW than those of twin-born lambs in the third class (*P*< 0.05). The 6 MW of twin-born lambs within first class of inbreeding was significantly (*P*< 0.05) lower than those of lambs in the second and third classes (*P*< 0.05). Twin-born lambs in the third class of inbreeding showed significant differences for 9 MW and YW with lambs in the second class (*P*< 0.05).

Table 6 shows the effects of inbreeding on body weight traits of the lambs according to the sex of lambs. Male lambs showed no significant differences in their WW, 6 MW and YW irrespective of the inbreeding coefficient. The BW of male lambs within third class of inbreeding was significantly lower than those of male lambs in the first and second classes (3.43 1.05 kg vs. 3.77 0.82 kg and 3.96 0.51 kg, respectively). Also, 9 MW of male lambs within first class of inbreeding was significantly lower than those of male lambs in the second and third classes (40.64 9.72 kg vs. 46.05 6.43 kg and 46.80 3.14 kg, respectively).

Female lambs showed significant differences in their body weights based on their inbreeding coefficient. The BW and WW of female lambs within third class of inbreeding were significantly (*P*< 0.05) lower than those of lambs in the first and second classes. The 6 MW of female lambs showed a significant difference (*P*< 0.05) between third class of inbreeding and first and second classes (38.50 1.00 kg vs. 34.09 7.26 kg and 35.66 5.69 kg; respectively). The 9 MW of female lambs showed a significant difference between all three classes of inbreeding (*P*< 0.05). On the other hand, the YW of female lambs showed significant difference between second and third classes of inbreeding (52.17 6.99 kg vs. 58.53 0.47 kg; *P*< 0.05).

Regression coefficients of body weights

Table 7 shows the regression coefficients of body weights on inbreeding of lambs for a change of 1% in inbreeding. The regression coefficients of BW, WW, 6 MW and YW on lamb inbreeding were estimated to be −6.34 0.69, −14.68 5.33, 48.00 9.43 and 98.65 15.65, respectively (*P*< 0.01). Therefore, BW and WW decreased, respectively, by 6.34 g and 14.68 g due to 1% increase in inbreeding and 6 MW and YW increased, respectively, by 48.00 g and 98.65 g due to 1% increase in inbreeding (*P*< 0.01). The regression coefficient of 9 MW on lamb inbreeding was not significant.

Discussion

Reported estimates of lamb inbreeding effects on growth performance traits showed the same trend by other authors. Similar to the current results, some reported a

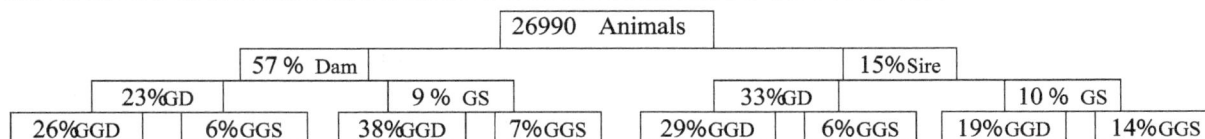

Figure 2 Pedigree completeness up to 3 generations back. *GD*: grand dam, *GS*: grand sire, *GGD*: great grand dam, *GGS*: great grand sire.

Table 5 Distribution of records for body weight traits in different inbreeding classes of animals grouped by the type of birth and born between 1994 and 2011

Birth type	Inbreeding class	Traits (kg)									
		BW		WW		6 MW		9 MW		YW	
		N	Mean SE	N	Mean SE	N	Mean SE	N	Mean SE	N	Mean SE
Single	F=0	1344	3.96 0.47 [a]	681	22.44 4.92 [a]	281	34.66 8.13 [a]	33	39.07 8.29 [b]	12	50.20 0.01 [a]
	0< F<0.05	6820	3.87 0.68 [a]	4476	21.98 4.59 [a]	3317	36.73 7.47 [a]	2198	45.78 6.66 [a]	1474	52.35 8.73 [a]
	F ≥ 0.05	11	3.85 0.65 [a]	6	22.37 3.36 [a]	6	36.22 5.51 [a]	6	46.57 3.36 [a]	4	52.60 3.99 [a]
Twin	F=0	192	3.33 0.46 [a]	121	22.61 4.19 [a]	67	34.41 5.83[b]	-	-	-	-
	0< F<0.05	1880	2.96 0.77 [b]	1427	22.55 3.92 [a]	1186	36.71 6.24 [ab]	897	44.76 5.77 [b]	524	53.65 5.28 [b]
	F ≥ 0.05	8	2.59 0.83 [c]	4	18.98 1.69 [b]	5	39.00 2.53 [a]	5	48.98 0.67 [a]	5	57.56 1.28 [a]

[a,b,c]Means with similar letters in each sub class within a column do not differ significantly at P<0.05. BW: birth weight, WW: weaning weight, 6 MW: 6-month weight, 9 MW: 9-month weight, YW: yearling weight. F: inbreeding coefficient. SE: standard error.

lower regression coefficient for BW due to increase in inbreeding, e.g. Ghavi Hossein-Zadeh [22] observed a reduction of 0.009 kg for 1% increase of inbreeding in Iranian Moghani sheep; Selvaggi et al. [23] found a mean value of 0.019 kg in Leccese sheep; MacKinnon et al. [10], Analla et al. [24], Van Wyk et al. [9], Ercanbrack and Knight [25], Khan et al. [26] and Mirza et al. [8] reported regression coefficients of −0.027, −0.013, −0.008, −0.010, −0.008 and −0.007 kg, respectively. Reasons of variation in inbreeding effects could be due to differences between the breeds in allele separation, amount of genetic variation in the base population, management, and diversity of the founders of the flocks examined [10].

Similar to the current results, Van Wyk et al. [16] and Selvaggi et al. [23] reported significant reduction in WW of lambs due to 1% increase in inbreeding in different breeds of sheep and inconsistent with the current result, Ghavi Hossein-Zadeh [22] and Lamberson and Thomas [11] reported no significant reduction in WW due to inbreeding. Sex of lambs was a significant effect in the current analysis of inbreeding; but Barczak et al. [7] and Ghavi Hossein-Zadeh [22] observed non-significant differences between males and females. Barczak et al. [7] reported positive inbreeding effects on fourth week weight in a multi-breed sheep population and Ghavi Hossein-Zadeh [22] reported positive inbreeding effects on 6 MW and YW in Moghani sheep population. There are several methodological and biological factors which determine the estimated inbreeding impact on the performance traits. It is well known that both negative and positive effects exist. Therefore, in a population, bad and good inbreeding effects are mixed [4].

The results of this study indicated a significant increase in 6WW and YW of lambs due to 1% increase in inbreeding, but Ghavi Hossein-Zadeh [22] reported a significant reduction in YW of male lambs (0.357 kg). The possible explanation for the strong inbreeding depression observed for 6 MW and YW in this study

was the higher heritability of this trait compared to other weight traits in Mehraban sheep [27].

The inbreeding level estimates are strongly determined by the two main factors: depth and completeness of pedigree and selection intensity. Selection intensity is often increased by the reproductive technologies being focused on a few superior animals (especially sires) and the application of advanced methods of genetic evaluation. Embryo transfer and artificial insemination technology currently allow the intensive use of the same sires, leading to increase in the relationship coefficient between animals, which help to the increase in inbreeding in this population. A high inbreeding level is observed for populations rebuilt from small number of founders [7], but on the other hand in this case the accuracy is strongly determined by the incompleteness of pedigrees [7]. Animal breeding emphasis on the genetic breeding values of the traits, used as criteria of sires and dams selection, can also raise the inbreeding coefficient, since relationship between animals tend to present similar genetic values, having as a consequence the selection of the most frequent relatives [3]. Average inbreeding estimates reported in this study were lower than reported estimates of Ghavi Hossein-Zadeh [22] in Moghani sheep (2.93%), Dorostkar et al. [28] in Moghani sheep (2.069%), Pedrosa et al. [3] in Santa Ins sheep in Brazil (2.33%). Van Wyk et al. [9] and Selvaggi et al. [23] reported high rates of inbreeding in Dormer sheep (16%) and Leccese sheep (8.1%), respectively. On the other hand, Eteqadi et al. [29] reported lower inbreeding (0.15%) in Guilan sheep. The lower inbreeding coefficient in the current sheep population compared with other studies could be due to the lack of designed mating programs and absence of selection, especially before 1999. The rapid increase in the rate of inbreeding in 1999 2000 could be resulted from the reduction in the number of sires. Similar to the current results, Eteqadi et al. [29], Ghavi Hossein-Zadeh [22], Dorostkar et al.

Table 6 Distribution of records for body weight traits in different inbreeding classes of animals grouped by the sex of lamb and born between 1994 and 2011

lamb sex	Inbreeding class	Traits (kg)										
		BW		WW		6 MW		9 MW		YW	
		N	Mean SE	N	Mean SE	N	Mean SE	N	Mean SE	N	Mean SE
	F= 0	730	3.96 0.51 [a]	373	22.48 4.69 [a]	152	35.30 8.37 [a]	19	40.64 9.72 [b]	12	50.20 0.01 [a]
Male	0< F<0.05	4264	3.77 0.82 [a]	2996	22.28 4.16 [a]	2265	36.81 6.12 [a]	1556	46.05 6.43 [a]	1042	53.16 6.48 [a]
	F ≥ 0.05	12	3.43 1.05 [b]	6	22.77 2.30 [a]	7	36.90 5.67 [a]	7	46.80 3.14 [a]	6	53.77 3.66 [a]
	F= 0	806	3.81 0.46 [a]	429	22.45 4.63 [a]	196	34.09 7.26 [b]	14	36.80 0.83 [c]	-	-
Female	0< F<0.05	4438	3.54 0.76 [a]	2917	21.95 4.26 [a]	2238	35.66 5.69 [b]	1539	44.97 6.32 [b]	956	52.17 6.99 [b]
	F ≥ 0.05	7	3.13 0.76 [b]	4	18.38 2.68 [b]	4	38.50 1.00 [a]	4	49.98 0.82 [a]	3	58.53 0.47 [a]

[a,b,c]Means with similar letters in each sub class within a column do not differ significantly at $P<0.05$. BW: birth weight, WW: 3-month weight, 6 MW: 6-month weight, 9 MW: 9-month weight, YW: yearling weight. F: inbreeding coefficient. SE: standard error.

[28], Pedrosa et al. [3] and Barczak et al. [7] reported positive trend for inbreeding over the years.

The generation interval was 2.15 years in the current population, and MacKinnon [10] reported the generation intervals of 2.65 and 4.28 years for different crossbred sheep. Van Wyk et al. [9], Pedrosa et al. [3], Ghavi Hossein-Zadeh [22] and Eteqadi et al. [29] reported generation intervals of 3.27, 3.70, 3.34 and 2.385 years for Elsenburg Dormer sheep, Santa Ins sheep, Moghani sheep and Guilan sheep, respectively. Lower estimates of generation interval would cause larger responses [30].

The f_e and f_g values are important parameters which can be used for management and control of small populations. Also, these parameters can increase the accuracy of changes in some parameters such as effective population size and inbreeding rate [31]. In a population, abundance of some forms of founder animals may be more than the others, in creating the next generation. This makes these animals have greater contributions than others in the population gene pool. The f_e parameter was calculated for correcting this item. The f_e value was 2965 in 2011 which was proportional to the increased number of animals (9905) in this year. This indicated the unequal contribution of founder animals in creating offspring. The most important limitation of f_e is converging genetic contribution of founder animals after several generations which will lead to remain f_e in a constant value.

The f_g parameter is an indicator for showing the unequal participation of founder animals and accidental loss of genes during transmission from parents to offspring. For this reason, the value of f_g is always lower than the f_e value and decreases rapidly over the time. As expected, the value of f_g was reduced over the studied years.

The G_e parameter is a factor to indicate the depth and quality of the pedigree. The G_e value was increased over the studied years. Therefore, this indicated the increase in pedigree information and its evolution over the years. The AC values were increased over the years. Animals AC predicts average inbreeding of future generation in a population. For this reason, this parameter can be used to calculate the effective population size in the future. High relative population size, meaning low variation in a population due to the reduction of variance between individuals, will lead to a decrease in the response to selection.

The effective size of population is described by the number of animals that mate in an ideal population and produce the same inbreeding increment of the population under study [32]. Evolutionary biologists have suggested that an effective population size in the range of 500 5000 is mandatory to secure evolutionary potential of natural

Table 7 Regression coefficients (SE) of body weight traits (in grams) on inbreeding of lambs for a change of 1% in inbreeding

Item	BW	WW	6 MW	9 MW	YW
Single	1.84 0.67**	11.45 6.45**	54.50 12.35**	22.00 19.58	232.65 27.96**
Twin	7.22 1.33**	28.63 9.16**	30.26 13.68*	17.81 16.52	19.59 15.30
Male	5.74 0.99**	9.23 7.24	32.18 12.87*	10.72 16.11	68.47 17.83**
Female	7.23 0.94**	21.66 7.85**	65.23 13.81**	7.23 22.62	171.28 29.97**
All	6.34 0.69**	14.68 5.33**	48.00 9.43**	4.09 13.12	98.65 15.65**

BW: birth weight, WW: 3-month weight, 6 MW: 6-month weight, 9 MW: 9-month weight, YW: yearling weight.*$P<0.05$. **$P<0.01$.

populations [33]. The reduction in effective population size, as a direct outcome of reduction in genetic diversity, associates with various unfavorable phenomena such as inbreeding depression in fitness-related traits and an increased change in response to selection [18]. One problem with the N_e value is that the value indicates the number of breeding animals needed to produce the average ΔF and does not quantify the cumulative decrease in allelic diversity or changes in breeding structure from year to year (10). Hence, the value obtained for N_e is not comparable to measures of f_e and f_g [10]. Most breeding programs may try to minimize accumulation of inbreeding and quantify the increase by calculating the change in inbreeding per generation (ΔF) [34] in order to decrease the possible negative effect of inbreeding on productive traits.

Conclusion

In conclusion, average inbreeding was 1.69% in Iranian Mehraban sheep and an increasing trend for inbreeding was observed over the years. Both positive and negative inbreeding effects were found in the current study. Different methods are proposed to maximize response to selection in an acceptable level of inbreeding such as balanced use of animals as parents of the next generation, limiting the size of families and creating sub-lines. Implementation of these methods and use of designed mating system can help to obtain the optimal response to the selection by least accumulation of inbreeding in Mehraban sheep flock in the future. Overall, avoidance from inbreeding is a main objective on the management of vulnerable species and breeds and this is especially true with respect to these new findings in Mehraban sheep.

Competing interests
The authors declare that they have no competing interests.

Authors contributions
RY participated in the acquisition of data, statistical and genetic analyses of data and drafted the manuscript. NGH designed and conceived this study and contributed to the statistical and genetic analyses of data and writing of the manuscript. AAS contributed to the conception of study and writing of the manuscript. All authors read and approved the final manuscript.

References
1. Barker JSF: Conservation and management of genetic diversity: a domestic animal perspective. Can J Forest Res 2001, 31:588 595.
2. Norberg E, Sørensen AC: Inbreeding trend and inbreeding depression in the Danish populations of Texel, Shropshire, and Oxford down. J Anim Sci 2007, 85:299 304.
3. Pedrosa VB, Santana ML, Oliveira J, Eler JP, Ferraz JBS: Population structure and inbreeding effects on growth traits of Santa Inê sheep in Brazil. Small Rumin Res 2010, 93:135 139.
4. Alvarez I, Royo LJ, Fernandez I, Gutierrez JP, Gomez E, Goyache F: Genetic relationships and admixture among sheep breeds from Northern Spain assessed using microsatellites. J Anim Sci 2004, 82:2246 52.
5. Canon J, Alexanderino P, Bessa I, Carleos C, Carretero Y, Dunner S, Ferran N, Garcia D, Jordana J, Lalo?D, Pereira A, Sanchez A, Moazami-Goudarzi K: Genetic diversity measures of local European cattle breeds for conservation purposes. Genet Sel Evol 2001, 33:311 32.
6. Rendo F, Iriondo M, Jugo BM, Mazon LI, Aguirre A, Vicario A, Estonba A: Tracking diversity and differentiation in six sheep breeds from the North Iberian Peninsula through DNA variation. Small Rumin Res 2004, 52:195 202.
7. Barczak E, Wolc A, Wøtowski J, Ślązarz P, Szwaczkowski T: Inbreeding and inbreeding depression on body weight in sheep. J Anim Feed Sci 2009, 18:42 50.
8. Negussie E, Abegaz S, Rege JOE: Genetic trend and effects of inbreeding on growth performance of tropical fat-tailed sheep. Proc 7th World Congress Genet Appl Livestock Prod Montpellier, France 2002, 25:1 4.
9. Van Wyk JB, Fair MD, Cloete SWP: The effect of inbreeding on the production and reproduction traits in the elsenburg dormer sheep stud. Livest Sci 2009, 120(3):218 224.
10. MacKinnon KM: Analysis of Inbreeding in a Closed Population of Crossbred Sheep. Virginia, USA: MSc Thesis, Virginia Polytechnic Institute and State University Blacksburg; 2003.
11. Lamberson WR, Thomas DL: Effects of inbreeding in sheep: A review. Anim Breed Abstr 1984, 52:287 297.
12. Zamani P, Mohammadi H: Comparison of different models for estimation of genetic parameters of early growth traits in the Mehraban sheep. J Anim Breed Genet 2008, 125:29 34.
13. Sargolzaei M, Iwaisaki HJ, Colleau J: CFC: A tool for monitoring genetic diversity. In Proceedings of 8th World Congress on Genetics Applied to Livestock Production. CD-ROM Communication 27-28. Belo Horizonte, Brazil. 13-18 August 2006.
14. Lacy RC: Analysis of founder representation in pedigrees: founder equivalents and founder genome equivalents. Zoo Biol 1989, 8:111 123.
15. Caballero A, Toro MA: Interrelations between effective population size and other pedigree tools for the management of conserved populations. Genet Res 2000, 75:331 343.
16. Woolliams JA, Mantysaari EA: Genetic contributions of Finnish Ayshire bulls over four generations. Anim Sci 1995, 61:177 187.
17. Aguilar I, Misztal I: Technical note: recursive algorithm for inbreeding coefficients assuming nonzero inbreeding of unknown parents. J Dairy Sci 2008, 91:1669 1672.
18. Falconer DS, Mackay TFC: Introduction to Quantitative Genetics. 4th edition. Harlow, Essex, UK: Longman Group LTD; 1996.
19. Gonzłez-Recio O, Lōpez de Maturana E, Gutiřrez JPJ: Inbreeding depression on female fertility and calving ease in Spanish dairy cattle. J Dairy Sci 2007, 90:5744 5752.
20. Maignel L, Boichard D, Verrier E: Genetic variability of French dairy breeds estimated from pedigree information. Interbull Bull 1996, 14:49 54.
21. Gutiřrez JP, Cervantes I, Molina A, Valera M, Goyache F: Individual increase in inbreeding allows estimating realised effective sizes from pedigrees. Genet Sel Evol 2008, 40:359 378.
22. Ghavi Hossein-Zadeh N: Inbreeding effects on body weight traits of Iranian Moghani sheep. Arch Tierz 2012, 55(2):171 178.
23. Selvaggi M, Dario C, Peretti V, Ciotola F, Carnicella D, Dario M: Inbreeding depression in Leccese sheep. Small Rumin Res 2010, 89:42 46.
24. Analla M, Montilla JM, Serradilla JM: Study of the variability of the response to inbreeding for meat production in merino sheep. J Anim Breed Genet 1999, 116(6):481 488.
25. Ercanbrack SK, Knight A: Effects of inbreeding on reproduction and wool production of Rambouillet, Targhee, and Columbia ewes. J Anim Sci 1991, 69:4734 4744.
26. Khan MS, Ahmad MD, Ahmad Z, Jadoon JK: Effect of inbreeding on performance traits of Rambouillet sheep. J Anim Plant Sci 1995, 5:299 304.
27. Gamasaee VA, Hafezian SH, Ahmadi A, Baneh H, Farhadi A, Mohamadi A: Estimation of genetic parameters for body weight at different ages in Mehraban sheep. Afri J Biotech 2010, 9:5218 5223.
28. Dorostkar M, Faraji Arough H, Shodja J, Rafat SA, Rokouei M, Esfandyari H: Inbreeding and inbreeding depression in Iranian Moghani sheep breed. J Agric Sci Technol 2012, 14:549 556.
29. Eteqadi B, Ghavi Hossein-Zadeh N, Shadparvar AA: Population structure and inbreeding effects on body weight traits of Guilan sheep in Iran. Small Rumin Res 2014, 3:45 51.
30. Ghavi Hossein-Zadeh N: Bayesian estimates of genetic changes for body weight traits of Moghani sheep using Gibbs sampling. Trop Anim Health Prod 2012, 44:531 536.

31. Martñez AT, Rencoret J, Marques G, Gutiřrez A, Ibarra D, Jimñez-Barbero J, del Rð JC: **Monolignol acylation and lignin structure in some nonwoody plants.** *J Phytochem* 2008, **69**:2831 2843.
32. Hill R: **Math. Proc. Cambridoe Phil.** *Soc* 1979, **85**:179.
33. Frankham R, Ballou JD, Briscoe DA: *Introduction to Conservation Genetics.* Cambridge, England: Cambridge University Press; 2002.
34. Boichard D, Maignel L, Verrier ? **The value of using probabilities of gene origin to measure genetic variability in a population.** *Genet Sel Evol* 1997, **29**:5 23.

15

Effects of *Prunella vulgaris* labiatae extract on specific and non-specific immune responses in tilapia (*Oreochromis niloticus*)

Kwan-Ha Park and Sanghoon Choi[*]

Abstract

We examined the effects of *Prunella vulgaris* Labiatae (*P. vulgaris* L.) on specific and non-specific immune responses of Nile tilapia, *Oreochromis niloticus*. The optimal concentration without toxicity of *P. vulgaris* was determined to 30-40 µg/ml *in vitro* and 120 µg/100 g of fish *in vivo*. *P. vulgaris* significantly elicited an antibody titer compared to FCA or β-glucan. β-glucan plus *P. vulgaris* group synergistically enhanced antibody production. No significant difference in antibody production was observed between *P. vulgaris* and *P. vulgaris* plus β-glucan group. A respiratory burst activity of head kidney (HK) leucocytes of tilapia administered with 300 or 500 µg *P. vulgaris* was significantly ($p < 0.05$) enhanced compared with the PBS-injected control group and FCA-treated group. Maximum increase in the NBT reduction value was observed in 500 µg *P. vulgaris* group but no significant difference was found between 300 and 500 µg *P. vulgaris* group. The level of serum lysozyme activity was significantly ($p < 0.05$) higher in the 300 and 500 µg *P. vulgaris* than 100 µg *P. vulgaris* and FCA group. The phagocytic activities of HK leucocytes from tilapia administered with 300 and 500 µg *P. vulgaris* were significantly ($p < 0.05$) higher than 100 µg *P. vulgaris* and the control group. *P. vulgaris* was revealed with a good immunoadjuvant evoking the specific and non-specific immune responses of tilapia.

Keywords: *Prunella vulgaris*, Respiratory burst activity, Lysozyme activity, Tilapia

Background

One of the most promising methods for controlling diseases in aquaculture is strengthening the defense mechanisms of fish through prophylactic administration of immunostimulants [1]. Immunostimulants are naturally occurring compounds that modulate the immune system by increasing host resistance to infectious pathogens, and they have been widely used in aquaculture [2-4]. Traditional disease control strategies employing antibiotics and chemical disinfectants are no longer recommended due to the emergence of bacterial resistance as well as concerns over the environment and wildlife protection. Although vaccination has been shown to be an effective prophylactic method for disease control in fish [5], there are some methodological problems related to high costs and stress [6]. Already, remarkable success has been achieved using immunostimulants as a more environmentally friendly approach to disease management [7-9].

Several compounds, including β-glucans, chitin, algal and mistletoe extracts, and bacterial polysaccharides, have been used to enhance immunity and disease resistance in a variety of fish species [7,10-12]. β-glucan administration has been reported to augment antibody production, complement activity, lysozyme activity, phagocytic activity, and respiratory bursts in channel catfish *Ictalurus punctatus* [13], Atlantic salmon *Salmo salar* [14], rainbow trout *Oncorhynchus mykiss* [15], gilthead rainbow trout *Sparus auratus* [16], and sea bass *Dicentrarchus labrax* [17].

Prunella vulgaris is a perennial herb that is used in traditional medicine for the clinical treatment of sore throat, fever, and accelerated wound healing [18,19]. The organic fraction of *P. vulgaris* exhibits antioxidative and antimicrobial activities [18], whereas aqueous extracts of *P. vulgaris* inhibit HIV-1 infection [20]. In aquaculture, *P. vulgaris* was reported to have efficacy as a dietary supplement, although only up-regulation of natural immunity was observed [21].

* Correspondence: shchoi@kunsan.ac.kr
Department of Aquatic Life Medicine, Kunsan National University, Gunsan, Geonbuk 573-400, South Korea

To further study the availability of *P. vulgaris* as an immunoadjuvant to elicit a vaccination effect, we measured the specific antibody titer following intraperitoneal injection of *P. vulgaris* with an antigen in tilapia as a fish model.

To overcome the disease problem affecting fish culture systems, the present study applied indigenous *P. vulgaris* as an appropriate immnoadjuvant in order to augment specific and non-specific immunities as well as disease resistance in fish.

Materials and methods

Fish

Nile tilapia, *Oreochromis niloticus*, fish weighing about 100-150 g each were obtained from a fish farm in Kunsan National University, Korea and acclimated for 2 weeks to laboratory conditions in 70 L glass aquaria containing re-circulated and aerated water at 23-23°C. They were acclimated to this environment for at least 2 weeks prior to use and fed daily using a commercial diet during the adaptation and experimental periods. The health status of the animals was checked daily by observing fish behavior, and there were no clinical symptoms such as abnormal swimming patterns or body color changes.

Reagents

Nitroblue tetrazolium (NBT), Percoll, hemocyanin (HC), 3-(4,5-dimethylthiazol-2yl)-2,5-diphenyl-2H-tetrazolium bromide (MTT), phorbol myristate acetate (PMA), and Minimum essential medium (MEM) were purchased from Sigma Chemicals CO. Hanks balanced salt solution (HBSS), fetal bovine serum (FBS), and antibiotic-antimycotic were obtained from Gibco BRL, Grand Island, NY. Sodium nitrite, sulfanylamide, and phosphoric acid were purchased from ICN Biomedicals. Bakers' yeast and *Saccharomyces cerevisiae* purchased from Oriental Yeast Co. Ltd., Tokyo, and thioglycollate broth was obtained from Difco Laboratories, Detroit, USA.

Extraction of *P. vulgaris*

P. vulgaris originating from South Korea was kindly donated by an herbal medicine company (Sanyacho-Nongwon, Namyangju, Korea). Extraction of *P. vulgaris* was performed according to Lee's method [22]. Briefly, the chopped flowers, stems, and leaves of *P. vulgaris* (100 g) were placed in distilled water (1,000 ml) and stirred at 4°C overnight. After centrifugation at 15,000 × g for 20 min, the supernatant was filtered through 0.2 μm pore-sized filters. Protein content of *P. vulgaris* extract was determined using a commercial protein assay kit (Bio-Rad Lab, USA) and stored at 4°C.

In vitro and *in vivo* toxicities of *P. vulgaris*

The *in vitro* toxicity of *P. vulgaris* was tested against EPC and CHSE-214 fish cell lines. Respective cell lines (1×10^6 cells) were dispensed into each well of a 24-well plate (Costar, USA), followed by administration of various concentrations of *P. vulgaris* extract ranging from 10 ng to 100 μg/ml. After incubation at 24°C for 3 days, MTT assay was performed according to the method of Daly et al. [23]. Briefly, the tissue culture plates were centrifuged at 500 × g for 10 min, after which the supernatant fluids were carefully removed without disturbing the cell pellet or formazan precipitate. The formazan crystals were then dissolved by addition of 200 μl of dimethyl sulphoxide (DMSO) (Sigma) to each well, followed by 25 μl of glycine buffer (0.1 M glycine, 0.1 M NaCl, pH 10.5). Contents of the wells were then thoroughly mixed with a multichannel pipette. After 10 min, formazan development was read at 595 nm using an ELISA reader (ASYS HITECH, Austria).

To determine whether or not *P. vulgaris* has serious toxicity *in vivo*, 100 and 1000 μg of *P. vulgaris*/100 g of fish were intraperitoneally (I.P.) injected into seven fish per group. Fish blood was collected 4 days after injection, and the concentrations of glutamic oxaloacetatic trams aminase (GOT), glutamic pyruvate transaminase (GPT), and c-creatin were determined on a Fuji Dry Chem System (Fuji Photo Film Co. Ltd, Japan).

Administration of *P. vulgaris* to elicit non-specific immune response in tilapia

Tilapia were divided into five groups of seven fishes per group. Fish in each group were I.P injected with 100, 300, and 500 μg of *P. vulgaris*/100 g of fish in 0.5 ml of phosphate-buffered saline (PBS). The remaining group of fish was injected with an equivalent volume of sterile PBS or 1:1 emulsified Freund's complete adjuvant (FCA) (Sigma) as a control. On day 4 post-injection, blood and head kidney leucocytes were obtained from each fish.

Antibody production upon administration of *P. vulgaris* plus other immunostimulants

The immunostimulating effect of *P. vulgaris* was compared with those of FCA and β-glucan based on elevated antibody production. HC was used as an antigen to evoke a specific antibody response. For I.P. injection, 300 μg of *P. vulgaris* and 100 μg of HC were mixed and administered in a volume of 200 μl. Tilapia were divided into six groups (five fish per group), after which HC was injected alone (control) or mixed and injected with 100 μl of other adjuvants. The total injection volume was adjusted to 200 μl in all experiments. *P. vulgaris* was I.P. injected at a concentration of 300 μg suspended in 200 μl of phosphate-buffered saline (PBS). In the FCA group, 100 μl of FCA was 1:1 emulsified with HC suspended in 100 μl of PBS. In the FCA plus *P. vulgaris* group, 100 μl of FCA was added to the *P. vulgaris* and HC mixture and then I.P. injected at a volume of 200 μl. The optimized concentration (50 μg) of β-glucan was injected with HC at a volume of 200 μl. In the β-glucan plus *P. vulgaris* group, 50 μg of β-glucan and 300 μg *P. vulgaris* were mixed together and I.P. injected at a volume of 200 μl. On day 30 post-injection, blood was harvested from the fishes in each group, followed by antibody titer assay using an ELISA reader.

Isolation of head kidney (HK) leucocytes

The method described by Santarem et al. [24] was followed with some modifications. The tilapia HK was dissected out by ventral incision, cut into small fragments, and then transferred into 5 ml of HBSS. Cell suspensions of the HK were obtained by teasing HK tissues with two slide glasses in HBSS in a Petri dish (Coring, USA). After sedimentation of tissue debris at 4°C for 1 min, the supernatants were removed. HK cell suspensions were then layered over a 34-51% Percoll gradient and centrifuged at 1000 × g for 40 min at 14°C. After centrifugation, leucocyte bands located above the 34-51% interfaces were collected using a Pasteur pipette and washed twice at 120 × g for 8 min in HBSS. The concentration of viable cells was determined by trypan blue exclusion.

Serum

Blood was collected from the dorsal aorta of tilapia. Blood was allowed to clot at 20°C for 30 min and then cooled at 0°C for 1 h. Serum was obtained by centrifugation at 1000 × g for 8 min. Sera were frozen at -20°C until used.

Reactive oxygen intermediates (ROI) production assay

ROI production by tilapia kidney cells after administration of *P. vulgaris* was assessed by monitoring reduction of NBT [25]. Leucocytes (1×10^5 cells) were washed once with HBSS at 60 × g for 3 min at 4°C and then incubated in 100 μl of complete media in the presence of PMA and 1 μg/ml of NBT. After 1 h of incubation at 25°C, excess NBT was washed out with PBS, and the leucocytes were fixed with 70% methanol. After discarding the methanol, the leucocytes were washed twice with PBS. The reduced formazan was then solubilized with 120 μl of KOH and 140 μg of DMSO, after which optical density values were read at 620 nm on an ELISA reader.

Lysozyme activity

Serum lysozyme activity was measured using a modified turbidimetric microtiter plate technique according to Ellis [6]. Briefly, a standard suspension of 0.15 mg/ml of *Micrococcus lysodeikticus* (Sigma) was prepared in 66 mM phosphate buffer (pH 6.0). Tilapia serum (50 μl) was then added to 1 ml of the bacterial suspension, after which the absorbance reduction was recorded at 0.5 and 4.5 min intervals at 450 nm on a spectrophotometer (SHIMADZU

UV-1600PC). One unit of lysozyme activity was defined as a reduction in absorbance of 0.001/min.

Phagocytic activity

Tilapia HK leucocytes were adjusted to 1×10^6 cells/ 200 μl/well in 5% FBS-MEM and dispensed in an 8-well slide chamber (Nunc, Denmark), followed by overnight incubation at 25°C. Following incubation, 1×10^7 cells/ml of zymosan (Sigma) was added. The mixture was incubated at 25°C for 1 h with occasional shaking, after which 50 μl of the mixture was smeared onto a glass slide, air-dried, and stained with Wright's solution. Phagocytic activity (PA) [26] was calculated by enumerating 500 leucocytes per fish under a microscope. PA = number of cells ingesting zymosans/number of cells observed × 100.

Statistical analysis

Statistical significance of the differences between the groups was calculated by applying Student's 2-tailed *t*-test.

Results and discussion

Prunella vulgaris is a perennial plant known for its self-healing properties in Western herbal medicine [27,28], and it traditionally has been used for treating various diseases such as an allergies and inflammation in East Asian countries [29]. In addition, *P. vulgaris* has been reported to have immunomodulatory effects such as activation of macrophages [27,28,30]. The effect of *P. vulgaris* on fish immunity in aquaculture has only been reported by Harikrishnan et al. [21]. Specifically, they investigated the dietary effects of *P. vulgaris* on the non-specific immune response as well as disease resistance against *Uronema marinum*. In the

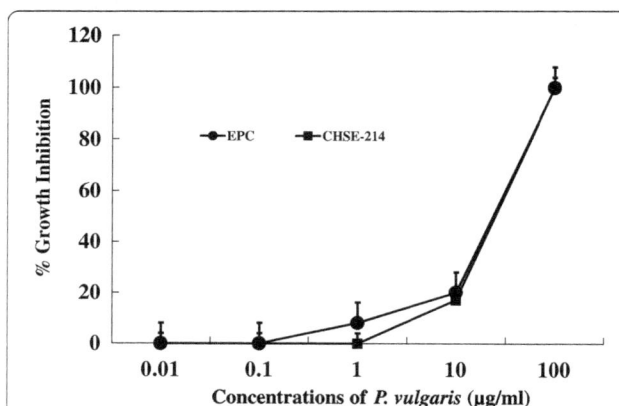

Figure 1 *In vitro* cytotoxicity of *P. vulgaris* on EPC and CHSE-214 fish cell lines. The cells (1×10^6) from each cell line were incubated with *P. vulgaris* ranging from 10 ng to 100 μg/ml and incubated at 24°C for 3 days. Formazan development was read at 595 nm using ELISA reader (ASYS HITECH, Austria). Error bars represent SD from the mean of triplicate wells. The result is a representative of three experiments.

Table 1 The levels of GOT, GPT and c-creatin in tilapia sera following injection of P. vulgaris

Standards for toxicity	Doses of P. vulgaris injected in fish		
	PBS	P100/100[1]	P1000/100
GOT	19 ± 3 mg/ML	23 ± 5 mg/ML	25 ± 3 mg/ML
GPT	40 ± 6 mg/ML	39 ± 5 mg/ML	42 ± 6 mg/ML
c-creatin	31 ± 4 mg/ML	28 ± 5 mg/ML	33 ± 7 mg/ML

[1]μg of P. vulgaris/g of fish.

present study, we tested the availability of *P. vulgaris* as a potent immunoadjuvant to achieve an enhanced vaccination effect. Furthermore, the *in vitro* and *in vivo* toxicities of *P. vulgaris* were investigated in tilapia as a fish model. Lastly, to determine the optimal amount of *P. vulgaris* extract that evokes an immune response in tilapia, we administered *P. vulgaris* extract by intraperitoneal (I.P.) injection.

First, we tested the *in vitro* and *in vivo* toxicities of *P. vulgaris* against transformed fish cell lines and tilapia. The half-killing concentrations of *P. vulgaris* against EPC and RTG-2 cells were 30 and 40 μg/ml, respectively (Figure 1). These *in vitro* toxicities were similar to a previous study in which Korean mistletoe showed negligible toxicity on mammalian cell lines [31]. Table 1 shows the levels of GOT, GPT, and c-creatin in blood from tilapia sensitized with *P. vulgaris*. There were no significant differences in toxicity between the groups administered 100 and 1000 μg of *P. vulgaris*/100 g of fish. Further, the GOT, GPT, and c-creatin levels of the PBS control group were within normal ranges, indicating non-toxicity. In contrast, a previous study showed that a 10-fold greater

concentration of mistletoe injected into eel significantly augmented toxicity [11]. The immunomodulatory effect of *P. vulgaris* was compared with those of FCA and β-glucan based on HC-specific antibody production. As shown in Figure 2, *P. vulgaris* induced significantly stronger antibody production than either FCA or β-glucan. Although β-glucan failed to elicit efficient antibody production, β-glucan plus *P. vulgaris* synergistically enhanced antibody production. However, no significant difference in toxicity between the *P. vulgaris* and *P. vulgaris* plus β-glucan groups was observed, indicating that *P. vulgaris* alone has strong immunoadjuvant activity.

Phagocytes produce respiratory bursts as a form of attack against invasive pathogens. As such, they are common measure of the defense ability against pathogens, although excessive accumulation of reactive oxygen intermediates (ROIs) is extremely toxic to host cells [32]. ROIs such as superoxide (O_2^-), hydrogen peroxide (H_2O_2), hydroxyl radical (OH), and singlet oxygen play important roles in the antimicrobial activity of phagocytic cells [33]. As shown in Figure 3, ROI production was significantly ($p < 0.05$) up-regulated in HK leucocytes from tilapia injected with 300 and 500 μg of *P. vulgaris*/100 g of fish compared to both the control and FCA groups, suggesting that ROIs are an indicator of *P. vulgaris*-induced non-specific immunity in tilapia. Although ROI production in the 100 μg of *P. vulgaris* group was higher than that in the control group, the difference was not significant. Maximum NBT reduction value was observed in the 500 μg of *P. vulgaris* group, but no significant difference was observed between the 300 and 500 μg of *P. vulgaris* groups. On the other hand, injection of more than 500 μg of

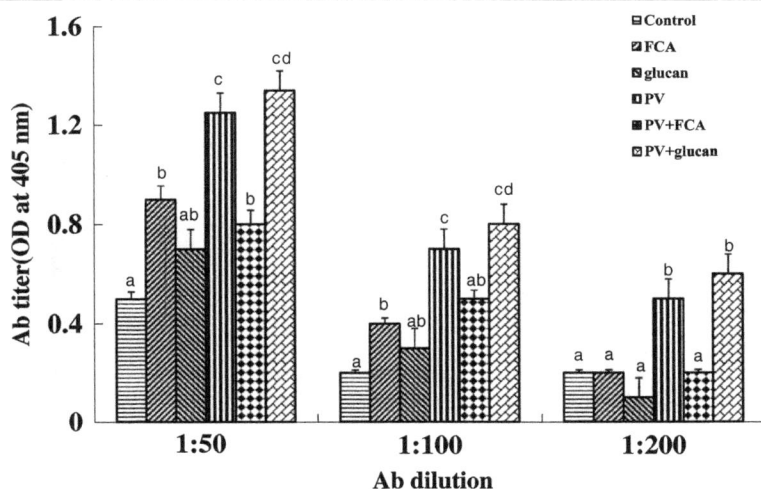

Figure 2 Antibody titer induced by P. vulgaris administered in tilapia. Six groups (5 fish/group) of fish were I.P. immunized with hemocyanin supplemented with *P. vulgaris* (PV), FCA, *P. vulgaris* + FCA (PV + FCA), β-glucan, *P. vulgaris* + β-glucan (PV + glucan) and PBS as a control, respectively. Data represent the mean + S.D. (n = 5). Statistical differences (p < 0.05) between groups are indicated by different letters over the bar.

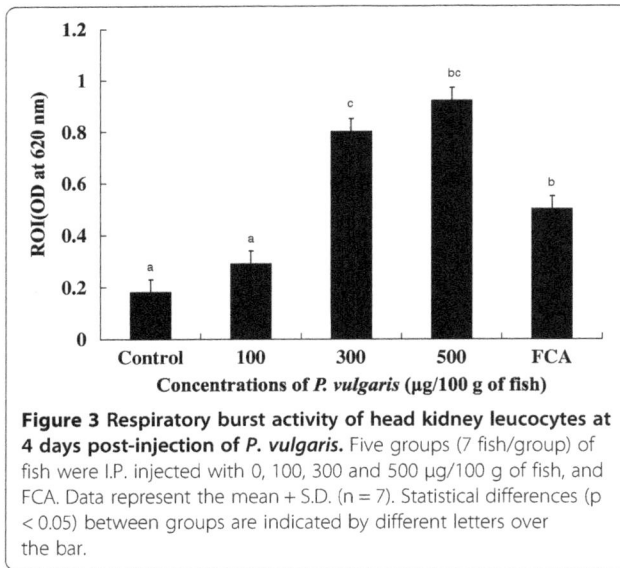

Figure 3 Respiratory burst activity of head kidney leucocytes at 4 days post-injection of *P. vulgaris*. Five groups (7 fish/group) of fish were I.P. injected with 0, 100, 300 and 500 μg/100 g of fish, and FCA. Data represent the mean + S.D. (n = 7). Statistical differences (p < 0.05) between groups are indicated by different letters over the bar.

P. vulgaris/100 g of fish failed to up-regulate ROI production (data not shown). Usually, the effect of immunostimulants is strongest at intermediate dosages with minimal activity and even toxicity at high doses [34,35]. This phenomenon has been established in fish through *in vivo* [36,37] and *in vitro* studies [38].

Lysozyme has both bactericidal as well as opsonin effects that activate the complement system and phagocytes to prevent infection and disease [39]. Figure 4 shows lysozyme activities in the sera of tilapia with or without *P. vulgaris* administration. Serum lysozyme activities were significantly (p < 0.05) higher in the 300 and 500 μg of *P. vulgaris* groups compared to the 100 μg of *P. vulgaris* and FCA groups. Further, there was no significant difference in

lysozyme activity between the 300 and 500 μg of *P. vulgaris* groups. Considering that serum from *P. vulgaris*-injected fish showed elevated lysozyme activity, *P. vulgaris* is likely to play a critical role in evoking lysozyme activity from tilapia kidney phagocytes. However, injection of 1000 μg of *P. vulgaris* reduced lysozyme activity compared to the 300 and 500 μg of *P. vulgaris* groups (data not shown), suggesting that an excess concentration of *P. vulgaris* interferes with lysozyme activity in tilapia.

Phagocytes are the first cells to recognize invading foreign bodies and are thus central to cell-mediated and humoral immunities [40]. To test whether or not *P. vulgaris* can influence phagocytic activity, tilapia kidney leucocytes sensitized with *P. vulgaris* (100, 300, and 500 μg of *P. vulgaris*) were incubated overnight with zymosans. In our study, instead of foreign pathogens, zymosans were treated to phagocytes from either *P. vulgaris*-injected tilapia or non-treated tilapia. As shown in Figure 5, the phagocytic activities of HK leucocytes isolated from tilapia injected with 300 and 500 μg of *P. vulgaris* were significantly higher compared to the 100 μg of *P. vulgaris* and PBS control groups. Further, was a significant difference (p < 0.05) between the 100 μg of *P. vulgaris* as well as 300 or 500 μg of *P. vulgaris* groups, but no significant difference between the 300 and 500 μg of *P. vulgaris* groups themselves. Lastly, excess injection of *P. vulgaris* (1000 μg) inhibited phagocytosis and respiratory bursts in HK leucocytes isolated from tilapia.

Although the materials used in the study were different, Gopalakannan and Aurl [41] and Luo et al. [42] also reported that fish treated with a high dosage of chitosan display significantly inhibited phagocytosis compared to low dosage. This result suggests that the high level of *P. vulgaris* directly induced phagocytosis, thereby exhausting the cells.

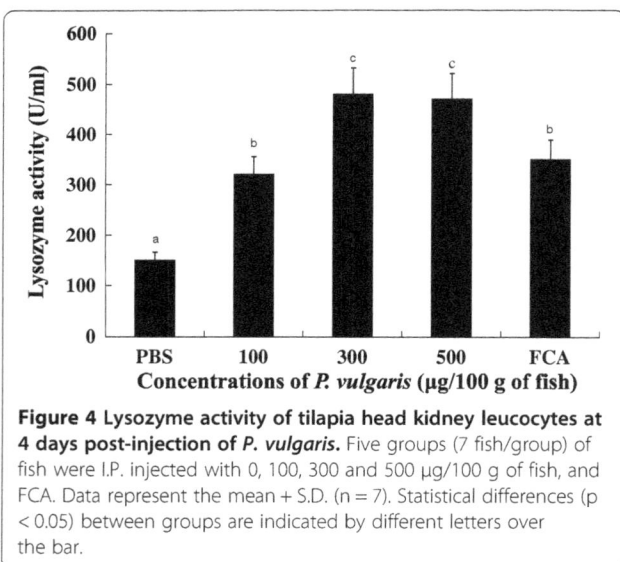

Figure 4 Lysozyme activity of tilapia head kidney leucocytes at 4 days post-injection of *P. vulgaris*. Five groups (7 fish/group) of fish were I.P. injected with 0, 100, 300 and 500 μg/100 g of fish, and FCA. Data represent the mean + S.D. (n = 7). Statistical differences (p < 0.05) between groups are indicated by different letters over the bar.

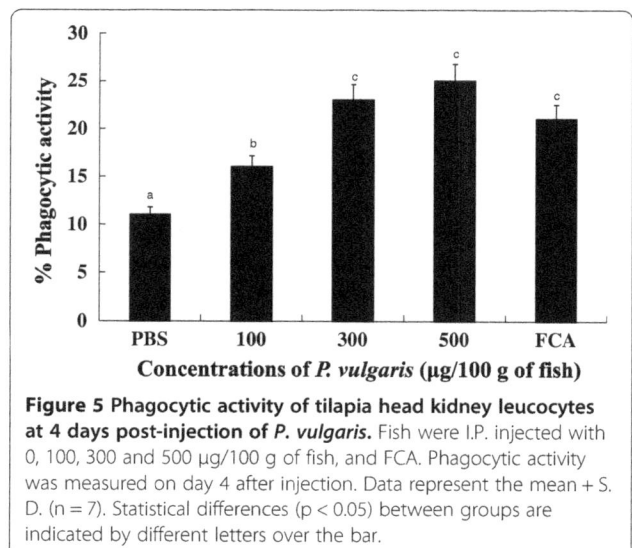

Figure 5 Phagocytic activity of tilapia head kidney leucocytes at 4 days post-injection of *P. vulgaris*. Fish were I.P. injected with 0, 100, 300 and 500 μg/100 g of fish, and FCA. Phagocytic activity was measured on day 4 after injection. Data represent the mean + S. D. (n = 7). Statistical differences (p < 0.05) between groups are indicated by different letters over the bar.

Conclusions

P. vulgaris elevated almost all non-specific immune parameters as well as specific humoral immunity. Therefore, *P. vulgaris* could be a promising immunomodulatory material for inducing specific and non-specific immune responses in fish. Further studies on using *P. vulgaris* as a dietary supplement in aquaculture are currently underway.

Competing interests
The authors declare that they have no competing interests.

Authors' contributions
KH collected *Prunella vulgaris* samples, wrote the manuscript and performed *in vitro* toxicity test as well as a statistical analysis. SH carried out the specific and non-specific immune-related experiments. Both authors read and approved the final manuscript.

Acknowledgments
This work was supported by a grant from the Fisheries Science Institute in Kunsan National University in 2013.

References

1. Robertsen B: **Modulation of the non-specific defense of fish by structurally conserved microbial polymers.** *Fish Shellfish Immunol* 1999, 9:269–90.
2. Ai QH, Mai KS, Zhang I, Tan BP, Zhang WB, Xu W: **Effects of dietary β-1,3 glucan on innate immune response of large yellow croaker, *Pseudosciaena crocea*.** *Fish Shellfish Immunol* 2007, 22:394–402.
3. Selvaraj V, Sampath K, Sekar V: **Administration of lipopolysaccharide increaseds specific and non-specific immune parameters and survival in carp infected with *Aeromonas hydrophila*.** *Aquaculture* 2009, 286:176–183.
4. Ringø E, Olsen RE, Gifstad TØ, Dalmo RA, Amlund H, Hemre GI, Bakke AM: **Prebiotics in aquaculture: a reviews.** *Aquacul Nutr* 2010, 16:117–136.
5. Midtlyng PJ, Reitan LJ, Lillehaug A, Ramstad A: **Protection, immune responses and side effects in Atlantic salmon (*Salmo salar* L.) vaccinated against furunculosis by different procedures.** *Fish Shellfish Immunol* 1996, 6:599–613.
6. Ellis AE: **Immunity to bacteria in fish.** *Fish Shellfish Immunol* 1999, 9:291–308.
7. Raa J: **The use of immunostimulatory substances in fish and shellfish farming.** *Rev Fish Sci* 1996, 4:229–288.
8. Sakai M: **Current research status of fish immunostimulants.** *Aquaculture* 1999, 172:63–92.
9. Peddie S, Zou J, Secombes J: **Immunostimulation in the rainbow trout (*Oncorhynchus mykiss*) following intraperitoneal administration of Ergosan.** *Vet Immunol Immunopathol* 2002, 86:101–113.
10. Anderson DP: **Environmental factors on fish health: immunological aspects.** In *The fish immune system: organism, pathogen, and environment.* Edited by Iwama G, Nakanishi T. San Diego, CA: Academic; 1996:289–310.
11. Yoon TJ, Park KH, Choi SH: **Korean mistletoe (*Viscum album* Coloratum) extract induces eel (*Anguilla japonica*) non-specific immunity.** *Immune Network* 2008, 8(4):124–129.
12. Park KH, Choi SH: **The effect of mistletoe, *Viscum album* coloratum, extract on innate immune response of Nile tilapia (*Oreochromis niloticus*).** *Fish Shellfish Immunol* 2012, 32:1016–1021.
13. Chen D, Ainsworth AJ: **Glucan administration potentiates immune defense mechanisms of channel catfish, *Ictalurus punctauts Rafinesque*.** *J Fish Dis* 1992, 15:295–304.
14. Engstad R, Robertsen B, Frivold E: **Yeast glucan induces increase in lysozyme and complement-mediated haemolytic activity in Atlantic salmon blood.** *Fish Shellfish Immunol* 1992, 2:287–297.
15. Verlhac V, Gabaudan J, Obach A, Schuep W, Hole R: **Influence of dietary glucan and vitamin D on non-specifici responses of rainbow trout (*Oncorhynchus mykiss*).** *Aquaculture* 1996, 144:123–133.
16. Castro R, Couso N, Obach A, Lamas J: **Effect of different beta-glucans on the respiratory burst of turbot (*Psetta maxima*) and gilthead rainbow trout (*Sparus aurata*) phagocytes.** *Fish Shellfish Immunol* 1999, 9:529–541.
17. Bagni M, Romano N, Finoia MG, Abelli L, Scapigliate G, Tiscar PG: **Short- and long-term effects of a dietary yeast β-glucan (Macrogard) and alginic acid (Ergosan)preparation on immune response in sea bass (*Dicentrarchus labrax*).** *Fish Shellfish Immunol* 2005, 18:311–325.
18. Psotova J, Kolar M, Sousek J, Svagera Z, Vicar J, Ulrichova J: **Biological activities of *Prunella vulgaris* extract.** *Phytother Res* 2003, 17:1082–1087.
19. Psotova J, Svobodova A, Kotarova H, Walterova D: **Photoprotective properties of *Prunella vulgaris* and rosmariinic acid on human keratinocytes.** *J Photochem Photobiol B* 2006, 84:167–174.
20. Oh C, Price J, Brindley MA, Widrlechner MP, Qu L, McCoy JA, Murphy P, Hauck D, Maury W: **Inhibition of HIV-1 infection by aqueous extracts of *Prunella vulgaris* L.** *Virol J* 2011, 23:8–12.
21. Harikrishnan R, Kim JS, Kim MC, Balasundaram C, Heo MS: ***Prunella vulgaris* enhances the non-specific immune response and disease resistance of *Paralichthys olivaceus* against *Uronema marinum*.** *Aquaculture* 2011, 318:61–66.
22. Lee JB, Kang TB, Choi SH, Lee U, Kim AJ, Jeong CJ, Lee HC, Cho YS, Won JG, Lim JC, Yoon TJ: **Effect of *Prunella vulgaris* Labiatae extract on inmate immune cells and anti-metastatic effect in mice.** *Food Sci Biotechnol* 2009, 18(1):218–222.
23. Daly JG, Moore AR, Olivier G: **A colorimetric assay for the quantivication of brook trout (*Salvelinus fontinalis*) lymphocyte mitogenesis.** *Fish Shellfish Immunol* 1995, 5:265–273.
24. Santarem M, Novoa B, Figueras A: **Effect of beta-glucans on the non-specific immune responses of turbot (*Scophthalmus* L.).** *Fish Shellfish Immunol* 1997, 7:429–437.
25. Secombes CJ, Chung S, Jeffries AH: **Superoxide anion production by rainbow trout macrophages detected by the reduction of ferricytochrome C.** *Dev Comp Immunol* 1988, 12:201–206.
26. Sahoo PK, Mukherjee SC: **Effect of dietary β-1,3 glulcan on immune responses and disease resistance of healthy and aflatoxin B1-induced immunocompromised rohu (*Labeo rohita* Hamilton).** *Fish Shellfish Immunol* 2001, 11:683–695.
27. Fang X, Chang RC, Yuen WH, Zee SY: **Immune modulatory effects of *Prunella vulgaris* L.** *Int J Mol Med* 2005, 15:491–496.
28. Fang X, Yu MM, Yuen WH, Xee SY, Chang RC: **Immune modulatory effects of *Prunella vulgaris* L. on monocytes/macrophages.** *Int J Mol Med* 2005, 16:1109–1116.
29. Kim SY, Kim SH, Shin HY, Lim JP, Chae BS, Park JS, Hong SG, Kim MS, Jo DG, Park WH, Shin TY: **Effects of *Prunella vulgaris* on mast cell-mediated allergic reaction and inflammatory cytokine production.** *Exp Biol Med* 2007, 232:921–926.
30. Han EH, Choi JH, Hwang YP, Park HJ, Choi CY, Chung YC, Seo JK, Jeong HG: **Immunostimulatory activity of aqueous extract isolated from *Prunella vulgaris*.** *Food Chem Toxicol* 2009, 47(1):62–69.
31. Yoon TJ, Yoo YC, Kang TB, Her E, Kim K, Azuma I, Kim JB: **Cellular and humoral adjuvant activity of lectins isolated from Korean mistletoe (*Viscum album colaratum*).** *Int Immunopharmacol* 2001, 1:881–889.
32. Salinas I, Cuesta A, Esteban MA, Meseguer J: **Dietary administration of *Lactobacillus delbrueckii* and *Bacillus subtilis*, single or combined, on gilthead seabream cellular innate immune responses.** *Fish Shellfish Immunol* 2005, 19:67–77.
33. Babior BM: **Oxidants from phagocytes: agents of defense and destruction.** *Bolld* 1984, 64:959–966.
34. Bliznakov EG, Adler AD: **Nonlinear response of the reticuloendothelial system upon stimulation.** *Pathol Microbiol* 1972, 38:393–410.
35. Gialdroni-Grassi G, Grassi C: **Bacterial products as immunomodulating agents.** *Int Arch Allergy Appl Immunol* 1985, 76(Supp.1):119–127.
36. Kenyon CJ, McKeever A, Oliver JA, Henderson IW: **Control of renal and adrenocortical function by the renin-angiotensin system in two euryhaline teleost fishes.** *Gen Comp Endocrinol* 1985, 58:93–100.
37. Anderson DP, Jeney G: **Immunostimulants added to injected *Aeromonas salmonicida* bacterin enhance the defense mechanisms and protection on rainbow trout (*Oncorhynchus mykiss*).** *Vet Immunol Immunopathol* 1992, 34:379–389.
38. Siwicki AK, Cossarini-Dunier M, Studnicka M, Demael A: **In vivo effect of the organophosphorus insecticide trichlorphon on immune response of carp (*Cyprinus carpio*). II. Effect of high doses of trichlorphon on nonspecific immune response.** *Ecotoxicol Environ Saf* 1990, 19:99–105.
39. Alexander JB, Ingram GA: **Noncellular non-specific defense mechanisms of fish.** *Ann Rev Fish Dis* 1992, 2:249–79.

40. Cavaillon JM: **Cytokines and macrophages.** *Biomed Pharmacother* 1994, **48:**445–453.

41. Gopalakannan A, Aurl V: **Immunomodulatory effects of dietary intake of chitin, chitosan and levamisole on the immune system of** *Cyprinus carpio* **and control of** *Aeromonas hydrophila* **infection in ponds.** *Aquaculture* 2006, **255:**179–187.

42. Luo L, Cai X, He C, Xue M, Wu X, Cao H: **Immune response, stress resistance and bacterial challenge in juvenile rainbow trouts** *Oncorhynchus mykiss* **fed diets containing chitosan-oligosaccharides.** *Curr Zool* 2009, **55:**416–22.

16

Hanwoo cattle: origin, domestication, breeding strategies and genomic selection

Seung-Hwan Lee[1†], Byoung-Ho Park[2†], Aditi Sharma[1†], Chang-Gwon Dang[1], Seung-Soo Lee[2], Tae-Jeong Choi[2], Yeon-Ho Choy[2], Hyeong-Cheol Kim[1], Ki-Jun Jeon[1], Si-Dong Kim[2], Seong-Heum Yeon[1*], Soo-Bong Park[2] and Hee-Seol Kang[1]

Abstract

Hanwoo (Korean cattle) is the native, taurine type of cattle breed of Korea and its history as a draft animal dates back to 5000 Years. In earlier times Hanwoo was used extensively for farming, transportation. Over the period of time, Hanwoo has changed to be meat type cattle. Full-scale production of Hanwoo as meat-type cattle has occurred since 1960s with the rapid growth of the Korean economy. Hanwoo is one of the most economically important species in Korea as it is a significant source of nutrition to the Korean people. Hanwoo beef is the most cherished food of Korea. One of the main goals of researchers is to increase the meat quality, quantity and taste of the beef. In this review we describe the origin, domestication of Hanwoo cattle and breeding program initiated from 1980's. Moreover the advent of technological advancement had provided us a platform to perform genome wide selection on economic traits and its implementation into traditional breeding programs.

Keywords: Hanwoo (Korean cattle), Origin, Domestication and breeding program

Introduction

Hanwoo (Korean cattle) is the native, taurine type, small sized cattle breed of Korea and its history as a draft animal dates back to 5000 Years. In earlier times Hanwoo was used extensively for farming, transportation and for religious sacrifice. Over the period of time Hanwoo cattle has evolved to be meat type cattle. Full-scale production of Hanwoo as meat-type cattle has occurred since 1960s with the rapid growth of the Korean economy. Despite its high price, beef from Hanwoo cattle is very popular among both Koreans and foreigners. Hanwoo beef is known for its marbled fat, tenderness, juiciness and characteristic flavor. There are four breeds of Hanwoo in Korea viz. brown Hanwoo, brindle Hanwoo, black Hanwoo and jeju black (Figure 1). Among all these colour variants brown is the most common one (Figure 1). National Institute of Animal Science along with the small stakeholders has taken an initiative for a better management and improvement of Hanwoo cattle. Hanwoo Experiment Station

(A Research Station of National Institute of Animal Science) in Pyeongchang-gun, Gangwon-do, Korea is dedicated exclusively for research on Hanwoo cattle. Along with Hanwoo experiment station there are other institutes and universities as well which are working towards the betterment of this native cattle breed.

Korea along with Hanwoo has a market for imported beef from Australia, USA, New Zealand, Canada and Mexico as well (KMTA, [1]). When Hanwoo beef was compared with the Australian Angus, it showed characteristic compositional and quality differences that should result from genetic and environmental differences between them [2]. Hanwoo is lower in cholesterol compared to other beef. It also has a higher omega 3 fatty acid count which makes it healthier than beef from other cattle. Taste-wise, it's very soft, juicy, and delicious. The marbling in Hanwoo steak is excellent as well and it has the right balance of meat and fat. Korean consumers decided their overall acceptability of Hanwoo beef are as follows: weights of tenderness 55%, juiciness 18%, and flavor-likeness 27% [3]. No wonder Korean consumers, despite the high price prefer Hanwoo beef to the imported one. In order to produce high quality Hanwoo beef along with specially designed breeding programs special care and feed strategies are deployed.

* Correspondence: yeonsu58@korea.kr
†Equal contributors
[1]Hanwoo Experiment Station, National Institute of Animal Science, RDA, Pyeong-Chang 232-950, Korea
Full list of author information is available at the end of the article

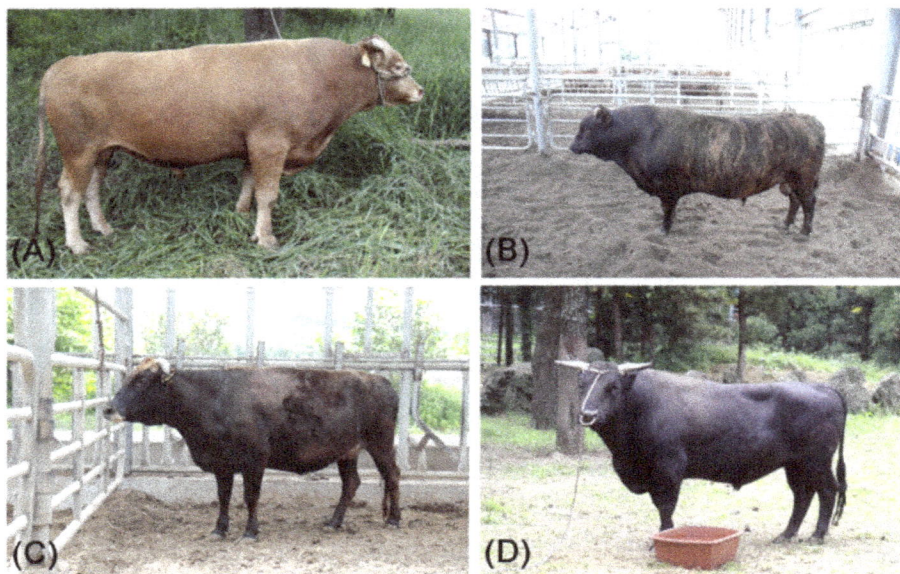

Figure 1 Pictures of four Hanwoo cattle. (**A**) Brown Hanwoo, (**B**) Brindle Hanwoo, (**C**) Black Hanwoo and (**D**) Jeju black Hanwoo.

The major goal of researchers is to increase both the quality (Marbling, tenderness and flavor) and the quantity (Carcass weight) of the meat to benefit the Hanwoo beef industry. Therefore, the current selection index used in Korea Proven Bulls (KPN) program is based on 1) carcass weight on marbling score 2) the comparison on carcass weight. In order to support the beef industry along with the conventional selection program, the genomic selection is also being incorporated in selection and breeding programs. In the present review we discussed the origin, domestication, breeding strategies and genomic selection in Hanwoo cattle.

Origin and Domestication

Origin and domestication of not just Hanwoo but all north-east asian cattle is a topic of discussion among researchers. Researchers are of divided view when it comes to ancestry and domestication of Hanwoo. There are some studies on the origin and ancestry of this cattle breed. Han S. W., [4] suggested Hanwoo to be a cross between zebu and taurine cattle, which migrated to Korea from Mongolia via north china while Lee & Pollack [5] proposed Hanwoo to have originated as a hybrid from auroch and zebu cattle. Yoon et al. [6] suggested an independent domestication event for Hanwoo. Mannen et al. [7] also suggested the independent mitochondrial origin in Asian cattle. Their study included Mongolian, Korean and Japanese cattle. In a recent study McTavish et al. [8] suggested that the asian cattle (Hanwoo and Japanese Black) to be of hybrid taurine–indicine origin. We carried out cattle diversity analysis which showed a clear differentiation of asian taurine from western taurine

cattle (Figure 2B-2C). Based on 50 K SNP chip data we computed a neighbor joining phylogenetic tree where we observed a clear partitioning for African taurine, European taurine, Asian taurine and Zebu cattle. The tree clearly clustered Korean cattle breeds on a separate node than the European taurine cattle breeds. This might suggest a separate domestication event for north-east asian cattle. Also, mitochondria based studies have identified a haplotype T4 that is observed in north east asian cattle breeds but not in other breeds (Mannen et al., [7]). The PCA plot also clearly separated Hanwoo and Wagyu (Japanese cattle) from Hereford and angus cattle thus supporting the inference from neighbor joining tree.

Population genetic parameters in Hanwoo

Linkage disequilibrium (LD) and effective population size (Ne) are important genetic parameters to understand population structure for optimal breeding program design. Lee et al., [9] performed a LD analysis and studied Ne for the entire Korean Hanwoo cattle genome, which was the first LD map and effective population size estimate ever calculated for this breed. A panel of 4,525 markers was used in the final LD analysis. The pairwise $r2$ statistic of SNPs up to 50 Mb apart across the genome was estimated. A mean value of $r2 = 0.23$ was observed in pairwise distances of <25 kb and dropped to 0.1 at 40 to 60 kb, which is similar to the average intermarker distance used in this study. The proportion of SNPs in useful LD ($r^2 \geq 0.25$) was 20% for the distance of 10 and 20 kb between SNPs. Analyses of past effective population size estimates based on direct estimates of recombination rates from SNP data demonstrated a

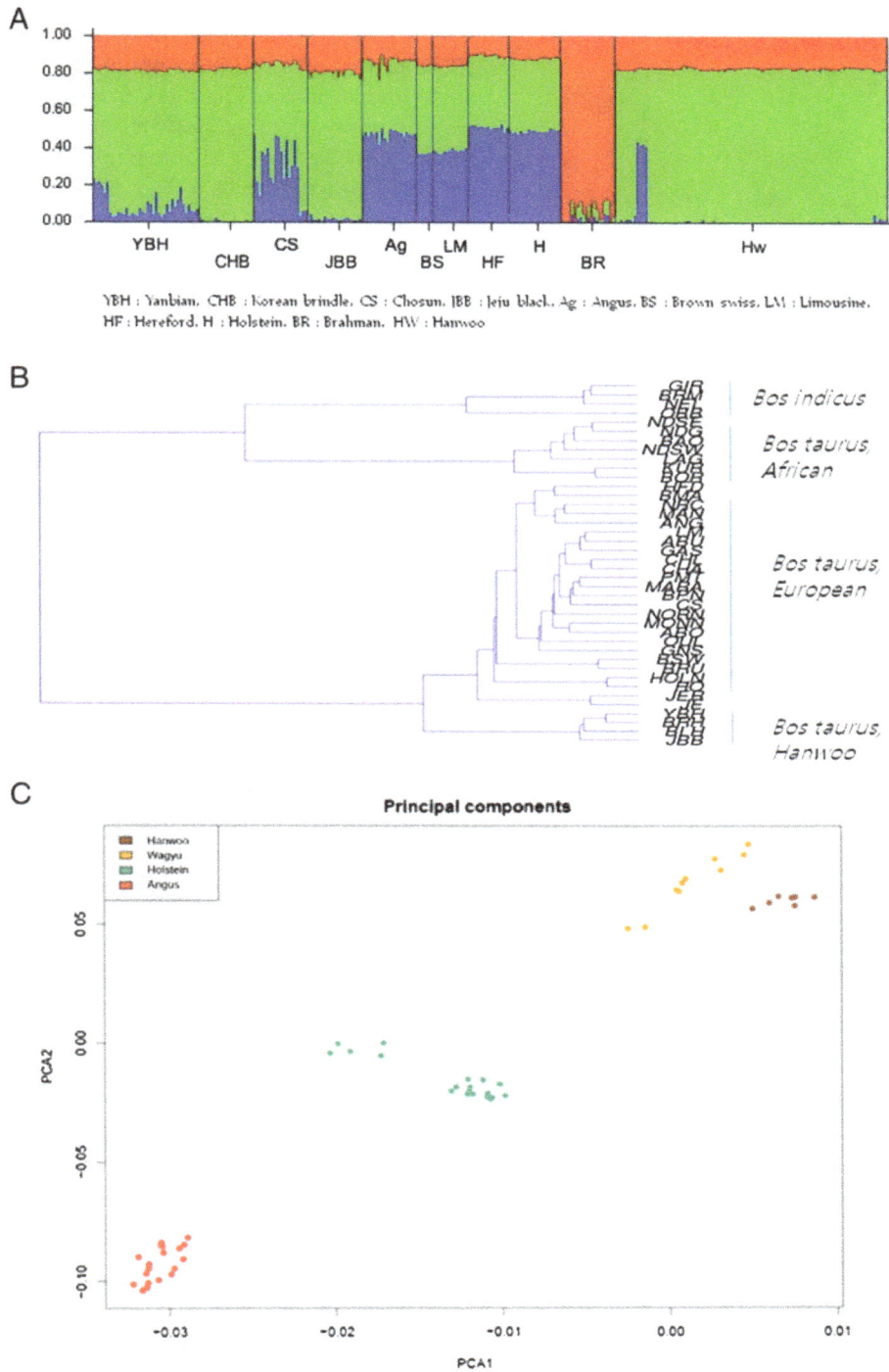

Figure 2 Comparison of Hanwoo compared to other taurine cattle. (A) Model-based population assignment for individuals based on the 50K SNP panel using STRUCTURE (K=3), YBH: Yanbian, CHB: Korean brindle, CS: Chosun, JBB: Jeju black, Ag: Angus, BS: Brown swiss, LM: Limousine, HF: Hereford, H: Holstein, BR: Brahman, HW: Hanwoo **(B)** A neighbor joining tree based on Pairwise Fst values, **(C)** Principal component analysis capturing a clear difference of north-east asian cattle.

decline in effective population size to $Ne = 98.1$ up to three generations ago (Figure 3). In this population genetic parameter (LD and Ne) study, the Ne in Hanwoo was larger overall than in North American Holstein cattle. In particular, the Ne of Hanwoo dramatically decreased 25 generations ago, this is when official Hanwoo

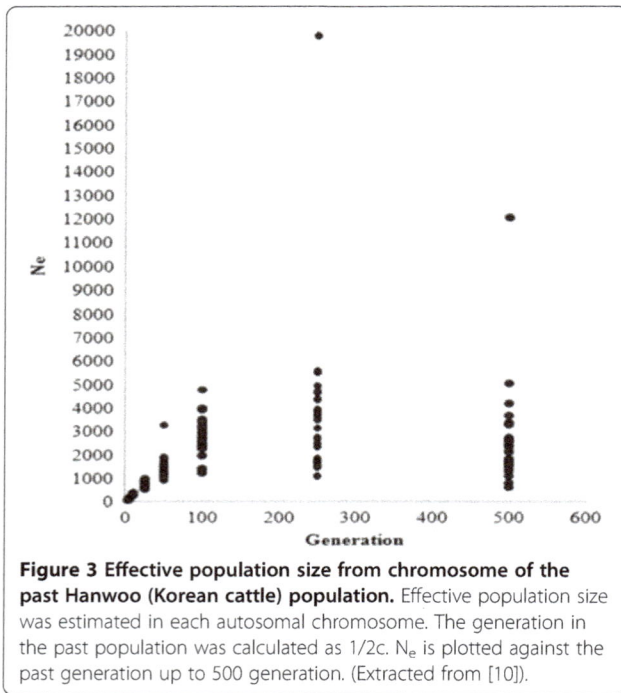

Figure 3 Effective population size from chromosome of the past Hanwoo (Korean cattle) population. Effective population size was estimated in each autosomal chromosome. The generation in the past population was calculated as 1/2c. N_e is plotted against the past generation up to 500 generation. (Extracted from [10]).

breeding programs were initiated. Effect of selection pressure on Ne in the Hanwoo population was evident from this study.

Hanwoo breeding program

Marbling score (MS), carcass weight (CWT), eye muscle area (EMA) and back fat thickness (BF) are the traits to impact Korean beef industry the most. Therefore the selection index for Korea Proven Bulls uses eye muscle area on marbling score and their comparison on CWT.

The first genetic breeding program called "Hanwoo-Gaeryang-Danji (HGD)" was initiated by Ministry of Agriculture and Forestry (MAF) in 1979. However, due to lack of awareness in the farmers to record the data and scarcity of the trained emulators, the program didn't meet its

objective. In 1999, a new program called "Hanwoo-Gaeryang-Nongga (HGN)" was introduced. Whereas HGD focused on a particular province HGN concentrated on individual farms. The cows from HGN program are used as major breeding stock for "Hanwoo Performance and Progeny Test (HPPT)" program (MAF, [11]). Prior to the HPPT program, bulls to be used for artificial insemination were selected solely on the basis of phenotypes. It was in 1987 that the first proven bull was produced using HPPT program. As shown in Figure 4, the HPPT program was a two stage selection program. The first stage was a performance test for the young bulls and the second stage was progeny test of selected young bulls. Young bull calves of current proven bulls were harvested at the age of 6 months from HGN belonging to HGD based on their phenotypic values and underwent a performance test up to age of 12 months. At this stage young bulls were selected based on a combination of their breeding values of weight at 12 months and average daily gain estimated from its performance test records. In the progeny test, cows from the HGN program were inseminated with the semen from young bulls and the male calves were harvested from farms and raised for the station progeny test. These steers were raised in a group until slaughter at 24 months. Carcass data comprised EMA, BF, MAR and CWT. Based on the progeny test results 20 bulls are selected for the AI program. Selection is based on a selection index which used weighted breeding values for BF, MAR and EMA.

Park et al. [12] carried out a study to determine the trend of improvement as well as the estimation of genetic parameters of the traits being used for seedstock selection based on the data collected from the past. In his study he concluded that the performance and progeny test based selection system of Korea is good enough to accommodate circumstances where fewer sires are used on many more cows. Although progeny tests take longer and cost more, they seem to be appropriate under the circumstances of the domestic market with its higher

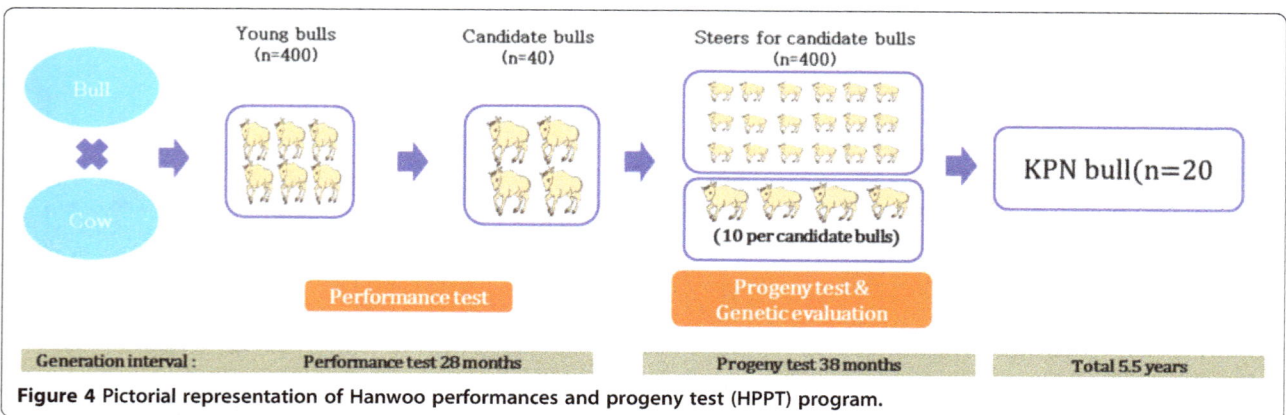

Figure 4 Pictorial representation of Hanwoo performances and progeny test (HPPT) program.

requirement for better meat quality. The study suggested accumulative data collection, genetic evaluation model development, revision of selection indices, as well as cooperation among farms, associations, National Agricultural Cooperative Federation, universities, research institutes and government agencies must be applied to the Hanwoo selection program. According to the Park's study, the current progeny tested evaluation runs every 6 months to select KPN in Korea. The breeding program has achieved a substantial genetic improvement for CWT and EMA. However there has been just a slightly negative genetic response for MS with an -0.036 of genetic response per year [12].

The heritability and genetic correlation are important genetic parameters in breeding program. In Hanwoo population, heritability and genetic correlation were estimated using carcass data from National Hanwoo progeny tested population (Figure 5A). A negative genetic correlation was observed between BF and EMA (-0.2) and MS and BF (-0.02) while positive correlation was observed in all other traits. A high positive correlation was observed between EMA and CWT (0.45) indicating that a greater EMA is associated with a higher production of CWT. The value of correlation between EMA and CWT in our study is more than that reported by and Koch et al. [13] and less as reported by Baik et al. [14] and Cundiff et al. [15]. High estimated heritabilities for BF (0.5), EMA (0.41), MS (0.4) and moderate for CWT (0.33) were observed. This indicates that in Hanwoo cattle a considerable genetic variability still exits which may be used for the improvement of carcass traits. Estimation of heritability, from genome-wide SNPs has recently attracted interest and offers several advantages over traditional pedigree-based methods. A comparison of pedigree and genotype based heritability

for carcass traits was thus drawn (Figure 5B). The amount of variation that could be accounted for by SNP genotypes was concordant with pedigree-based heritabilities and varied from very low for CWT to medium-high for BF.

We know conventional breeding methods rely on physical characteristics or phenotypes to calculate the breeding values (BV) of animals. The traditional methods are often not so accurate, inefficient and time consuming as most traits like meat quality are hard to measure and evident only when animal reaches maturity leading to a delay in verifying breeding results. With the advent of genome wide SNP panels we can overcome the drawbacks of the conventional breeding methods. Genome wide SNP panels will allow us to accurately and cost & time effectively determine the genomic estimated breeding values (GEBV) even in young animals and thus select animals at young age.

Genetic architecture for carcass trait in Hanwoo

Korean beef industry lay emphasis on meat yield and quality traits like carcass weight, marbling, intramuscular fat, eye muscle area. To identify the significant associations of SNPs with these traits can directly affect the selection of animals for a breeding program.

Genome-wide association study (GWAS) is a process for inspection and screening of detectable common genetic variants (single-nucleotide polymorphisms) in individuals to identify the variant(s) associated with the trait under study. There are several GWAS studies that are successfully conducted in livestock species. A recent study on Hanwoo cattle conducted by lee et al. (2013) identified 6 highly significant SNPs on chromosome 14 to be associated with carcass weight in Hanwoo cattle (Figure 6). This genome-wide association study was conducted to identify

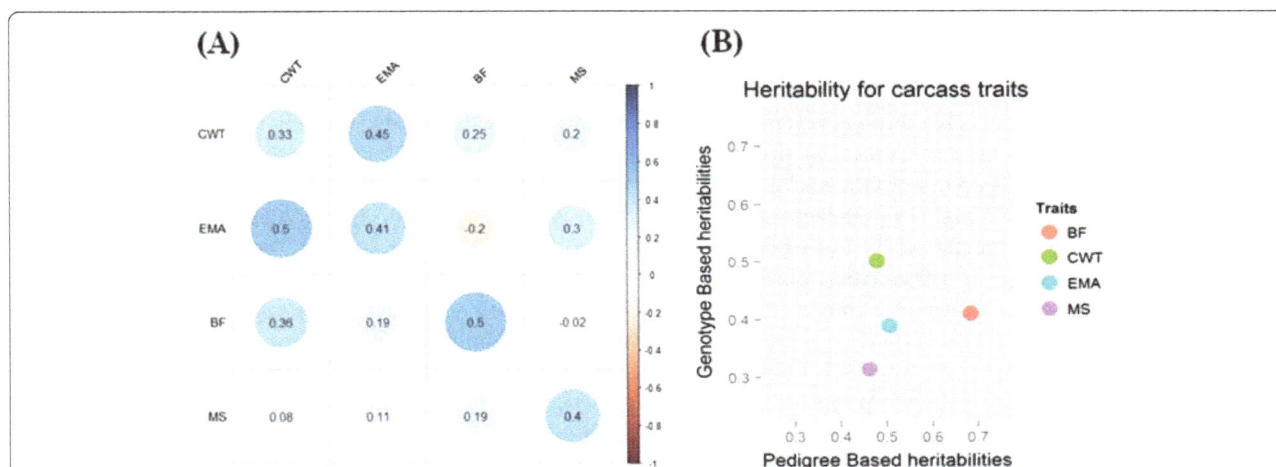

Figure 5 Heritability and genetic correlation for carcass trait in Hanwoo. Data are taken from a report for Hanwoo genetic improvement meeting (2013). (A) Along the diagonal (From upper left) are heritability values, upper triangle shows variance for genetic correlation and lower triangle shows residual variance. **(B)** Estimate of heritability from a pedigree and genotype for carcass traits in Hanwoo.

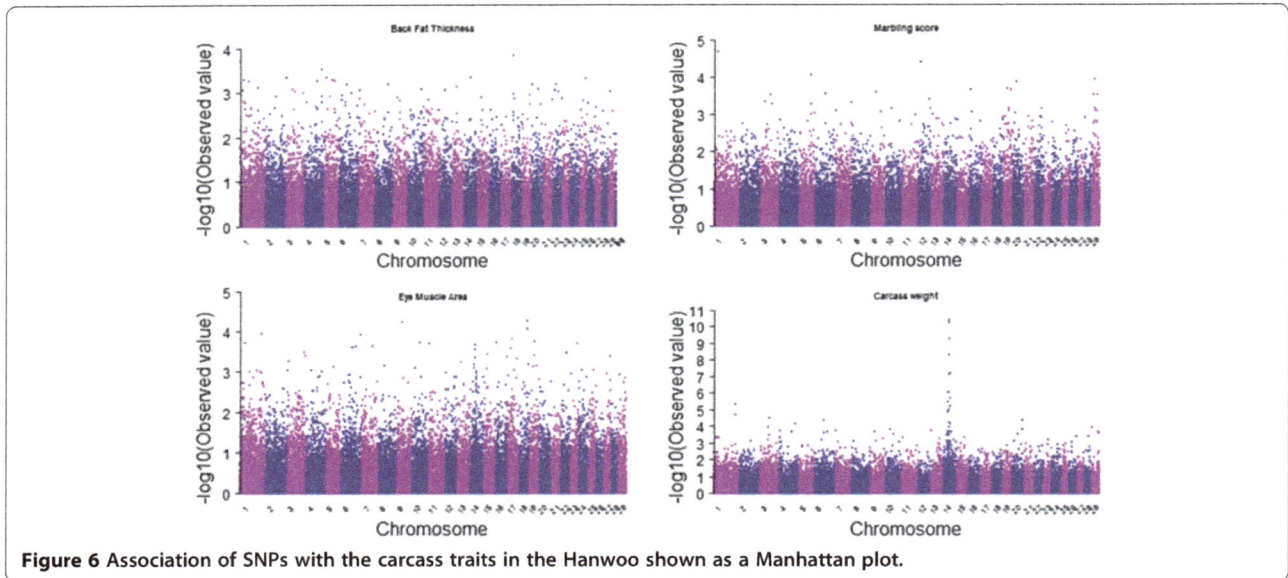

Figure 6 Association of SNPs with the carcass traits in the Hanwoo shown as a Manhattan plot.

major loci that are significantly associated with carcass weight, and their effects, so as to provide an insight into the genetic architecture of carcass weight in Hanwoo. This genome-wide association study identified a major chromosome region ranging from 23 Mb to 25 Mb on chromosome 14 as being associated with carcass weight in Hanwoo. The most significant SNP detected was *BTB-01280026* ($P = 4.02 \times 10^{-11}$), located in the 25 Mb region on Bos taurus autosome 14 (BTA14). The most significant SNPs accounted for 6.73% to 10.55% of additive genetic variance, which is quite a large proportion of the total additive genetic variance. The most significant SNP (*BTB-01280026*; $P = 4.02 \times 10^{-11}$) had 16.96 kg of allele substitution effect, and the second most significant SNP (*Hapmap27934-BTC-065223*; $P = 4.04 \times 10^{-11}$) had 18.06 kg of effect on carcass weight, which correspond to 44% and 47%, respectively, of the phenotypic standard deviation for carcass weight in Hanwoo cattle. Our results demonstrated that carcass weight was affected by a major Quantitative Trait Locus (QTL) with a large effect and by many SNPs with small effects that are normally distributed. As evident from manhattan plots of GWAS for four carcass traits shown in Figure 6 a highly significant association was observed for carcass weight on BTA14 while for other three traits there was no such association observed which shows the effect of many genes on a trait across all chromosomes.

In order to determine if polygenic inheritance is a general phenomenon for carcass traits in Hanwoo population, we directly estimated the proportion of phenotypic variance explained by the common SNPs on each chromosomes. Proportion of variance attibuted to each chromosome averaged across four carcass traits against chromosome length (Figure 7) was calculated using genotype relationship matrix (GRM) on each chromosomes which supported the results shown in manhattan plots of GWAS. While for all other traits we can see most of the chromosomes having marginal effect but for CWT where Chromosome 14 is clearly showing a high effect on the trait. By calculating the proportion of the genome represented by each chromosome (not including the length of sex chromosomes), we tested for a correlation between the variance explained by each chromosome relative to its size. It was observed that several of the smaller chromosomes contributed less to the overall variance than several of the larger chromosomes, however trend was not significant.

Also there are reports on GWAS studies on other traits of economical importance such as MS, IMF, EMA and sensory panel. Yi et al. [16] carried out a GWAS study to identify QTL for growth and carcass quality traits using high-density SNP panels. The data set comprised of 61 sires, their 486 steers, and the 54,001 SNP markers on 29 bovine autosomal chromosomes. Traits to be analyzed in this study were six growth and carcass quality traits including weaning weight (WWT), 365-d yearling weight (YWT), CWT after slaughter, BF, EMA, and MS. A total of 16(0), 18(4), 20(13), 11(23), 10(13) and 19(1) SNPs were detected at the 5% chromosome (genome)-wise level for the traits, WWT, YWT, CWT, BF, EMA and MS, respectively. Among the 148 SNPs, 91 SNPs had dominance effects, suggesting that dominance inheritance mode be considered in genetic improvement for growth and caracass quality in Hanwoo. Thirty five QTL regions on 17 Bos taurus chromosomes (i.e. BTA 3, 4, 5, 6, 7, 11, 12, 13, 14, 15, 16, 17, 18, 20, 23, 26, and 28) were detected. Strong evidence for the QTL influencing CWT were detected on BTA14. Also, the QTL for WWT, YWT, BF, and EMA were detected on BTA20.

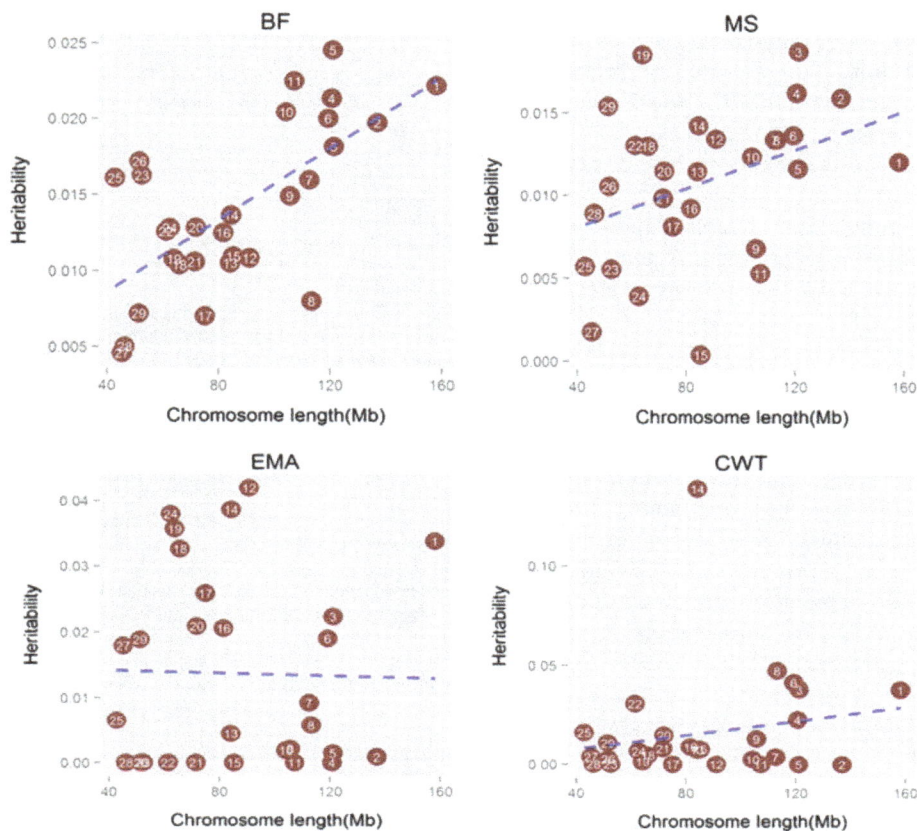

Figure 7 Proportion of variance attributed to each chromosome averaged across four carcass traits against chromosome length.

The GWAS studies could thus greatly help in developing the selection strategies for breeding programs.

Genomic strategy for breeding scheme

Genomic selection (GS) employs selection of an individual based on the genomic breeding value assessed through evaluating all the genetic markers located throughout the genome of that individual. Thus, the underlying principle of GS is to exploit the linkage disequilibrium of the QTL with one or more genetic marker(s) [17]. Molecular markers can be used to predict GBV of breeding animals by exploiting population-wide linkage disequilibrium between QTL and genetic markers spanned over the genome. The key–factor behind the success of genomic selection is to utilize the next generation sequencing approaches and the associated bioinformatics tools to identify the SNPs. Simulation studies in some domesticated species like, beef cattle, swine and chicken (Meuwissen et al., [18]) have suggested that the breeding values can be predicted with high accuracy using genetic markers alone but its validation is required especially in samples of the population different from that in which the effect of the markers was estimated. In genomic selection, the estimation of genomic breeding values is predicted to sum up all loci that are estimated based on phenotypes and genotypes in training population. This is particularly useful for traits that are very difficult to measure, such as marbling. The accuracy of these genomic predictions depends on the genetic architecture of the complex traits. For example, number of loci affecting the trait and distribution of their effects. In case of Hanwoo, MAS (Marker Associated Selection) and GS will provide a way to predict phenotype of marbling score as a preselection method that can be used in performance test [19].

Strategies for genomic selection program in Hanwoo

Application of genomic selection in Hanwoo preselection process such as performance test and then in progeny test in the breeding program might raise the genetic gain for marbling. A comparison of accuracy of GEBV and traditional EBV is tabulated in Table 1. In

Table 1 Accuracy of GEBV and traditional EBV estimated by 50 K SNP chip for Hanwoo cattle

Traits	BLUP	GBLUP	Difference
Eye muscle area (cm²)	0.11(0.08)	0.29(0.07)	0.18
Back fat thickness (mm)	0.11(0.08)	0.30(0.11)	0.19
Marbling score (1 ~ 9)	0.11(0.08)	0.27(0.12)	0.16

order to estimate the genomic breeding value, well organized reference population (Training data set) is to be build. The two most important features of reference population are 1) to representation of the entire Hanwoo population (all the variants) and the sample size. For the same, current progeny tested steers, selected proven bulls and candidate bulls from the progeny test program will be a reasonable population. Reference population must contain minimum of 5,000 animals. A genetic gain of 30% can be observed if we have a large reference population.

Conclusion

Recent advances in molecular biotechnology facilitate not only detection of genes that contribute to genetic variation of quantitative traits but also incorporation of genomic information into a conventional animal breeding program. The incorporated molecular information into GEBV may achieve an improvement of EBV and selection accuracy in cattle populations. In conclusion, we suggest incorporating molecular information into the conventional breeding programs to achieve better results in comparatively shorter time. Including molecular information will be a step further to achieve higher standards of Hanwoo beef.

Competing interests
The authors declare that they have no competing interests.

Authors' contributions
Conceived and designed the experiments: SHL BHP AS CGD SSL. Performed the experiments: TJC YHC HCK KJJ SDK. Analyzed the data: SHL BHP AS CGD TJC. Contributed reagents/materials/analysis tools: SHL SHY SBP HSK. Wrote the paper: SHL BHP AS CGD. Read and commented on the earlier drafts of this manuscript: SHL AS SHY SBP HSK. All authors read and approved the final manuscript.

Acknowledgements
This study was supported by awards from the AGENDA project (Grant no. PJ907008062013) and Molecular Breeding program (PJ0081882013) of Next Generation BIOGREEN21 project in the National Institute of Animal Science, RDA.

Author details
[1]Hanwoo Experiment Station, National Institute of Animal Science, RDA, Pyeong-Chang 232-950, Korea. [2]Animal Genetic and Breeding Division, National Institute of Animal Science, Cheon-An, Korea.

References
1. Korea Meat Trade Association (KMTA): *Information and Data of Agricultural Statistics Of Korea.* 2011. Accessed on January 12, 2011. http://kmta.or.kr/html/sub6-1.html?scode=6.
2. Jo C, Cho SH, Chang J, Nam KC: **Keys to production and processing of Hanwoo beef: a perspective of tradition and science.** *Anim Front* 2012, **2**(4):32–38.
3. Cho SH, Kim J, Park BY, Seong PN, Kang GH, Kim JH, Jung SG, Im SK, Kim DH: **Assessment of meat quality properties and development of a palatability prediction model for Korean Hanwoo steer beef.** *Meat Sci* 2010, **86**:236–242.
4. Han SW: *The Breed of Cattles*: Sun-Jin publishing; 1996:148–160.
5. Lee C, Pollak EJ: **Genetic antagonism between body weight and milk production in beef cattle.** *J Anim Sci* 2002, **80**:316–321.
6. Yoon DH, Park EW, Lee SH, Lee HK, Oh SJ, Cheong IC, Hong KC: **Assessment of genetic diversity and relationships between Korean cattle and other cattle breeds by microsatellite loci.** *J Anim Sci Technol (Kor)* 2005, **47**(3):341–354.
7. Mannen H, Kohno M, Nagata Y, Tsuji S, Bradley DG, Yeo JS, Nyamsamba D, Zagdsuren Y, Yokohama M, Nomura K, Amano T: **Independent mitochondrial origin and historical genetic differentiation in North Eastern Asian cattle.** *Mol Phylogenet Evol* 2004, **32**:539–544.
8. McTavish EJ, Decker JE, Schnabel RD, Taylor JF, Hillis DM: *New World Cattle Show Ancestry From Multiple Independent Domestication Events.* 2013. PNAS Early Edition.
9. Lee SH, Cho YM, Lim D, Kim HC, Choi BH, Park HS, Kim OH, Kim S, Kim TH, Yoon D, Hong SK: **Linkage disequilibrium and effective population size in Hanwoo Korean Cattle.** *Asian Australas J Anim Sci* 2011, **24**(12):1660–1665.
10. Lee SH, Julius VW, Lee SH, Lim DJ, Park EW, Gondro C, Yoon D, Oh SJ, Kim OH, Gibson J, Thompson J: **Genome wide QTL mapping to identify candidate genes for carcass traits in Hanwoo (Korean Cattle).** *Genes Genom* 2012, **34**(1):43–49.
11. Ministry of Agriculture and Fishery (MAF): *Code of Practice For Development Programs In Agriculture And Forestry.* 1999:40.
12. Park B, Choi T, Kim S, Oh S-H: **National genetic evaluation (system) of Hanwoo (Korean Native Cattle).** *Asian-Aust J Anim Sci* 2013, **26**(2):151–156.
13. Koch RM, Cundiff LV, Gregory EK: **Heritabilities and genetic, environmental and phenotypic correlations of carcass traits in a population of diverse biological types and their implications in selection programs.** *J Anim Sci* 1982, **55**:1319–1324.
14. Baik DH, Hoque1 MA, Choe HS: **Estimation of genetic and environmental parameters of carcass traits in hanwoo (korean native cattle) populations.** *Asian-Aust J Anim Sci* 2002, **15**(11):1523–1526.
15. Cundiff LV, Gregory EK, Koch RM, Dickerson GE: **Genetic relationships among growth and carcass traits of beef cattle.** *J Anim Sci* 1971, **33**:550–555.
16. Yi L: *Genome-Wide Association Study to Identify Qtl For Growth And Carcass Quality Traits In Korean Native Cattle, Hanwoo.* 2012.
17. Goddard ME, Hayes BJ: **Mapping genes for complex traits in domestic animals and their use in breeding programmes.** *Nat Rev Genet* 2009, **10**:381–391.
18. Meuwissen T, Hayes B, Goddard M: **Prediction of total genetic value using genome-wide dense marker maps.** *Genetics* 2001, **157**:1819–1829.
19. Lee SH, van der Werf JHJ, Hayes BJ, Goddard ME, Visscher PM: **Predicting unobserved phenotypes for complex traits from whole-genome SNP data.** *PLoS Genet.* 2008, **4**:e1000231.

17

Changes in expression of insulin signaling pathway genes by dietary fat source in growing-finishing pigs

Seung-Chang Kim[1], Hong-Chul Jang[1], Sung-Dae Lee[2], Hyun-Jung Jung[2], Jun-Cheol Park[2], Seung-Hwan Lee[1], Tae-Hun Kim[1] and Bong-Hwan Choi[1*]

Abstract

This study investigated changes in gene expression by dietary fat source, *i.e.*, beef tallow, soybean oil, olive oil, and coconut oil (each 3% in feed), in both male and female growing-finishing pigs. Real-time PCR was conducted on seven genes (insulin receptor; INSR, insulin receptor substrate; IRS, phosphatidylinositol (3,4,5)-triphosphate; PIP3, 3-phosphoinositide-dependent protein kinase-1; PDK1, protein kinase B; Akt, forkhead box protein O1; FOXO1 and cGMP-inhibited 3', 5'-cyclic phosphodiesterase; PDE3) located upstream of the insulin signaling pathway in the *longissimus dorsi* muscle (LM) of pigs. The INSR, IRS, PIP3, and PDE3 genes showed significantly differential expression in barrow pigs. Expression of the PIP3 and FOXO1 genes was significantly different among the four dietary groups in gilt pigs. In particular, the PIP3 gene showed the opposite expression pattern between barrow and gilt pigs. These results show that dietary fat source affected patterns of gene expression according to animal gender. Further, the results indicate that the type of dietary fat affects insulin signaling-related gene expression in the LM of pigs. These results can be applied to livestock production by promoting the use of discriminatory feed supplies.

Keywords: Dietary fat, Gene expression, Growing-finishing pig, Insulin signaling pathway

Background

Fat supplementation with high energy value is important for growing-finishing pigs. Addition of dietary fat has been shown to improve feed efficiency during the post-weaning period [1-5]. Dietary fat type affects fatty acid composition in the LM [6]. Intramuscular fat (IMF) deposition and back fat thickness (BF) are the most important candidate traits for understanding the interactions between nutrition and gene expression in pigs [7].

The insulin signaling pathway has a well established relationship with fat metabolism. Therefore, genes related to the insulin signaling pathway have long been the subject of major research. Insulin is the major hormone for fatty acid synthesis, glycolysis, and glycogenesis, and it suppresses β-oxidation, gluconeogenesis, glycogenolysis, and apoptosis by controlling critical energy functions such as

glucose and lipid metabolism [8-10]. Insulin activates insulin receptor (IR), which is a tyrosine kinase that phosphorylates and recruits different substrate adaptors such as the IRS family of proteins. Phosphorylated IRS then displays binding sites for numerous signaling partners. Among them, PI3K plays a major role in insulin function, mainly via activation of the Akt/PKB pathway. Activated Akt induces anti-lipolysis through activation of PDE3 as well as regulates gluconeogenesis and glycogenolysis through inhibition of forkhead box protein O1 (FOXO1). Activation of PDE3 decreases the concentration of cAMP, which in turn reduces protein kinase A (PKA) activity [11]. PKA is responsible for activation of lipase, which induces lipolysis as well as other physiological pathways [12]. Inhibition of FOXO1 decreases transcription of glucose 6-phosphates, which consequently reduces rates of gluconeogenesis and glycogenolysis [13]. Protein phosphorylation is controlled by the opposing and coordinated activities of protein kinases and phosphatases catalyzing protein phosphorylation and dephosphorylation, respectively [14]. This reversible phosphorylation of proteins is a

* Correspondence: bhchoi@korea.kr
[1]Animal Genomics & Bioinformatics Division, National Institute of Animal Science, Rural Development Administration, Chuksan-gil 77, Kwonsun-gu, Suwon, Korea
Full list of author information is available at the end of the article

major mechanism responsible for the regulation of cellular functions, including metabolism, signal transduction, cell division, and memory [15].

The Akt/PI3K signaling pathway is crucial to cell growth and survival. As such, current research has attempted to develop anti-cancer drugs based on the Akt/PI3K signaling pathway. Further, many studies have focused on mechanisms related to glucose uptake via Glut4 as well as protein synthesis via mTOR in the insulin signaling pathway. However, little is known about other pathway mechanisms. In this study, we investigated genes located upstream of the insulin signaling pathway related to glycolysis and anti-lipolysis in growing-finishing pigs.

Methods

Animals and diets

A total of 72 crossbred pigs (Landrace × Large White × Duroc) consisting of 36 gilt and barrow pigs each were used. The animals had an average body weight of 71 ± 1 kg, were about 130 days of age, and were divided according to gender. The pigs were randomly allocated into 24 pens (320 × 150 cm with solid concrete flooring) in a confined pig house, with three pigs per pen and six replicate pens per treatment. Treatment groups consisted of the same numbers of gilts and barrows. Each pen was equipped with a nipple water bottle and a stainless steel feeder, and pigs were given free access to feed and water throughout. Animals received care in accordance with the standard guideline for the Care and Use of Laboratory Animals provided by the National Institute of Animal Science Animal Care Committee, and the experiment was conducted with approval from the animal ethics committee and Operation rule of animal experiment ethics in the National Institute of Animal Science (approval number: 2009–076).

The ingredients and chemical compositions of the growing and finishing diets used in this experiment are shown in Table 1. All other nutrient requirements met or exceeded NRC recommendations for growing and finishing pigs (NRC, 1998). Dietary fat sources used in the present study were beef tallow, coconut oil, olive oil, and soybean oil, which were added to feed at a concentration of 3.0%. For this, fat sources were melted at approximately 50°C, after which they were diluted to approximately 10%. The 10% fat diets were then formulated to 3.0% fat diets. Growing diet was administered to crossbred pigs for an experimental period of 14 ± 3 days, whereas finishing diet was administered to crossbred pigs for an experimental period of 28 ± 3 days.

Slaughtering and sampling

Pigs with a live weight of 102 ± 3 kg were transported to a standard abattoir near the experimental station. The

Table 1 Composition of experiment diets, as-fed basis

Items	Growing	Finishing
Ingredients, %		
Corn grain	62.38	57.64
Soybean meal	22.00	14.00
Wheat	10.00	11.00
Wheat bran	0.00	12.00
Fat source[1]	3.00	3.00
L-lysine	0.06	0.06
Limestone	0.65	1.10
Tricalcium phosphate	1.11	0.30
Sodium chloride	0.30	0.30
Vitamin + mineral premix[2]	0.40	0.40
Antibiotics	0.10	0.00
Chemical composition[3]		
DE, kcal/kg	3,500	3,400
Crude protein, %	15.44	13.42
Crude fat, %	5.50	5.67
Crude fiber, %	3.45	3.94
Lysine,%	0.82	0.66
Methionine + Cystine, %	0.52	0.47
Calcium, %	0.69	0.60
Phosphorus, %	0.54	0.47

[1]Fat source : Beef tallow, soybean oil, olive oil, coconut oil.
[2]Vitamin and mineral contents per kilogram of diet provided by premix: Vitamin A, 2,000,000 IU; Vitamin D_3, 400,000 IU; Vitamin E, 2,500 IU; Vitamin K_3, 100 mg; Vitamin B_1, 100 mg; Vitamin B_2, 300 mg; Vitamin B_{12}, 1,200mcg; Niacin, 2,000 mg; d-Pantothenicalcium, 1,000 mg; Folic acid, 200 mg; Biotin, 20 mg; Choline chloride, 25,000 mg; Mn, 12,000 mg; Zn, 15,000 mg; Fe, 4,000 mg; Cu, 500 mg; I, 250 mg; Co, 100 mg; Mg, 2,000 mg; B.H.T., 5,00 mg.
[3]Chemical composition was calculated from ingredient proportion.

pigs were then slaughtered at 12 h after feed restriction. Briefly, pigs were stunned electrically (300 V for 3 s) with a pair of stunning tongs, shackled by the right leg, and exsanguinated while hanging. The carcasses were placed in a dehairer at 62°C for 5 min, and remaining hair was removed using a knife and flame. The carcasses were eviscerated and split before being placed in a chiller set at 4°C for 12 h. Immediately, 24 LM samples were taken from animals in the four dietary groups, frozen in liquid nitrogen, and stored at -80°C until preparation of total RNA.

RNA isolation and cDNA synthesis

The tissue was powdered with liquid nitrogen, and total RNA was extracted from 10 mg of muscle tissue using 1 mL of TRIzol® reagent (Invitrogen, Inc., USA). RNA quality was confirmed by examining 28S and 18S rRNA bands on 1.5% agarose gels stained with ethidium bromide. Total RNA was purified from all samples using an RNeasy MinElute cleanup kit (Qiagen, USA).

Complementary DNA (cDNA) synthesis was performed by reverse transcription using SuperScript™ II reverse transcriptase (Invitrogen, USA) as follows. Aliquots (4 μL) of total RNA were preincubated with 50 ng (1 μL) of random primer mix (Promega, USA) and 2.5 mM (1 μL) dNTP mix at 65°C for 5 min. The tubes were placed on ice, after which 4 μL of 5× first-stand buffer (250 mM Tris–HCl, pH 8.3, 375 mM KCl, 15 mM MgCl$_2$), 2 μL of 0.1 M DTT, and 40 units (0.5 μL) of RNase inhibitor (Promega, USA) were added, followed by incubation at 42°C for 2 min. After addition of 200 units (1 μL) of SuperScript™ II reverse transcriptase (Invitrogen, USA), incubation was continued at 42°C for 50 min. Reverse transcriptase activity was terminated by incubation at 70°C for 15 min. The resulting cDNA was stored at –20°C until used in quantitative real-time PCR (qRT-PCR).

Quantitative real-time polymerase chain reaction (qRT-PCR) analysis

To validate the seven differentially expressed genes (DEGs) related to insulin signaling based on the KEGG database in the LM of pigs, we performed qRT-PCR using Power SYBR Green PCR Master Mix (Applied Biosystems, USA) and the ABI 7500 Real-Time PCR system (Applied Biosystems USA). All primer sets were designed using the Primer3 program (http://bioinfo.ut.ee/primer3-0.4.0/) to amplify products ranging from 100 to 200 base pairs (Table 2). The β-actin gene (GenBank Acc. No. AY550069) was used as an internal control. qRT-PCR was performed in a total volume of 20 μL containing 2 μL of cDNA (0.1 μg/μL), 10 μL of 2× SYBR® Green PCR Master Mix (Applied Biosystems, USA), and 1 μL each of 10 pM forward and reverse primers. The amplification reaction was initiated by incubation for 2 min at 50°C, followed by 40 cycles of 95°C for 10 min, 95°C for 10 s, and 60°C for 1 min. After 36 cycles, a final extension step was performed at 72°C for 1 min. qRT-PCR for each gene was repeated three times. Following amplification, melting curve analysis was performed to verify the specificity of the reactions. The endpoint used in real-time RT-PCR quantification (Ct) was defined as the PCR threshold cycle number. The ΔCt value was determined by subtracting the β-actin Ct value for each sample from the target Ct value. Finally, we transformed the expression level to the $2^{-\Delta Ct}$ value for further analysis.

Statistical analysis

To identify DEGs among the dietary fat groups, statistical analysis was performed by analysis of variance (ANOVA) using the MIXED procedure with the R statistical package (http://www.R-project.org) for animals nested within age as the random effect. We also examined the least square means (LSM) to test the significance of differences among the groups using Duncan's multiple range test. The following statistical model was used to estimate the effects of dietary fat type on individual gene expression:

$$Y_{ij} = \mu + FED_i + DAY_j + e_{ij}$$

Where Y_{ij} is the target gene intensity ($2^{-\Delta Ct}$), μ is the overall mean, FED_i is the fixed effect of the ith dietary type, and DAY_j is animals nested within age as a random effect.

Table 2 List of insulin signaling pathway primers for qPCR

Gene Symbol	Primer sequences Forward / Reverse	Product Size (bp)
INSR	F:5'-TTCACTGGCAATCGCATTGAGCTG-3'	137 bp
	R:5'-TCATGGGTCACAGGGCCAATGATA-3'	
IRS	F:5'-AGGAAGTTTGGCAGGTGATCCTGA-3'	200 bp
	R:5'-ACGGCCCACTTCGATGAAGAAGAA-3'	
PIP3	F:5'-CTTTGCAGAGCTTGACCCAGAT-3'	100 bp
	R:5'-GAGCTTGTGGGCTTGCCTTCATTT-3'	
PDK1	F:5'-GGAAACCCTTGGCACCAGTTTGTA-3'	183 bp
	R:5'-TCGGAGTTCTTGTGACCACGGAAT-3'	
Akt	F:5'-AGAAGCTCTTCGAGCTCATCCTCA-3'	148 bp
	R:5'-TGCATGATCTCCTTGGCATCCTCA-3'	
FOXO1	F:5'-TCCCACACAGTGTCAAGACAACGA-3'	118 bp
	R:5'-ACTGCTTCTCTCAGTTCCTGCTGT-3'	
PDE3	F:5'-CCTGCAGAACCACAAGATGTGGAA-3'	190 bp
	R:5'-TCACTGGTTTGGCTTTGGTGTTGG-3'	

INSR, insulin receptor; IRS, insulin receptor substrate; PIP3, phosphatidylinositol 3-kinase; PDK1, 3-phosphoinositide-dependent protein kinase-1; Akt, protein kinase B; FOXO1, forkhead box protein O1; PDE3, cGMP-inhibited 3',5'-cyclic phosphodiesterase.

Results and discussion

Animal fat sources such as beef tallow have low digestibility, which can be improved by mixing animal fat with various vegetable oils to increase meat quality via elevation of lipase activity [16-19]. Therefore, gene expression in the LM can be manipulated according to dietary nutrients.

DNA microarray analysis has previously revealed that dietary fat type influences LM gene expression profiles. In particular, expression of insulin signaling pathway-related genes has been shown to be significantly enriched in differential gene expression sets [6]. These changes also suggest significant changes in other insulin signaling pathway genes. Thus, we compared differential gene expression in the LM of three barrows and gilts for each dietary fat type. Various genes linked to insulin signaling pathway genes as well as differentially expressed genes identified through microarray analysis were confirmed by RT-PCR.

The insulin signaling pathway involves a number of genes that control glucose storage and uptake, protein synthesis, and regulation of lipid synthesis in pigs. Insulin inhibits lipid metabolism by activating a cAMP-specific phosphodiesterase in adipocytes, thereby reducing cellular cAMP concentrations [20]. As the insulin signaling pathway is closely related to fat metabolism [21], the seven genes were subjected to qRT-PCR to determine whether or not dietary fat type influences their expression in pigs. The seven genes were insulin receptor (INSR), insulin receptor substrate (IRS), phosphatidylinositol (3,4,5)-triphosphate (PIP3), 3-phosphoinositide-dependent protein kinase-1 (PDK1), protein kinase B (Akt), forkhead box protein O1 (FOXO1), and cGMP-inhibited 3′, 5′-cyclic phosphodiesterase (PDE3).

Expression levels of the seven DEGs were measured using the ΔCt method, and the results are shown in Tables 3 and 4. RT-PCR analysis determined the expression profiles of the insulin signaling pathway genes in growing-finishing pigs. In particular, the INSR, IRS, PIP3, and PDE3 genes showed significantly differential expression according to dietary oil composition in barrow pigs (Figure 1). INSR plays a key role in the regulation of glucose homeostasis, a functional process that may result in a range of clinical manifestations, including diabetes and cancer, under degenerate conditions [22]. IRS plays a key role in transmitting signals from insulin and insulin-like growth factor-1 (IGF-1) receptors to the intracellular Akt/PI3K and Erk/MAPK pathways [23]. PIP3 functions to activate downstream signaling components such as protein kinase Akt, which activates downstream anabolic signaling pathways required for cell growth and survival [24]. Especially, PDE3 undergoes phosphorylation and short-term activation in response to insulin as well as agents that increase cAMP in adipocytes, hepatocytes, and platelets [11].

Table 3 ANOVA table for each gene associated with feeding groups in barrows

Gene	Source	Df	Sum Sq	Mean Sq	F value	Pr(>F)
INSR	Feed	3	2.084	0.695	4.014	<0.05*
	Residuals	20	3.461	0.173		
IRS	Feed	3	1.450	0.483	3.868	<0.05*
	Residuals	20	2.499	0.125		
PIP3	Feed	3	0.422	0.141	4.370	<0.05*
	Residuals	20	0.643	0.032		
PDK1	Feed	3	0.100	0.033	0.396	0.758
	Residuals	20	1.683	0.084		
Akt	Feed	3	0.209	0.070	1.021	0.405
	Residuals	20	1.365	0.068		
FOXO1	Feed	3	1.527	0.509	1.880	0.166
	Residuals	20	5.416	0.271		
PDE3	Feed	3	0.927	0.309	5.213	<0.01**
	Residuals	20	1.186	0.059		

INSR, insulin receptor; IRS, insulin receptor substrate; PIP3, phosphatidylinositol (3,4,5)-triphosphate; PDK1, 3-phosphoinositide-dependent protein kinase-1; Akt, protein kinase B; FOXO1, forkhead box protein O1; PDE3, cGMP-inhibited 3′,5′-cyclic phosphodiesterase. *, ** Significant differences (P < 0.05 and 0.01) among the feeding groups determined using the mixed ANOVA module. DF = Degrees of freedom, Sum Sq = Sum of square, Mean Sq = Mean of square.

The INSR and IRS genes showed opposite gene expression patterns in barrows. Beef tallow was the only dietary fat source associated with low INSR expression as well as increased IRS expression. The PIP3 and PDE3 genes showed similar reduced expression levels in pigs

Table 4 ANOVA table for each gene associated with feeding groups in gilts

Gene	Source	Df	Sum Sq	Mean Sq	F value	Pr(>F)
INSR	Feed	3	0.457	0.153	0.923	0.448
	Residuals	20	3.304	0.165		
IRS	Feed	3	0.197	0.066	0.249	0.861
	Residuals	20	5.268	0.263		
PIP3	Feed	3	0.970	0.323	4.644	<0.05*
	Residuals	20	1.393	0.070		
PDK1	Feed	3	0.386	0.129	1.258	0.316
	Residuals	20	2.043	0.102		
Akt	Feed	3	0.308	0.103	1.380	0.278
	Residuals	20	1.486	0.074		
FOXO1	Feed	3	2.712	0.904	7.830	<0.01**
	Residuals	20	2.309	0.115		
PDE3	Feed	3	0.300	0.100	0.530	0.667
	Residuals	20	3.770	0.189		

INSR, insulin receptor; IRS, insulin receptor substrate; PIP3, phosphatidylinositol (3,4,5)-triphosphate; PDK1, 3-phosphoinositide-dependent protein kinase-1; Akt, protein kinase B; FOXO1, forkhead box protein O1; PDE3, cGMP-inhibited 3′,5′-cyclic phosphodiesterase. *, ** Significant differences (P < 0.05 and 0.01) among the feeding groups determined using the mixed ANOVA module. DF = Degrees of freedom, Sum Sq = Sum of square, Mean Sq = Mean of square.

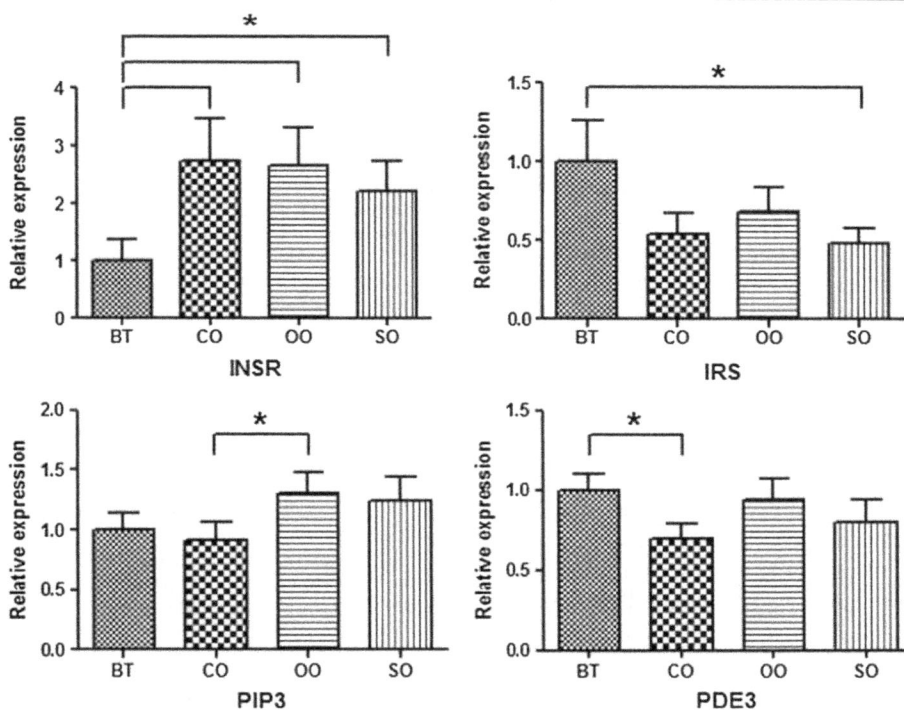

Figure 1 Expression patterns of differentially expressed genes in four dietary oil groups in barrows, as determined by qRT-PCR. Dietary fat sources were BT (beef tallow), CO (coconut oil), OO (olive oil), and SO (soybean oil), which were added at concentrations of 3.0% in feed. Experiments were performed using the LM from three barrows, and data are expressed as mean ± SD. Asterisks show statistically significant values (*$P < 0.05$).

treated with coconut oil. In particular, INSR gene expression was up-regulated 2-fold in pigs treated with other oils compared to beef tallow.

The PIP3 and FOXO1 genes showed significantly different expression levels among the four dietary oil groups in gilts (Figure 2). Specifically, the PIP3 and FOXO1 genes showed relatively high expression in gilts with coconut oil compared to other oils. Gene expression of FOXO1, which negatively regulates adipogenesis, was two times higher in gilts [25]. FOXO1 belongs to

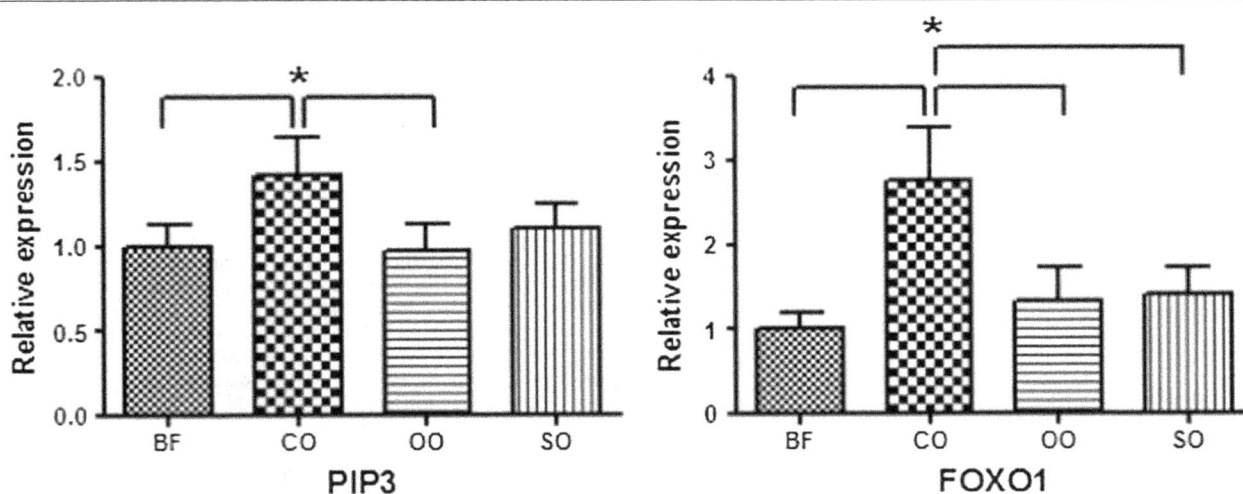

Figure 2 Expression patterns of differentially expressed genes in four dietary oil groups in gilts, as determined by qRT-PCR. Dietary fat sources were BT (beef tallow), CO (coconut oil), OO (olive oil), and SO (soybean oil), which were added to concentrations of 3.0% in feed. Experiments were performed using the LM from three gilts, and data are expressed as mean ± SD. Asterisks show statistically significant values (*$P < 0.05$).

the forkhead family of transcription factors that increases transcription of glucose-6-phosphatase, resulting in elevated rates of gluconeogenesis and glycogenolysis [26]. Compared with other oils, activation of FOXO1 in pigs treated with coconut oil had a negative effect on intramuscular fat accumulation.

Barrows and gilts showed opposite patterns of PIP3 gene expression. These results show that dietary fat type affected patterns of gene expression according to animal gender. It is known that expression levels of genes are linked to gender differences and variations in fatty acids [27,28]. These results can be applied to livestock production by promoting the use of discriminatory feed supplies.

Conclusion

DNA microarray analysis has previously revealed that dietary fat type alters LM gene expression profiles. These changes also suggest significant changes in other insulin signaling pathway genes. Thus, we compared differential gene expression in the LM of three barrows and gilts for each dietary fat type. Various genes linked to insulin signaling pathway genes as well as differentially expressed genes identified through microarray analysis were confirmed by RT-PCR. Results show that seven genes (INSR, IRS, PIP3, PDK1, Akt, PDE3, and FOXO1) were located upstream of the insulin signaling pathway. In barrows, the INSR, IRS, PIP3, and PDE3 genes showed significantly differential expression according to dietary oil type. The PIP3 and FOXO1 genes showed significant differences in gene expression among the four dietary oil groups in gilts. In particular, barrow and gilt pigs showed opposite patterns of PIP3 expression. Therefore, dietary fat type affected patterns of gene expression by gender difference. These results can be applied to livestock production by promoting the use of discriminatory feed supplies.

Competing interests
The authors declare that they have no competing interests.

Authors' contributions
SCK, THK and BHC participated in the design of the study and total organization. SCK and HCJ made the laboratory work. SDL and HJJ made feeding and animal management work. JCP made sampling work. SHL performed the statistical analysis. All authors helped to draft the manuscript, and all authors read and approved the final manuscript.

Acknowledgements
This study was supported by the Agenda (PJ0087112013) and 2013 PostDoctoral Fellowship Program of National Institute of Animal Science, Rural Development Administration, Republic of Korea.

Author details
[1]Animal Genomics & Bioinformatics Division, National Institute of Animal Science, Rural Development Administration, Chuksan-gil 77, Kwonsun-gu, Suwon, Korea. [2]Swine Science Division, National Institute of Animal Science, Rural Development Administration, Cheon-an, Chungnam 330-801, Korea.

References
1. Cera KR, Mahan DC, Reinhart GA: Effects of dietary dried whey and corn oil on weanling pig performance, fat digestibility and nitrogen utilization. *J Anim Sci* 1988, **66**:1438–1445.
2. Howard KA, Forsyth DM, Cline TR: The effect of an adaptation period to soybean oil additions in the diets of young pigs. *J Anim Sci* 1990, **68**:678–683.
3. Li DF, Thaler RC, Nelssen JL, Harmon DL, Allee GL, Weeden TL: Effect of fat sources and combinations on starter pig performance, nutrient digestibility and intestinal morphology. *J Anim Sci* 1990, **68**:3694–3704.
4. Overland M, Tokach MD, Cornelius SG, Pettigrew JE, Rust JW: Lecithin in swine diets: I. Weanling pigs. *J Anim Sci* 1993, **71**:1187–1193.
5. Tokach MD, Pettigrew JE, Johnston LJ, Overland M, Rust JW, Cornelius SG: Effect of adding fat and(or) milk products to the weanling pig diet on performance in the nursery and subsequent grow-finish stages. *J Anim Sci* 1995, **73**:3358–3368.
6. Park JC, Kim SC, Lee SD, Jang HC, Kim NK, Lee SH, Jung HJ, Kim IC, Seong HH, Choi BH: Effects of dietary fat types on growth performance, pork quality, and gene expression in growing-finishing pigs. *Asian-Aust J Anim Sci* 2012, **25**:1759–1767.
7. Yin J, Li D: Nutrigenomics Approach - A strategy for identification of nutrition responsive genes influencing meat edible quality traits in swine. *Asian-Aust J Anim Sci* 2009, **22**:605–610.
8. Saltiel AR, Kahn CR: Insulin signalling and the regulation of glucose and lipid metabolism. *Nature* 2001, **414**:799–806.
9. Fritsche L, Weigert C, Haring H-U, Lehmann R: How insulin receptor substrate proteins regulate the metabolic capacity of the liver-implications for health and disease. *Curr Med Chem* 2008, **15**:1316–1329.
10. Izuchi R, Nakai Y, Takahashi H, Ushiama S, Okada S, Misaka T, Abe K: Hepatic gene expression of the insulin signaling pathway is altered by administration of persimmon peel extract: a DNA microarray study using type 2 diabetic goto-kakizaki rats. *J Agric Food Chem* 2011, **59**:3320–3329.
11. Degerman E, Belfrage P, Manganiello VC: Structure, Localization, and Regulation of cGMP-inhibited Phosphodiesterase (PDE3). *J Biol Chem* 1997, **272**:6823–6826.
12. Maurice DH, Palmer D, Tilley DG, Dunkerley HA, Netherton SJ, Raymond DR, Elbatarny HS, Jimmo SL: Cyclic nucleotide phosphodiesterase activity, expression, and targeting in cells of the cardiovascular system. *Mol Pharmacol* 2003, **64**:533–546.
13. Daitoku H, Fukamizu A: FOXO transcription factors in the regulatory networks of longevity. *J Biochem (Tokyo)* 2007, **141**:769–774.
14. Huang T, Xiong Y-Z, Lei M-G, Xu D-Q, Deng C-Y: Identification of a Differentially Expressed Gene PPP1CB between Porcine Longissimus dorsi of Meishan and Large Whitex Meishan Hybrids. *Acta Biochim Biophys Sin* 2006, **38**:450–456.
15. Cohen PTW: Novel protein serine/threonine phosphatases: Variety is the spice of life. *Trends Biochem Sci* 1997, **22**:245–251.
16. Powles J, Wiseman J, Cole DJA, Hardy B: Effect of chemical structure of fats upon their apparent digestible energy value when given to growing/finishing pigs. *Ani Sci* 1993, **57**:137–146.
17. Powles J, Wiseman J, Cole DJA, Hardy B: Effect of chemical structure of fats upon their apparent digestible energy value when given to young pigs. *Ani Sci* 1994, **58**:411–417.
18. Mountzouris KC, Fegeros K, Papadopoulos G: Utilization of fats based on the composition of sow milk fat in the diet of weanling pigs. *Anim Feed Sci Technol* 1999, **77**:115–124.
19. Cho JH, Kim HJ, Chen YJ, Yoo JS, Min BJ, Kim JD, Kim IH: the effect of soybean oil, tallow and coconut oil supplementation on growth performance, serum lipid changes and nutrient digestibility in weaned pigs. *J Anim Sci Technol* 2007, **49**:33–40.
20. Kitamura T, Kitamura Y, Kuroda S, Hino Y, Ando M, Kotani K, Konishi H, Matsuzaki H, Kikkawa U, Ogawa W, Kasuga M: Insulin-induced phosphorylation and activation of cyclic nucleotide phosphodiesterase 3b by the serine-threonine kinase akt. *Mol Cell Biol* 1999, **19**:6286–6296.
21. Wong RHF, Sul HS: Insulin signaling in fatty acid and fat synthesis: a transcriptional perspective. *Curr Opin Pharmacol* 2010, **10**:684–691.
22. Ward CW, Lawrence MC: Ligand-induced activation of the insulin receptor: a multi-step process involving structural changes in both the ligand and the receptor. *BioEssays* 2009, **31**:422–434.

23. Sun X, Rothenberg P, Kahn C, Backer J, Araki E, Wilden P, Cahill D, Goldstein B, White M: **Structure of the insulin receptor substrate IRS-1 defines a unique signal transduction protein.** *Nature* 1991, **352**:73–77.

24. Whitman M, Downes CP, Keeler M, Keller T, Cantley L: **Type I phosphatidylinositol kinase makes a novel inositol phospholipid, phosphatidylinositol-3-phosphate.** *Nature* 1988, **332**:644–646.

25. Farmer SR: **The Forkhead Transcription Factor Foxo1: A Possible Link between Obesity and Insulin Resistance.** *Mol Cell* 2003, **11**:6–8.

26. Nakae J, Kitamura T, Silver DL, Accili D: **The forkhead transcription factor Foxo1 (Fkhr) confers insulin sensitivity onto glucose-6-phosphatase expression.** *J Clin Invest* 2001, **108**:1359–1367.

27. Zhang S, Knight TJ, Stalder KJ, Goodwin RN, Lonergan SM, Beitz DC: **Effects of breed, sex, and halothane genotype on fatty acid composition of pork longissimus muscle.** *J Anim Sci* 2007, **85**:583–591.

28. Miersch C, Doring F: **Sex differences in carbohydrate metabolism are linked to gene expression in Caenorhabditis elegans.** *PLoS One* 2012, **7**:e44748.

18

Changes in physicochemical and microbiological properties of isoflavone-treated dry-cured sausage from sulfur-fed pork during storage

Ji-Han Kim[1], Chang-Won Pyun[1], Go-Eun Hong[1], Soo-Ki Kim[2], Cheul-Young Yang[3] and Chi-Ho Lee[1*]

Abstract

This study was performed to investigate the physicochemical and microbiological properties of isoflavone-treated dry cured sausage from sulfur fed pork (0.3%) during storage at 15°C for 45 days. Groups were divided into three treatments: dry-cured sausages produced with pork fed general diet as the control group (CON), sulfur-fed pork (SUL) and isoflavone-(0.25%) treated sulfur-fed pork (ISO). Moisture content in all groups decreased dramatically from 55–57% to 10–11% during storage, whereas crude protein, crude fat, and ash content increased ($P < 0.05$). The ISO group showed excellent antioxidant effect compared to CON during storage. Redness and lightness of ISO was higher than that of CON during storage. VBN in the ISO group was significantly lower than that in the CON and SUL treatments during 30 and 45 days of storage ($P < 0.05$). A total plate count of ISO was significantly lower than that of CON at 45 days ($P < 0.05$). In this study, adding isoflavone to meat products indicated prevention of lipid oxidation and improved color stability in meat products.

Keywords: Isoflavone, Dry-cured sausage, Sulfur, Lipid oxidation, Storage

Background

Functional foods have beneficial effects besides merely supplying nutrition. Addition of functional materials can extend shelf-life and prevent rancidity of products as well as modulate various body functions [1]. Traditional fermented foods have been shown to possess many of these beneficial effects, and their extracts are often applied to other foods after extraction and purification [2].

Dry-cured pork sausage is made by thorough mixing with lean pork back fat and other non-meat ingredients, including salt, nitrite, spices, and a starter. Each additive is useful for preserving flavor, color, and water-holding capacity in meat products. On average, dry-cured pork sausage is 70–80% lean and 20–30% lard, which includes 43% saturated fatty acids (SFAs), 47% monounsaturated fatty acids, and 10% polyunsaturated fatty acids. Half of the fatty acids in pork fat are SFAs, which significantly contribute to cardiovascular disease [3].

Lipid oxidation, which has a negative influence on meat product quality [4], can be inhibited by natural or artificial antioxidants [5]. Isoflavones, which are abundantly available in soybeans, are natural antioxidative materials. Isoflavones reduce blood cholesterol and LDL-cholesterol oxidation through antioxidant and free radical scavenging activities [6]. Further, lipid oxidation must be considered during the drying period of dry-cured sausage due to its high lipid content. Therefore, many researchers have investigated ways to prevent lipid oxidation in order to increase the quality and safety of these meat products.

The aim of this study was to investigate the effects of isoflavones on the physicochemical and microbiological properties of dry-cured sausage during the drying period.

Methods

Formula and chemical composition of diet

A total of 90 three-way crossbred pigs (Landrace, Duroc, and Yorkshire) from Yang-ju Federation of Livestock Cooperatives in the Republic of Korea were used. Experimental protocol was approved by the animal care committee of Konkuk University of Seoul, Republic of Korea. Detailed feeding and rearing procedures as well as the

* Correspondence: leech@konkuk.ac.kr
[1]Konkuk University of Food Science & Technology, Seoul 143-701, Republic of Korea
Full list of author information is available at the end of the article

Table 1 Common components of the meat mixture

	Ingredients	Percentage (%)		
		CON	SUL	ISO
Materials	Lean pork	75	75	75
	Pork back fat	25	25	25
	Salt	2.8	2.8	2.8
	Black pepper	0.25	0.25	0.25
Pickle	Rosemary	0.05	0.05	0.05
	Sodium nitrite	0.02	0.02	0.02
	Glucose	1	1	1
	Isoflavone			0.25
Starter culture	*Staphylococcus carnosus* M17	0.25	0.25	0.25
	Pediococcus pentosaceus ATCC 33314			

CON and SUL groups were divided from the level of processed sulfur. Sulfur levels in diets were controlled by crude fiber content. (SUL: 0.3% sulfur).

composition of the feed were reported previously [7]. The pigs, which weighed 110 kg each at shipment, were divided into three groups based on the level of dietary processed sulfur (0%, 0.3%) fed for 3 months before shipment. Processed sulfur was used as received from Ebatha Co., Ltd. The control (CON) was not supplied processed sulfur, whereas the SUL group was supplied with processed sulfur at 3 g/kg feed for 3 months before shipment.

Weight, average daily feed intake, weight gain, feed efficiency, and carcass grade

Dry-cured sausage preparation

Pork meat and fat were trimmed off with a sterile knife and refrigerated overnight at about 4°C. The composition of the meat mixture is shown in Table 1. The basic formula was 75% lean pork and 25% back fat with curing ingredients. The trimmed meat was ground using a 2.7 mm plate meat mincer and mixed with other curing ingredients. Starter culture at 0.25% (*Staphylococcus carnosus* M17: *Pediococcus pentosaceus* ATCC 33314 = 1:1) was added into the mixed sample. This original mixture was then split into batches, after which 0.25% isoflavone powder obtained from SOLGAR® (Seoul, Republic of Korea) was added to the ISO group. Finally, the minced meat was filled into collagen casings (150 mm long, 30 mm diameter), and all samples were ripened for 45 days at $15 \pm 2°C$ and a relative humidity of $80 \pm 3\%$ in a chamber. Sampling was performed by randomly choosing each sausage group after 0, 15, 30, and 45 days for physicochemical and microbiological analyses [8].

Physicochemical analysis of dry-cured sausage

Proximate compositions (moisture, crude fat, crude protein, and ash) of the sausage samples were determined by the AOAC method [9]. Water activities (a_w) of the samples were determined using a water activity measuring device

Table 2 Effect of isoflavone on the change in moisture, crude protein, crude fat, and ash contents in dry-cured pork sausage during storage (%)

Proximate composition (%)	Storage period (days)	Treatments				P-value	
		CON	SUL	ISO	SEM	Treatment	Storage
Crude protein	0	17.2$^{NS\ B}$	16.9^B	17.4^B	0.45	0.76	
	15	31.0$^{NS\ A}$	31.6^A	31.9^A	2.63	0.95	***
	30	38.3$^{NS\ C}$	38.5^A	37.6^C	2.37	0.77	
	45	36.8$^{a\ C}$	34.2$^{ab\ A}$	32.6$^{b\ A}$	1.48	*	
Crude fat	0	19.8$^{a\ C}$	16.3^{bC}	17.9$^{ab\ C}$	0.93	0.1	
	15	38.4$^{NS\ AB}$	37.2^B	38.6^B	0.46	0.08	***
	30	39.9$^{NS\ A}$	38.4^B	38.3^B	1.43	0.18	
	45	45.3^{NSA}	43.3^A	44.8^A	1.94	0.44	
Moisture	0	55.7$^{b\ A}$	57.3$^{a\ A}$	57.97$^{a\ A}$	0.47	*	
	15	18.2$^{NS\ B}$	18.1^B	18.03^B	0.76	0.95	***
	30	11.2$^{NS\ C}$	12.1^C	12.40^C	2.94	0.73	
	45	11.5$^{NS\ C}$	11.1^C	10.44^C	0.51	0.24	
Ash	0	3.15$^{b\ B}$	3.64$^{a\ C}$	3.27$^{b\ C}$	0.03	***	
	15	6.32$^{NS\ A}$	6.23^B	6.46^B	0.29	0.86	***
	30	6.15$^{b\ A}$	7.22$^{a\ A}$	5.92$^{c\ B}$	0.33	0.07	
	45	6.65$^{NS\ A}$	7.52^A	7.35^A	0.66	0.44	

1) CON, Dry-cured sausage processed with 100% commercial diet fed pork; SUL, Dry-cured sausage processed with 0.3% sulfur fed pork; ISO, 0.25% isoflavone + dry-cured sausage processed with 0.3% sulfur fed pork. 2)^{NS}; Not significant; a-c means within a row with different letters are significantly different, P < 0.05; A-D means within a column with different letters are significantly different, P < 0.05. Significance level of treatment is given by ***(P <0.001); **(P<0.01); *(P<0.05)
*All values are mean ± standard deviation (n = 3).

Table 3 Effect of isoflavone on the pH and water activity (a_w) change in dry-cured pork sausage during the ripening period

	Storage period (days)	Treatments				P-value	
		CON	SUL	ISO	SEM	Treatment	Storage
pH	0	6.00[b A]	6.03[ab A]	6.06[a A]	0.012	*	
	15	5.91[NS B]	5.92[B]	5.99[B]	0.008	0.064	***
	30	6.17[NS C]	6.16[C]	6.10[C]	0.038	0.369	
	45	6.23[b D]	6.26[a D]	6.23[b D]	0.027	0.23	
a_w	0	0.97[NS A]	0.97[A]	0.99[A]	0.001	0.123	
	15	0.69[b B]	0.72[a B]	0.72[a B]	0.009	0.096	***
	30	0.69[b B]	0.71[a B]	0.71[a B]	0.005	*	
	45	0.65[c C]	0.68[a C]	0.66[b C]	0.002	***	

1) CON, Dry-cured sausage processed with 100% commercial diet fed pork; SUL, Dry-cured sausage processed with 0.3% sulfur fed pork; ISO, 0.25% isoflavone + dry-cured sausage processed with 0.3% sulfur fed pork. 2)[NS]; Not significant; a-c means within a row with different letters are significantly different, $P < 0.05$; A-D means within a column with different letters are significantly different, $P < 0.05$. Significance level of treatment is given by ***($P < 0.001$); **($P<0.01$); *($P<0.05$) *All values are mean ± standard deviation (n = 3).

(Aquaspector, AQS-31, NAGY, Gaeurfelden, Germany) The a_w values were determined in triplicate in order to optimize the weights of samples at 25°C until equilibrium was reached.

The pH was measured using a pH meter (pH 900, Precisa Co, Deitikon, Switzerland) in a slurry made by homogenizing 2 g of sample with 18 mL of distilled water for 90 sec using a Bag mixer 400 (Interscience Co, St Nom la Bretêche, France).

Thiobarbituric acid (TBA) values of dry-cured sausage stored for different times were determined by following a modified method of Witte et al. [10]. The absorbance of the supernatant was measured at 532 nm using a spectrophotometer (Optizen 2120UV, Mecasys, Seoul, Korea). Results are expressed as mg malonaldehyde (MDA)/kg sample. Volatile basic nitrogen (VBN) was determined by the micro-diffusion method of Conway [11] for dry-cured sausage stored for different times. Results were expressed as VBN value % mg (mg/100 g meat). The color values of dry-cured sausages were expressed as Hunter L-, a-, and b-values using a Handy colorimeter (NR-300, Nippon Denshoku, Tokyo, Japan). All experiments were performed in triplicate.

Microbiological analysis
Sample (2 g) and 0.85% NaCl (18 mL) in sterile deionized water were transferred to a sterile stomacher bag and homogenized for 90 sec using a bag mixer (Interscience Co, France). A 10^{-1} dilution was then used for subsequent serial dilutions. An aliquot (0.1 mL) of the appropriate dilution of sample was spread in triplicate onto agar plates. After serial dilution, the solution was inoculated onto Petrifirm Aerobic Count Plates (3 M, Korea) and cultured for 48 h at 35°C, after which the colony number was converted into a log value. The samples were treated as described above and inoculated onto Petrifirm *E. coli*

O157:H7 Count Plates (3 M, Korea) for 48 h at 35°C. Lactic acid bacteria were treated as described above, inoculated onto MRS agar (OXOID, England), and then cultured for 24 h at 35°C. *Staphylococcus aureus* bacteria were treated as described above, inoculated onto Baird-Parker agar (OXOID, England), and then cultured for 24 h at 35°C. *Salmonella* bacteria were treated as described above, inoculated onto MacConkey agar (Difco, USA), and then cultured for 24 h at 35°C.

Statistical analysis
Analysis of variance (ANOVA) was conducted to determine significant differences among the groups and storage times. All analyses were performed on all variables using the General Linear Model (GLM) procedure of SAS version 9.2 (SAS Institute Inc., Cary, NC, USA). All analyses were conducted in triplicate, and significant differences were detected using Duncan's multiple range test (p < 0.05).

Table 4 Effect of isoflavone on the change in thiobarbituric acid values in dry-cured pork sausage during the ripening period (Malondialdehyde mg/kg)

Storage (days)	Treatment			SEM	P-value	
	CON	SUL	ISO		Treatment	Storage
0	0.14[a C]	0.09[b C]	0.11[ab C]	0.01	*	
15	0.18[a B]	0.16[b B]	0.15[b B]	0.007	*	***
30	0.23[a A]	0.21[ab A]	0.20[b A]	0.031	0.19	
45	0.20[a A]	0.16[b B]	0.11[c C]	0.022	0.19	

1) CON, Dry-cured sausage processed with 100% commercial diet fed pork; SUL, Dry-cured sausage processed with 0.3% sulfur fed pork; ISO, 0.25% isoflavone + dry-cured sausage processed with 0.3% sulfur fed pork. 2)[NS]; Not significant; a-c means within a row with different letters are significantly different, $P < 0.05$; A-D means within a column with different letters are significantly different, $P < 0.05$. Significance level of treatment is given by ***($P < 0.001$); **($P<0.01$); *($P<0.05$)* All values are mean ± standard deviation (n = 3).

Table 5 Effect of isoflavone on the change in volatile basic nitrogen (VBN) in dry-cured pork sausage during the ripening period (mg %)

Storage (days)	Treatment				P-value	
	CON	SUL	ISO	SEM	Treatment	Storage
0	8.2[NS C]	8.9[B]	8.2[B]	0.304	0.22	
15	10.9[NS B]	11.0[A]	10.8[A]	0.459	0.93	
30	11.8[a A]	11.4[a A]	10.1[b A]	0.296	*	***
45	9.0[a C]	8.6[ab B]	7.8[b B]	0.216	*	

1) CON, Dry-cured sausage processed with 100% commercial diet fed pork; SUL, Dry-cured sausage processed with 0.3% sulfur fed pork; ISO, 0.25% isoflavone + dry-cured sausage processed with 0.3% sulfur fed pork. 2)[NS]; Not significant; a-c means within a row with different letters are significantly different, $P < 0.05$; A-D means within a column with different letters are significantly different, $P < 0.05$. Significance level of treatment is given by ***($P < 0.001$); **($P<0.01$); *($P<0.05$) *All values are mean ± standard deviation (n = 3).

Result and discussion

Physicochemical qualities

The initial moisture content of all dry-cured ham samples was 54.71–57.97%, which declined to 10.4–11.56% in the final product (Table 2). In the initial phase, moisture contents of the ISO and SUL groups were significantly higher than that of the CON ($p < 0.05$). This result could be due to the higher water-holding capacity of sulfur-fed pigs [12]. The reduction of moisture content was accompanied by elevation of the other components during the drying period. Crude fat contents of the SUL and ISO groups were lower than that of the CON during storage. Promotion of lipid metabolism and reduction of intramuscular fat have been observed in animals fed dietary sulfur [13]. Crude protein contents of the CON, SUL, and ISO groups increased significantly ($p < 0.05$) until day 30, followed by a decrease. The processing time of dry-cured ham depends on the product. Specifically, longer processing time results in more intense muscle proteases activities, leading to extensive protein breakdown [14].

pH value and water activity (a_w)

The pH values of all groups decreased during 15 days of storage but increased during the experimental period (Table 3). The pH value of the ISO group was significantly higher than that of the CON group at the beginning of the drying process ($p < 0.05$), possibly due to inhibition of lactic acid production caused by addition of isoflavones [15]. Reduction of the pH values of all samples during drying can be mainly attributed to lactic acid bacterial growth in the initial phase [16]. Many researchers have indicated that production of lactic acid in tissue leads to pH reduction in meat, resulting in less water [17]. After 15 days, elevation of the pH value could also be due to the accumulation of non-protein nitrogen or amino acid catabolic products [18].

The a_w of all three groups decreased significantly during 45 days of storage ($p < 0.05$, Table 3). The a_w of the SUL

Table 6 Effect of isoflavone on the change in lightness coordinate (L*), redness coordinate (a*), and yellowness coordinate (b*) in dry-cured pork sausage during the ripening period

L* (Lightness) Storage (days)	Treatment				P-value	
	CON	SUL	ISO	SEM	Treatment	Storage
0	62.1[b A]	64.7[a A]	64.2[a A]	0.447	*	
15	48.6[NS D]	47.2[C]	49.4[D]	0.633	0.130	
30	49.1[NSC]	50.4[B]	51.4[C]	0.692	0.134	***
45	51.2[b B]	52.8[ab B]	54.4[a B]	0.806	0.073	
a* (redness) Storage (days)	CON	SUL	ISO			
0	6.1[NS C]	6.8[D]	6.5[C]	0.361	0.435	
15	12.6[NS A]	11.5[A]	12.0[A]	0.567	0.453	
30	10.2[NS A]	10.6[B]	11.0[A]	0.457	0.518	***
45	8.1[NS B]	9.1[C]	9.1[B]	0.555	0.412	
b* (yellowness) Storage (days)	CON	SUL	ISO			
0	13.2[b C]	16.9[a B]	14.1[b D]	0.470	**	
15	17.6[a A]	15.4[b A]	18.5[a A]	0.481	**	
30	15.7[b B]	15.4[b A]	17.3[a B]	0.337	*	***
45	14.8[b B]	15.3[ab A]	16.8[a C]	0.436	*	

1) CON, Dry-cured sausage processed with 100% commercial diet fed pork; SUL, Dry-cured sausage processed with 0.3% sulfur fed pork; ISO, 0.25% isoflavone + dry-cured sausage processed with 0.3% sulfur fed pork. 2)[NS]; Not significant; a-c means within a row with different letters are significantly different, $P < 0.05$; A-D means within a column with different letters are significantly different, $P < 0.05$. Significance level of treatment is given by ***($P < 0.001$); **($P<0.01$); *($P<0.05$) *All values are mean ± standard deviation (n = 3).

Table 7 Effect of isoflavone on the change in microbial counts in dry-cured pork sausage during the ripening period (log CFU/g)

Composition	Storage (days)	Treatment			SEM	P-value	
		CON	SUL	ISO		Treatment	Storage
Total aerobes	0	5.8[NS B]	5.8[B]	5.8[B]	0.044	**	
	15	7.3[a A]	7.2[ab A]	7.1[b A]	0.027	*	
	30	4.6[NS C]	4.7[C]	4.3[C]	0.179	0.33	***
	45	4.7[a C]	4.7[a C]	4.3[b C]	0.179	0.97	
Lactic acid bacteria	0	5.6[NS B]	5.3[B]	5.4[B]	0.018	**	
	15	5.8[b A]	5.9[b A]	6.0[a A]	0.104	**	
	30	4.7[NS C]	5.1[C]	4.6[C]	0.184	0.13	***
	45	4.5[NS C]	4.5[C]	4.5[C]	0.097	0.75	
E. coli 0157:H7	0	-	-	-			
	15	-	-	-			
	30	-	-	-			
	45	-	-	-			
Staphylococcus aureus	0	-	-	-			
	15	-	-	-			
	30	-	-	-			
	45	-	-	-			
Salmonella spp.	0	-	-	-			
	15	-	-	-			
	30	-	-	-			
	45	-	-	-			

1) CON, Dry-cured sausage processed with 100% commercial diet fed pork; SUL, Dry-cured sausage processed with 0.3% sulfur fed pork; ISO, 0.25% isoflavone + dry-cured sausage processed with 0.3% sulfur fed pork. 2)[NS]; Not significant; a-c means within a row with different letters are significantly different, P < 0.05; A-D means within a column with different letters are significantly different, P < 0.05. *All values are mean ± standard deviation (n = 3).
[a,b] means with different superscripts in the same row are significantly different (p < 0.05).
Significance level of treatment is given by ***(P <0.001); **(P<0.01); *(P<0.05). *All values are mean ± standard deviation (n = 3).

and ISO groups were significantly higher than that of the CON during storage (p < 0.05). Meat products can be classified as intermediate moisture meats when their a_w values range from 0.60 to 0.90 [19].

Thiobarbituric acid reactive substances (TBARS)
TBA value of dry-cured pork sausage is shown in Table 4. TBA values of all samples increased significantly during storage (p < 0.05), which corroborates the results of a previous study [20]. The ISO and SUL groups showed significantly lower TBA values than that of the CON (p < 0.05). This result could be attributed to the antioxidant effects of sulfur-containing compounds and isoflavones [12,21].

Song et al [22] reported that dietary sulfur administered to pigs causes an increase in sulfur-containing antioxidants. Addition of isoflavones reduced the TBA value compared to that of the CON and SUL groups. According to Jiang et al. [23], this result could be due to addition of isoflavones, which increases pH and prevents lipid oxidation. Maximum MDA content values were observed in all samples on day 30, after which values decreased. This result

could be attributed to decomposition of MDA by bacteria, which can selectively breakdown and utilize carbonyl compounds such as MDA [24,25]. The reduced TBA value by day 45 could be due to lipid oxidation products such as organic acids and alcohols, which did not form colored compounds when TBA reacted with MDA [26].

Volatile basic nitrogen (VBN)
The VBN values of all dry-cured sausages increased significantly (p < 0.05) during storage (Table 5). The range of VBN values is 7–18 mg % in Chinese style dry-cured sausage during the ripening period [27]. The VBN value represents the total growth of aerobic bacteria in meat products [28]. In our results, the VBN value of the ISO sample was significantly lower than that of the CON during days 30–45. This result can be attributed to addition of isoflavones, which have antimicrobial effects [28]. The VBN value may also be an indicator of spoilage [21,29]. VBN values decreased in all samples on the last day of the ripening process compared to day 30, suggesting inhibition of VBN by lactic acid bacteria [30].

Color

Changes in the color of dry-cured pork sausage during storage are shown in Table 6. The rate of meat product discoloration is related to oxidation and reduction of metmyoglobin in meat as well as the effects of its three derivatives (myoglobin, deoxymyoglobin, oxymyoglobin) on L^*, a^*, and b^* values [31,32].

The lightness (L*) values of all samples were between 62–64 and decreased (p < 0.05) to about 15 during the ripening period. This reduction of the L-value resulted in dark coloration due to browning reaction in all dry-cured hams. It was previously reported that metmyoglobin generated by pigment oxidation reduces lightness value while pigment oxidation shows a positive relationship with lipid oxidation during the ripening period [33,34]. Therefore, the high lightness value of the ISO group could be due to prevention of pigment oxidation during storage. The redness (a*) values of all samples increased significantly compared to that of the initial meat product (p < 0.05), but a* values decreased slightly in all samples after 15 days. Similarly, Kayaardı and Gök [33] reported that a-values of dry-cured ham increase during the initial ripening period, after which they decrease during further ripening. Additionally, pH and redness have a proportional relationship in meat products. A similar result was observed by Fu et al [35]. However, no significant differences were observed among the three groups. Redness (a*) was not affected by isoflavone or sulfur content.

The yellowness (b*) value of the meat was between 13–18 during ripening. All b* values decreased slightly during the ripening period. This result corroborates the yellowness of Spanish sausage, which decreases during the ripening period [36]. No significant differences were observed among the groups during fermentation.

Microbiological assessment

Microbiological analyses of the dry-cured sausages during the 45-day storage period are shown in Table 7. Total plate counts increased during the initial phase and then decreased until day 45. Total aerobic bacterial counts were significantly lower in dry-cured sausage containing isoflavones compared to other sausages during the entire ripening process. This result may be due to the antimicrobial activity of isoflavonoids, as many researchers have reported that isoflavones are representative flavonoids with antibacterial activity [37,38]. The number of lactic acid bacteria increased by day 15 and then tended to decrease in all dry-cured sausages until day 45. Visible lactic acid bacteria counts in the ISO group were significantly higher than those in the other samples by day 30. Other studies have also found that conversion of isoflavone glycosides to aglycones has beneficial effects on the growth of *Lactobacillus* [39]. The results of the microbial analysis during sausage fermentation can be a measure of the quality of meat products such as pH and physicochemical properties [40].

Conclusion

The results demonstrated the antioxidative and antimicrobial effects of isoflavones on dry-cured sausage during ripening period (15°C). The combination of isoflavones and sulfur-fed pork showed the best results in terms of a_w and TBA values, which are the main concerns in dry-cured meat products during storage. These beneficial effects are related to inhibition of lipid oxidation and product spoilage. Isoflavones may have commercial potential for extending the shelf-life of meat products without other additives such as nitrites. Further research could demonstrate the antimicrobial effect of isoflavones on lactic acid bacteria during meat fermentation by optimizing the quantities of isoflavones used in this study.

Competing interests
The authors declare that they have no competing interests.

Authors' contributions
All authors helped to draft the manuscript, as well as read and approve the final manuscript.

Acknowledgment
This research was supported by the Technology Commercialization Support Program, IPET (Korea Institute of Planning and Evaluation for Technology in Food, Agriculture, Forestry and Fisheries), Ministry of Agriculture, Food and Rural Affairs, Republic of Korea.

Author details
[1]Konkuk University of Food Science & Technology, Seoul 143-701, Republic of Korea. [2]Konkuk University of Animal Science & Technology, Seoul 143-791, Republic of Korea. [3]Eulji University of Food Technology & Services, Sung-nam 461-713, Republic of Korea.

References
1. Marcel BR: **Concepts and strategy of functional food science: the European perspective.** *Am J Clin Nutr* 2000, **71**:1660–1664.
2. Kotilqinen L, Rajalahti R, Ragasa C, Pehu E: *Health enhancing foods: opportunities for strengthening the sector in developing countries*, Agriculture and Rural Development Discussion; 2006:30.
3. INRA: *Repertoire general des aliments. Table de composition de corps grras*. Paris, France: INRA; 1987.
4. Eburne RC, Prentice G: **Modified atmosphere- packed ready-to-cook and ready-to eat meat products.** In *Shelf Life Evaluation of Foods*. Edited by Man CMD, Jones AA; 1996:156–178.
5. Gray JJ, Gomaa EA, Buckley DJ: **Oxidative quality and shelf life of meats.** *Meat Sci* 1996, **43**:111–123.
6. Ruiz LMB, Mohan AR, Paganga G, Miller NJ, Bolwell GP, Rice ECA: **Antioxidant activity of phytoestrogenic isoflavones.** *Free Radic Res* 1997, **26**:63–70.
7. Kim JH, Lee HR, Pyun CW, Kim SK, Lee CH: **Changes in physicochemical, microbiological and sensory properties of dry-cured ham in processed sulfur-fed pigs.** *J Food Process* 2014. in press.
8. Han KH, Park JK, Lee CH: **Manufacture and product evaluation of fermented sausages inoculated with freeze-dried kimchi powder and starter culture.** *Korean J Food Sci Ani Resour* 2006, **26**(4):486–490.
9. AOAC: *Official Methods of Analysis*. 17th edition. Gaithersburg, MD: Association of Official Analytical Chemists; 2002.
10. Witte VC, Krause GF, Bailey ME: **A new extraction method for determining 2- thiobarbituric acid values of pork.** *Food Technol* 1970, **8**:326.

11. Conway EJ: *Microdiffusion Analysis and Volumetric Error.* London: Crosby Lockwood and Son Ltd; 1950.

12. Lee JI, Min HK, Lee JW, Jeong DJ, Ha YJ, Kwack SC, Park JS: **Changes on the quality of loin from pigs supplemented with dietary sulfonyl methane during cold storage.** *Korean J Food Sci Ani Resour* 2009, **29**:229–237.

13. Park JH, Ryu MS, Lee YE, Song GS, Ryu KS: **A comparision of fattening performance, physic-chemical properties of breast meat, vaccine titers in cross bred meat type hybrid chicks fed sulfur.** *Korean J Poult Sci* 2003, **30**(3):211–217.

14. Fidel T: **Proteolysis and lipolysis in flavour development of dry-cured meat products.** *Meat Sci* 1998, **49**:101–110.

15. Lee YB, Hargus GL, Webb JE, Rickansrud DA, Hagberg EC: **Effect of electrical stunning on postmortem biochemical changes and tenderness in broiler breast muscle.** *J Food Sci* 1979, **44**:1121–1122.

16. Bloukas JG, Paneras ED, Fournitzis GC: **Effect of replacing pork back fat with olive oil on processing and quality characteristics of fermented sausages.** *Meat Sci* 1997, **45**:133–144.

17. Puolanne E, Peltonen J: **The effects of high salt and low pH on the water-holding of meat.** *Meat Sci* 2013, **93**:167–170.

18. Maurice R, Didier A: **Culture method to study fungal growth in solid fermentation.** *Eur J Appl Microbiol* 1980, **9**:199–209.

19. Leistner L, Rodel W: **The stability of intermediate moisture foods with respect to microorganisms.** In *Intermediate Moisture Foods.* Edited by Davies R, Birch G, Parke K. London: Elsevier Applied Science; 1976:120–137.

20. Fanco I, Prieto B, Cruz JM, López M, Carballo J: **Study of the biochemical changes during the processing Androlla, a Spanish dry-cured pork sausage.** *Meat Sci* 2002, **78**:339–345.

21. Kumar P, Kumar S, Tripathi MK, Mehta N, Ranjan R, Bhat ZF, Singh PK: **Flavonoids in the development of functional meat products: a review.** *Vet World* 2013, **8**:573–578.

22. Song R, Chen C, Wang L, Johnston LJ, Kerr BJ, Weber TE, Shurson GC: **High sulfur content in corn dried distillers grains with soluble protects against oxidized lipids by increasing sulfur-containing antioxidants in nursery pigs.** *J Anim Sci* 2013, **91**(6):2715–2728.

23. Jiang ZY, Jiang SQ, Lin YC, Xi PB, Yu DQ, Wu TX: **Effects of soybean isoflavone on growth performance. Meat quality and antioxidation in male broilers.** *Poult Sci* 2007, **86**:1356–1362.

24. Smith LJ, Alfod JA: **Action of microorganisms on the peroxides and carbonyls of rancid fat.** *J Food Sci* 1968, **33**(1):93–97.

25. Moerck KE, Ball HR: **Lipid autoxidation in mechanically deboned chicken meat.** *J Food Sci* 1974, **39**(5):876–879.

26. Stapelfedt H, Bjorn H, Skovgaard IM, Skibsted LH, Bertelsen G: **Warmed-over-flavour in cooked sliced beef. Chemical analysis in relation to sensory evaluation.** *Z Lebensm, Unters Forsch* 1992, **195**:203–208.

27. Lin KW, Lin SN: **Effects of sodium lactate and trisodium phosphate on the physicochemical properties and shelf life of low-fat Chinese style sausage.** *Meat Sci* 2002, **60**:147–154.

28. Jung S, Choe JH, Kim BN, Yun HJ, Kruk ZA, Jo C: **Effect of dietary mixture of gallic acid and linoleic acid on antioxidative potential and quality of breast meat from broilers.** *Meat Sci* 2010, **86**:520–526.

29. Chen WS, Liu DC, Chen MT: **Determination of quality changes throughout processing steps in Chinese-style pork jerky.** *J Anim Sci* 2004, **17**:700–704.

30. Yin LJ, Pan CL, Jiang ST: **Effect of lactic acid bacterial fermentation on the characteristics of minced mackerel.** *J Food Sci* 2002, **67**:786–792.

31. Quervedo R, Valencia E, Cuevas G, Roneros B, Pedreschi F, Bastias JM: **Color changes in the surface of fresh cut meat: A fractal kinetic application.** *Food Res Int* 2013, **54**:1430–1436.

32. Lindahl G, Enfält AC, Seth GV, Joseli Å, Ingela HV, Andersen HJ, Braunschweig M, Andersson L, Lundström K: **A second mutant allele (V1991) at the PRKAG3 (RN) locus- ii. Effect on colour characteristics of pork loin.** *Meat Sci* 2004, **66**:621–627.

33. Kayaardı S, Gök V: **Effect of replacing beef fat with olive oil on quality characteristics of Turkish soudjouk (sucuk).** *Meat Sci* 2003, **66**:249–257.

34. Faustman C, Sun Q, Mancini R, Suman SP: **Myoglobin and lipid oxidation interactions: mechanistic bases and control.** *Meat Sci* 2010, **86**:86–94.

35. Fu AH, Molins RA, Sebranek JG: **Storage quality characteristics of beef rib eye steaks packaged in modified atmospheres.** *J Food Sci* 1992, **57**:283–287.

36. Perez AJA, Sayes BME, Fernandez-Lopez J, Aranda-Catala V: **Physicochemical characteristics of Spanish type dry-cured sausage.** *Food Res Int* 1999, **32**:599–607.

37. Hong HK, Landauer MR, Foriska MA, Ledney GD: **Antibacterial activity of the soy isoflavone genistein.** *J Basci Microbiol* 2006, **46**(4):329–335.

38. Narayana KR, Reddy SM, Chaluvadi MR, Krishna DR: **Bioflavonids classification, pharmacological, chemical effects and therapeutic potential.** *Indian J Pharmacol* 2001, **33**:2–16.

39. Pham TT, Shah NP: **Effect of lactulose on biotransformation of isoflavone glycosides to aglycones in soymilk by lactobacilli.** *J Food Sci* 2008, **73**(3):M158–M165.

40. Lücke FK: **Utilization of microbes to process and preserve meat.** *Meat Sci* 2000, **56**:105–115.

19

A method using artificial neural networks to morphologically assess mouse blastocyst quality

Felipe Delestro Matos[1,2]*, José Celso Rocha[1] and Marcelo Fábio Gouveia Nogueira[2]

Abstract

Background: Morphologically classifying embryos is important for numerous laboratory techniques, which range from basic methods to methods for assisted reproduction. However, the standard method currently used for classification is subjective and depends on an embryologist's prior training. Thus, our work was aimed at developing software to classify morphological quality for blastocysts based on digital images.

Methods: The developed methodology is suitable for the assistance of the embryologist on the task of analyzing blastocysts.
The software uses artificial neural network techniques as a machine learning technique. These networks analyze both visual variables extracted from an image and biological features for an embryo.

Results: After the training process the final accuracy of the system using this method was 95%. To aid the end-users in operating this system, we developed a graphical user interface that can be used to produce a quality assessment based on a previously trained artificial neural network.

Conclusions: This process has a high potential for applicability because it can be adapted to additional species with greater economic appeal (human beings and cattle). Based on an objective assessment (without personal bias from the embryologist) and with high reproducibility between samples or different clinics and laboratories, this method will facilitate such classification in the future as an alternative practice for assessing embryo morphologies.

Keywords: Embryology, Quality, Assessment, Artificial neural networks, Mice, Software, Blastocyst

Background

Since the first techniques for multiple ovulation embryo transfer (MOET) and in vitro fertilization (IVF) were successfully developed in mammals, a clear, direct relationship between embryo quality and gestation rate following embryo transfer to recipient females has been established. Embryos that are morphologically classified as high-quality yield higher gestation rates [1,2]. Thus, the field requires a system that can standardize the elements used to categorize embryos into different quality grades, which is an indirect indication of viability.

Currently, a four-grade system is used for cattle: excellent, good, fair and poor [3,4]. This system is based on visual analyses (subjective and qualitative) of embryo morphology, which are commonly performed through optical microscopy (stereomicroscopy). The technique depends on an embryologist's experience and accuracy in analyzing and categorizing samples from the most obvious variables to the nuances that indicate an embryo is more or less apt to develop. For this classic embryo morphology analysis, the variables are not measured objectively; thus, the method is subjective and has limited reproducibility [5]. As a result, the same embryo measured by different experts may be classified with different quality grades. Such inconsistency is typical for adjacent grades, such as good and excellent embryos [6].

Various alternative methods have been developed to solve the subjectivity problem in embryo morphological analyses [7-10]. The most significant such methods include in vitro embryo culture [7], blastomere membrane integrity analysis [7], embryo metabolism analysis [7], cellular respiration measurements [8], electron microscopy

* Correspondence: delestro@gmail.com
[1]Laboratory of Applied Mathematics (Laboratório de Matemática Aplicada - MaAp), School of Sciences and Letters (Faculdade de Ciências e Letras – FCL) São Paulo State University (Universidade Estadual Paulista – Unesp), Assis, Brazil
[2]Laboratory of Embryo Micromanipulation (Laboratório de Micromanipulação Embrionária - LaMEm), FCL/Unesp, Assis, Brazil

analysis [9] and zona pellucida birefringence indices [10]. However, no method has provided a definitive solution for measuring embryo quality, and it is necessary to develop such fast, non-invasive and objective methods [3,7]. In addition, such methods can be prohibitively expensive for widespread use. Thus, despite its subjectivity and limited reproducibility, visual morphology analysis persists for embryo quality determinations.

Herein, we validate a method for morphological analysis that is more precise, wherein information is extracted from two-dimensional digital embryo images and the images are subsequently analyzed using software. The software (Blasto4Q) is based on artificial neural networks (ANNs), which is an artificial intelligence technique that solves non-linear problems with interconnected variables [11-13]. ANNs have been applied to various areas, including administrative aids [14] and stock market index predictions [15]. An ANN is a system that solves problems by simulating biological neurons. The neurons in an ANN (also, "perceptrons") must receive training data to learn and generalize output based on an input dataset. Once it is properly trained, an ANN can generate predictions without a pre-established classification [11,12,14,16]. Therefore, an ANN is an intelligent system that can solve a complex problem based on assisted learning.

Methods

The embryos used herein were products from other ongoing projects in the Laboratory of Embryo Micromanipulation (Laboratório de Micromanipulação Embrionária - LaMEm, UNESP, Assis). The experiments in this laboratory were developed for applied embryology and embryo micromanipulation. Mouse (Mus musculus) embryos from the Swiss-Webster and C57BL/6 EGFP strains were used. The embryos originated from projects that were approved by the Ethical Commission on Animal Use (Comissão de Ética no Uso de Animais – CEUA) in the School of Sciences and Letters of Assis (protocols 007/2010, 015/2011 and 026/2011).

The embryo images were collected using an Eclipse Ti inverted microscope (Nikon, Japan) coupled to a Digital Sight (Nikon, Japan) camera, which was controlled by the NIS - Elements Advanced Research 3.0 (Nikon, Japan) software. The image may contain one or several embryos, as long as the individual embryo shape is not affected. The magnification of the image capture, as well the resolution of the image file can be chosen by the user, as the software uses on ratios between values.

For the purpose of organizing the database, each embryo was labeled using a code that included one number and letter; the number identified the image in the database, and the letter identified the embryo in the image. Such labeling was performed using the GIMP 2.6.11 software.

All the measurements on the image were made using the software ImageJ 1.45 s. First the previously captured image is loaded on the interface. The user then must use the "Straight Lines" and "Polygon" tools to assess the proportions from the embryo, as indicated on Figure 1.

The data were collected from the embryos using the ImageJ 1.45 s software. The following data were the basis for calculating the following variables: smallest embryo diameter (ED1), largest embryo diameter (ED2), smallest zona pellucida diameter (ZPD1), largest zona pellucida diameter (ZPD2), embryo area (EA), zona pellucida area (ZPA), dead cell area (DCA), live cell area (LCA), embryo color density (ECD), total color density (CDtotal) and zona pellucida color density (ZPCD).

When the perivitelline space was absent (expanded blastocyst), the first two vectors, which determine the embryo and zona pellucida areas, were sufficient to discern the zona pellucida color density using the following ratio (notably, the area vector also yields the mean color density for the area).

$$Mz = \frac{(Mez \times Aez) - (Me \times Ae)}{(Aez \times Ae)}$$

Mz is the mean color for the zona pellucida, Mez is the mean color for the zona pellucida and embryo, Aez is the area that comprises the zona pellucida and embryo, Me is the embryo mean color, and Ae is the embryo area.

Compared with manually selecting the zona pellucida area, this formula yielded a more rapid and efficient process for determining the zona pellucida color density (ZPCD); this formula was incorporated into Blasto4Q.

The ANN creation process, the algorithm that determines the best architecture and the graphical user interface were developed using MATLAB R2011a and the Artificial Neural Network Toolbox [17]. To better understand such processes and their adaptation to the overarching problem and its particularities, the standard models for constructing an ANN (ready and available in the "toolbox") were not used, and the ANN was constructed using the MATLAB metalanguage.

Results

Embryo collection and image capture

Training the ANN herein required a database with embryo images that can be properly classified and analyzed. The animals used were superovulated Swiss-Webster mice, and the structures were harvested (eggs and viable or degenerate embryos) 3.5 days after copulation (consistent with the blastocyst stage; see the Methods section for more details). After the embryos were collected, those with viable cells were grouped and photographed using a digital image capture system. We only used embryos that

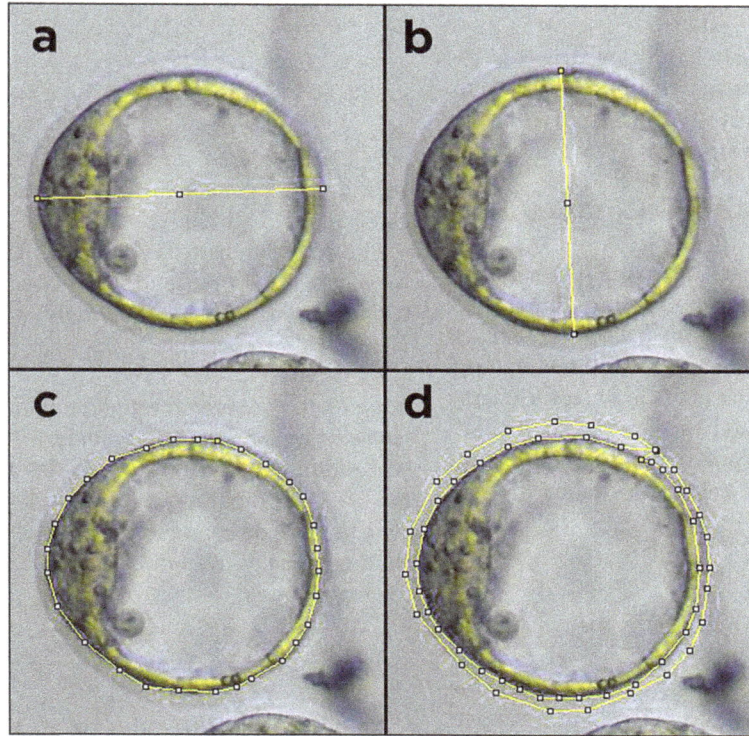

Figure 1 Example measurements from ImageJ, which represent the points required to determine the mean embryo diameter (images a and b), embryo area (image c) and zona pellucida area (image d).

were viable during the blastocyst stage (including the early blastocyst, blastocyst and expanded blastocyst stages) and images in which the blastocyst was largely in focus. Thus, the ANN was trained using 98 images.

Embryo classification

The conventional morphological classification system 3 was used to classify the selected images as excellent, good, fair or poor grade. Of the 98 embryos, 40 (40.8%) were classified as excellent, 46 (46.9%) were good, 8 (8.2%) were fair, and 4 (4.1%) were poor. These data were used to train the ANN, which generated 4 distinct outputs, one for each embryo quality grade.

Definitions for the ANN variables

We determined the features that were desirable for assessing embryo quality because such features should be discerned using only static two-dimensional images. Thus, we used the biological aspects of embryo morphology, experience from the quality-assessing embryologist and computational techniques for image processing; the following 12 variables were isolated.

Stage of embryonic development (SED)

The embryo's development stage is critical for the ANN to correctly manage the additional variables given the morphological differences in embryos throughout development before they are implanted (from the zygote to blastocyst stages). For the ANN used herein, this variable indicates whether an embryo is at an early blastocyst, blastocyst or expanded blastocyst stage.

Days after copulation (DAC)

The DAC variable is used to compare the SED and the ideal stage for the time elapsed since fertilization. Depending on the DAC value, the embryo should be at a specific stage. Thus, this variable is used to characterize consistency between the level of embryo development assessed with the ideal development level.

Ratio between developmental stage and group mean (RGM)

The developmental stage of an embryo relative to other embryos in the same harvest must be considered. For example, an embryo's stage may be delayed compared with its DAC value, which the ANN may penalize. However, if the other embryos are similarly delayed, such penalization may be reduced or eliminated.

Thus, the RGM was determined using the following formula.

$$RGM = \frac{SED}{SEDgroup}$$

SEDgroup is the mean embryonic stage for the remaining embryos in the same harvest.

Therefore, values greater than 1 indicate that the embryo is at a more advanced stage relative to its group, while values less than 1 indicate that an embryo is at a delayed stage relative to its group.

Ratio (RMD) for the mean embryo diameter (ED) and mean zona pellucida diameter (ZPD)

Information on embryo morphology is key to generating data using the ANN. The data generated are dimensionless to avoid scaling problems. Thus, distances can be measured using pixels, micrometers or millimeters, and a ratio for the embryo and zona pellucida measurements is used. This metric was selected because the zona pellucida diameter and embryo diameter ratio is highly consistent. The following formula defines this ratio:

$$RMD = \frac{ED}{ZPD}$$

ED is the embryo diameter, and ZPD is the zona pellucida diameter.

Both the ED and ZPD were determined using the means for the largest and smallest embryo and zona pellucida diameters, respectively.

Ratio for the live cell area and total area (RLC)

This ratio is calculated to determine the proportion of live cell area in an embryo's (LCA) total area, which is defined by the outer border for the zona pellucida (ZPA).

$$RLC = \frac{LCA}{ZPA}$$

Ratio (RDC) between the dead cell area (DCA) and live cell area (LCA)

This ratio was created so that the ANN considers dead cells in the embryo for quality analysis.

$$RDC = \frac{DCA}{LCA}$$

Greater RDC values indicate a larger proportion of dead cells in the embryo, which negatively impacts its quality.

Ratio (RCD) between the embryo color density (ECD) and zona pellucida color density (ZPCD)

Embryo color is another important factor for analysis because it is directly affected by cell density and viability. This variably is highly dependent on the conditions used to photograph the image, including both illumination and the camera control software settings. However, using the ratio between the embryo color (ECD)

and zona pellucida color (ZPCD) compensates for such variations.

$$RCD = \frac{ECD}{ZPCD}$$

RCD values less than 1 indicate that an embryo is lighter than its zona pellucida, while values greater than 1 indicate a darker embryo.

The color intensity (ECD or ZPCD) is measured as the mean brightness value for each pixel in a particular area. This value ranges from 0 (completely black) to 255 (completely white).

Ratio (RER2) between the embryo roundness (ER) and zona pellucida roundness (ZPR) squared

An additional factor that may be a good indicator for quality is comparing an embryo's roundness with the typical level of roundness at a blastocyst embryo stage. Mathematically, roundness is determined using the following formula:

$$Roundness = 4 \times \pi \times \frac{Area}{Perimeter^2}$$

An ideal circle has the value 1. As the value approaches 0, its shape is less similar to a circle.

Thus, roundness is defined as follows:

$$RER = \frac{ER}{ZPR}$$

ER is the embryo roundness, and ZPR is the zona pellucida roundness.

Because the zona pellucida is stable and round, a value near 1 indicates a round embryo, while values near 0 indicate low roundness.

However, in practice, the values are always near 1; thus, it is difficult for the ANN to assign different features to round or less round embryos. To solve this problem, RER2 (ratio of roundness squared) was used to numerically emphasize small differences in roundness. Notably, rounder embryos trend towards 1 because it is the upper limit.

Sharp edges macro (EDG)

The input variables must numerically indicate the visual morphological features for an embryo. However, the aforementioned variables cannot represent an embryo's roughness or granularity. Thus, a macro (series of automatic operations) was developed for the ImageJ software (further details in the Methods section) to identify and count the contrast regions in the embryo to numerically represent this visual feature. The macro is referred to as "Sharp Edges" because it uses the basic operations sharpen and find edges.

Ratios for blastocoel area, color density and roundness (RBA, RBCD and RBR)

Because only blastocysts were used, blastocoel features were also included as input variables. Blastocoel area, color density and roundness were used, and a ratio was established with the respective variable for the embryo. Blastocoel roundness was squared for the aforementioned rationale regarding embryo roundness.

$$RBA = \frac{BA\,(Blastocoel\,area)}{EA\,(Embryo\,area)}$$

$$RDCB = \frac{BCD\,(Blastocoel\,color\,density)}{ECD\,(Embryo\,color\,density)}$$

$$RBR = \left(\frac{BR\,(Blastocoel\,roundness)}{ER\,(Embryo\,roundness)}\right)^2$$

Data extraction and standardization

The data required to calculate the above-described variables must be extracted from the embryo images; thus, the ImageJ software was used. ImageJ is a free multifunctional image processing software that facilitates measurements using selected points in an image [18]. Figure 1 demonstrates how such selections were generated.

However, the output data from ImageJ cannot be directly used. Such data are the basis for calculating the variables that will be used to train the ANN and for quality analyses using Blasto4Q.

For example, the four measurements used to calculate the variable RMD included the largest and smallest diameters for both the embryo and zona pellucida. Similar procedures were used to collect the input variables (see the Methods section for further details).

Developing the ANN architecture

The structure of an ANN includes various elements, such as the numbers of neuron layers and neurons in each layer as well as their transfer functions and the network training function. Although it is important to correctly establish such factors to optimally develop an ANN, there is not a standard protocol to determine the best architecture [19].

To address this problem, an algorithm was developed that automatically tests various combinations and structures to determine the best result. The flowchart for this algorithm is shown in Figure 2.

This algorithm was executed using Matlab software with the stop condition 10,000 cycles; the range 5 to 20 neurons for the first and second layers; and randomly selected tansig, logsig and purelin transfer functions as well as trainlm, trainscg and traingdx training functions. The error was calculated using a confusion matrix (incorrect classification percentage). The ANN was developed

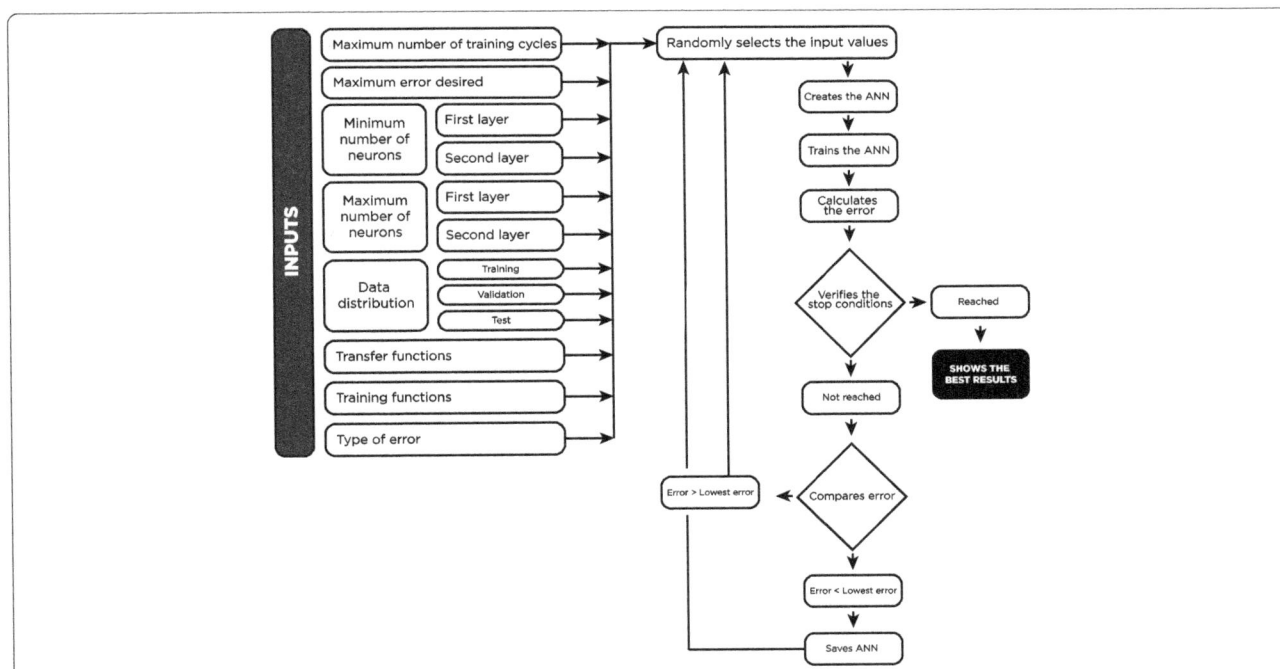

Figure 2 Flowchart for the algorithm used to construct the ANN architecture. Initially, the variables used to create the various ANN configurations were included. The program randomly selects values for each variable and generates an ANN, which is trained using the database; the data are divided into training, validation and test sets in accordance with the initial program selection. This process is repeated; the program compares the error for a network with previous networks, and the ANN with the lowest error is saved. When it encounters a stop variable, the program ends the cycle and displays the best result.

Table 1 Results from the algorithm used to determine the best ANN architecture

Error from training	12.07%
Error from validation	30.00%
Error from test	25.00%
Error from all data	18.37%
Mean Square Error (MSE)	0.023206
Training Data	60%
Validation Data	20%
Test Data	20%
Number of neurons in the first layer	18
Number of neurons in the second layer	13
Function in the first layer	purelin
Function in the second layer	logsig
Training function	trainscg

for the backpropagation algorithm. Table 1 shows the best results generated.

Thus, herein, the best ANN architecture was a network that included 18 neurons in the first layer with a purelin transfer function (linear function) and 13 neurons in the second layer with a logsig transfer function (logistic function). The algorithm selected trainscg (Scaled Conjugate Gradient Algorithm) for the training function [20]. The data used to train, validate and test the ANN are summarized in Figure 3. Each confusion matrix shows the relationship between the real data (template) and the data simulated by the network.

Validation

The data processed by the ANN were divided into three classes: training (60% of the data), validation (20%) and test (20%). Each dataset was randomly generated each time the ANN was trained. The training data were

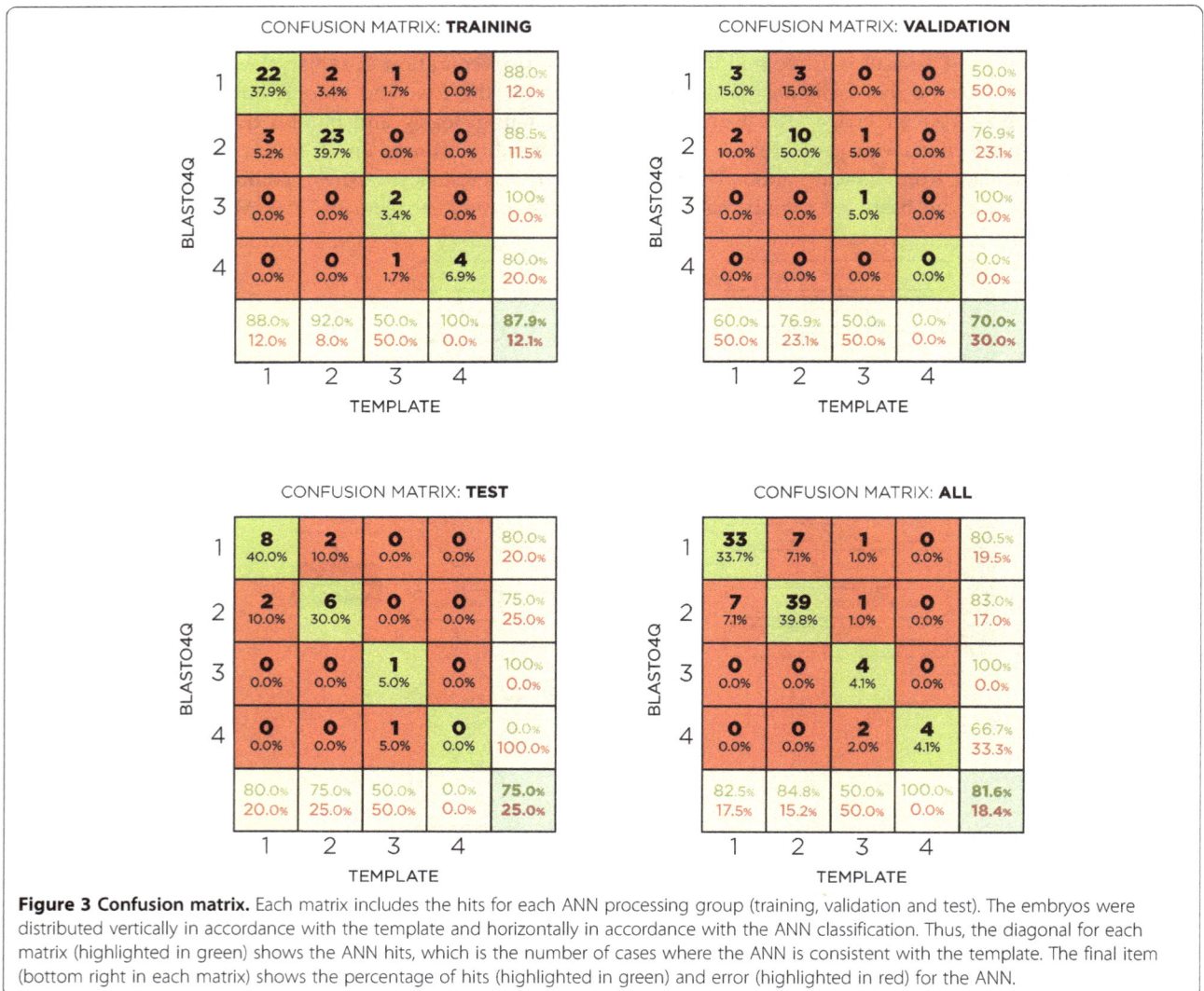

Figure 3 Confusion matrix. Each matrix includes the hits for each ANN processing group (training, validation and test). The embryos were distributed vertically in accordance with the template and horizontally in accordance with the ANN classification. Thus, the diagonal for each matrix (highlighted in green) shows the ANN hits, which is the number of cases where the ANN is consistent with the template. The final item (bottom right in each matrix) shows the percentage of hits (highlighted in green) and error (highlighted in red) for the ANN.

effectively used to teach the ANN. The validation data were used to avoid overfitting by the ANN; data may be overfit when a network is excessively trained with a dataset and incorporates input noise [11]. The final test dataset was not provided to the ANN during the training phase; thus, the test data were used to verify whether the ANN was effectively trained because the network is used to classify a "novel" dataset (without prior access to the data) after training and validation.

Table 2 shows the ANN results for the test data. In this table, the ID column identifies the embryo in the database. The error column was calculated by subtracting the ANN-assigned quality from the template-generated quality (embryologist assessment), where 0 is a hit, +1 indicates that the ANN assigned a lower quality score than the template, and −1 indicates that the ANN assigned a higher quality score than the template.

The ANN provided a correct prediction (its analysis was equal to the template) in 75% of the cases (15 hits from 20 test samples). Of the 5 incorrectly classified cases, the ANN assigned a quality score 1 grade above the template (for example, a grade 2 embryo was classified as grade 1) in 3 cases and at a lower grade in 2 cases (for example, a grade 1 embryo was classified as grade 2).

The 75% hit rate indicates the cases where the ANN generated the same value as the template used to train the ANN. However, embryo classification can vary depending on the evaluator, who can categorize the same embryo at two or three different but adjacent grades [6]. In reality, this is common, primarily where the image does not have a good focal plane for analyses or the embryo assessed is at the border between two quality grades, such as excellent and good.

Therefore, another analysis was performed using the test data and the same embryologist that performed the initial classification, who reassessed the embryos blind to both their prior classification and the ANN's classification. In the same analysis, the evaluator indicated the possible quality grades that may correctly classify the embryo (see Table 3).

These results were compared with the original template and ANN classifications, which are shown in Table 3. Comparing the reassessment by the embryologist, the ANN assessment and the original assessment showed that, while the ANN incorrectly classified five cases (25%), seven cases (35%) were incorrectly classified upon reassessment. This finding demonstrates that the human factor (assessment by an embryologist) was responsible for limited reproducibility of the assessments. Moreover, the three classifications were within the possible correct grades in almost each case (except for the image 027B, for which the reassessment was inconsistent with the original template and ANN classification; it was correct in the original assessment but was incorrectly classified upon reassessment). The image 029E presented an interesting case, for which fair was the only possible grade, and each assessment was consistent.

The 75% hit rate indicates the cases where the ANN is consistent with the template. However, the blind reassessment test presents the possible correct grades for the test embryos. A new analysis was performed using these possible grades for the error analysis (Table 4), not the original template, from which we generated a 95% hit rate with only 1 incorrect classification.

Blasto4Q

The software Blasto4Q was developed as the final result herein. This software is fully functional and can be installed on any computer running the Windows operating system (the compatibility test was conducted using Windows 7; other versions of Windows have not been tested). Matlab Compiler Runtime (MCR) and Java Virtual Machine (JVM) must be installed by the end-user; they can be acquired from their respective companies.

Creation and development of this system would be meaningless without a practical, fast and efficient way for the end-user to apply the program. Thus, a graphical user interface (GUI) was developed, as shown in Figure 4.

Table 2 Results from the ANN for the test data

ID	ANN outputs				Quality ANN	Quality template	Error
	1	2	3	4			
003D	0.11	0.84	0.03	0.01	2	1	1
004E	0.45	0.64	0.00	0.00	2	2	0
004 F	0.83	0.10	0.01	0.01	1	2	−1
004G	0.86	0.09	0.01	0.01	1	2	−1
008C	0.53	0.40	0.11	0.00	1	1	0
008D	0.21	0.62	0.05	0.00	2	2	0
011A	0.75	0.27	0.02	0.00	1	1	0
012C	0.47	0.48	0.11	0.00	2	2	0
013G	0.84	0.13	0.01	0.00	1	1	0
015A	0.55	0.47	0.05	0.00	1	1	0
016B	0.94	0.04	0.02	0.00	1	1	0
016D	0.37	0.63	0.02	0.01	2	2	0
016E	0.74	0.07	0.21	0.00	1	1	0
017 F	0.21	0.85	0.02	0.00	2	1	1
024C	0.22	0.79	0.03	0.00	2	2	0
027B	0.63	0.04	0.34	0.00	1	1	0
027 J	0.00	0.20	0.18	0.90	4	3	1
028I	1.00	0.00	0.00	0.01	1	1	0
029 F	0.00	0.58	0.58	0.15	2	2	0
029E	0.00	0.27	0.51	0.14	3	3	0

Table 3 Comparison for the original assessment, reassessment and ANN results

ID	Original assessment	Reassessment	Possible grades			ANN Result	Reassessment error	ANN error
003D	1	1	1	2		2	0	1
004E	2	2	1	2	3	2	0	0
004 F	2	1	1	2		1	−1	−1
004G	2	2	1	2		1	0	−1
008C	1	1	1	2		1	0	0
008D	2	2		2	3	2	0	0
011A	1	1	1	2		1	0	0
012C	2	2		2	3	2	0	0
013G	1	1	1	2		1	0	0
015A	1	2	1	2		1	1	0
016B	1	1	1	2		1	0	0
016D	2	2	1	2		2	0	0
016E	1	2	1	2		1	1	0
017 F	1	2	1	2		2	1	1
024C	2	2	1	2		2	0	0
027B	1	2		2	3	1	1	0
027 J	3	3		3	4	4	0	1
028I	1	2	1	2		1	1	0
029 F	2	3		2	3	2	1	0
029E	3	3		3		3	0	0

Table 4 All possible quality grades considered in a single embryo assessment

ID	Possible scores				Blasto4Q	Classification
003D	1	2			2	Correct
004E	1	2	3		2	Correct
004F	1	2			1	Correct
004G	1	2			1	Correct
008C	1	2			1	Correct
008D		2	3		2	Correct
011A	1	2			1	Correct
012C		2	3		2	Correct
013G	1	2			1	Correct
015A	1	2			1	Correct
016B	1	2			1	Correct
016D	1	2			2	Correct
016E	1	2			1	Correct
017F	1	2			2	Correct
024C	1	2			2	Correct
027B		2	3		1	Incorrect
027J			3	4	4	Correct
028I	1	2			1	Correct
029F		2	3		2	Correct
029E			3		3	Correct

Blasto4Q provides results in the following three ways (see Figure 4).

a) A bar chart that represents each ANN output (the 4 quality grades); the height of each bar is determined by the output value magnitude.
b) A quality index using the network's highest output value; the possible results are excellent, good, fair and poor.
c) A descriptive vector that is output vector for the ANN; this vector represents the values for the four neurons in the output layer ("Excellent", "Good", "Fair" and "Poor").

Discussion

Based on the work of this study, we established an alternative method to classify blastocyst morphology in mice. We used an ANN based on data from static, two-dimensional digital images and combined with a graphical user interface to generate a proposed method for a quantitative, objective and highly reproducible assessment. Embryo morphological classification is important for numerous laboratory techniques, which range from basic methods to assisted reproduction applications. Success rates (gestation) for associated biotechniques (cryopreservation, biopsy, embryo splitting and microinjection, among others) can be inferred using this technique, and the embryos

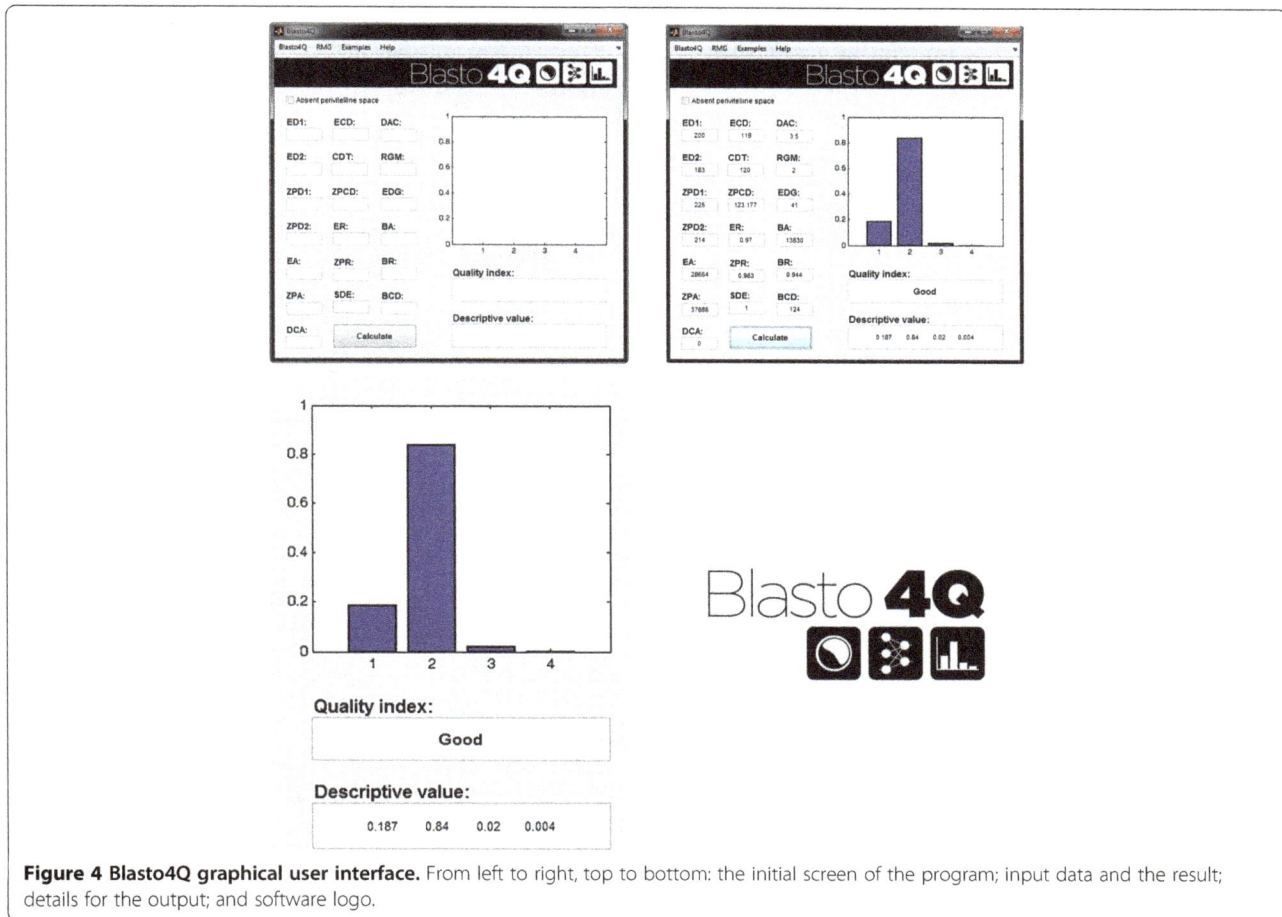

Figure 4 Blasto4Q graphical user interface. From left to right, top to bottom: the initial screen of the program; input data and the result; details for the output; and software logo.

used for scientific experiments can be standardized. However, the method used to classify mammalian embryos (e.g., from human beings, horse, cattle, rats and mice) has always been based on a subjective assessment by an evaluator. Despite the standards for quality grades and the morphological characteristics that characterize each grade, inconsistencies are common for classification by different evaluators, even if they are experienced embryologists [6].

Alternative models for morphological classification with greater objectivity have been developed [7-10]. However, such methods should be fast, low-cost, high-resolution and non-invasive [3,7]. In particular, it is necessary to exercise extreme caution to minimize iatrogenic damage to the embryo by the technique (e.g., through prolonged exposure to non-ideal conditions or excessive handling). Thus, conventional morphological assessments are widely used despite the limited subjectivity (due to the problem with reproducibility and accuracy). Blasto4Q is a reliable morphological analysis technique because the result will always be the same for a given input after the ANN is trained (objectivity and reproducibility).

The Blasto4Q software facilitates rapid assessment with minimal interference in embryonic development because only a single digital image of the embryo is necessary, which requires a microscopy system (inverted or not) coupled to a digital image capture system. The embryos are stored under favorable conditions inside CO_2 incubators when the analyses are performed (data collection and simulation through Blasto4Q). Additionally, the analyses performed by the software are more detailed and produce more data than from an embryologist. Although embryologists may have experience distinguishing among subgrades, one standard grade must be assigned (e.g., an embryo that is grade "1.5" would be either grade 1 or grade 2, depending on the evaluator's analysis). Because a result from Blasto4Q is the descriptive vector, each embryo can be given an "identity" or values that represent the probability that an embryo will be classified in each of the four possible grades.

The blastocyst stage was used for preliminary tests because it is important for commercial in vitro bovine embryo production and due to its growing relevance in assisted human reproduction compared with pre-compaction embryonic stages. Different embryologists can provide a template for ANN training in accordance with the specific classifications used for embryo morphological assessment. Thus, this process has a high potential for applicability because it can be adapted to additional species with greater economic appeal (human beings and cattle). Based on an objective assessment (without personal bias from the embryologist) and with high reproducibility between samples or different clinics and laboratories, this method will facilitate such classification in the future as an alternative practice for assessing embryo morphologies.

Conclusions

This process has a high potential for applicability because it can be adapted to additional species with greater economic appeal (human beings and cattle). Based on an objective assessment (without personal bias from the embryologist) and with high reproducibility between samples or different clinics and laboratories, this method will facilitate such classification in the future as an alternative practice for assessing embryo morphologies.

Competing interests
The authors declare that they have no competing interests.

Authors' contributions
MFGN and FDM determined the ANN variables. MFGN classified the embryos. JCR and FDM developed the ANN architecture. FDM designed the graphical user interface and drafted the manuscript. Each author reviewed the manuscript. All authors read and approved the final manuscript.

Acknowledgments
The authors thank the São Paulo Research Foundation (FAPESP) for supporting this study, which was conducted under protocols 2011/06179-7 and 2006/06491-2, and the computational resources, facilities, equipment and assistance provided by the School of Sciences and Letters of UNESP, Assis Campus. The authors also thank the members of the Laboratory of Embryo Micromanipulation (Laboratório de Micromanipulação Embrionária-LaMEm, UNESP/Assis), without which it would have been impossible to develop the embryo image database. The opinions, hypotheses, conclusions or recommendations expressed in this material are the responsibility of the authors and do not necessarily reflect the views of FAPESP.

References
1. Tervit HR, Cooper MW, Goold PG, Haszard GM: **Non-surgical embryo transfer in cattle.** *Theriogenology* 1980, **13**:63–71.
2. Schneider HJ Jr, Castleberry RS, Griffin JL: **Commercial aspects of bovine embryo transfer.** *Theriogenology* 1980, **13**:73–85.
3. Lindner GM, Wright RW Jr: **Bovine embryo morphology and evaluation.** *Theriogenology* 1983, **20**:407–416.
4. Wright RW Jr, Ellington J: **Morphological and physiological differences between in vivo- and in vitro-produced preimplantation embryos from livestock species.** *Theriogenology* 1995, **44**:1167–1189.
5. Benyei B, Komlosi I, Pecsi A, Pollott G, Marcos CH: **The effect of internal and external factors on bovine embryo transfer results in a tropical environment.** *Anim Reprod Sci* 2006, **93**:268–279.
6. Farin PW, Britt JH, Shaw DW, Slenning BD: **Agreement among evaluators of bovine embryos produced in vivo or in vitro.** *Theriogenology* 1995, **44**:339–349.
7. Overström EW: **In vitro assessment of embryo viability.** *Theriogenology* 1996, **45**:3–16.
8. Hoshi H: **In vitro production of bovine embryos and their application for embryo transfer.** *Theriogenology* 2003, **59**:675–685.
9. López-Damián EP, Galina CS, Merchant H, Cedillo-Peláez C, Aspron M: **Assessment of Bos taurus embryos comparing stereoscopic microscopy and transmission electron microscopy.** *J Cell Animal Biol* 2008, **2**:072–078.
10. Held E, Mertens EM, Mohammadi-Sangcheshmeh A, Salilew-Wondim D, Besenfelder U, Havlicek V, Herrler A, Tesfaye D, Schellander K, Hölker M: **Zona pellucida birefringence correlates with developmental capacity of bovine oocytes classified by maturational environment, COC morphology and G6PDH activity.** *Reprod. Fertil.* 2011, **24**:568–579.
11. Guoqiang ZB, Patuwo E, Hu MY: **Forecasting with artificial neural networks: The state of the art.** *Int J Forecast* 1998, **14**:35–62.
12. Goethals PLM, Dedecker AP, Gabriels W, Lek S, Pauw N: **Applications of artificial neural networks predicting macroinvertebrates in freshwaters.** *Aquat Ecol* 2007, **41**:491–508.
13. Li EY: **Artificial neural networks and their business applications.** *Inform Manag* 1994, **27**:303–313.
14. Rocha JC, Matos FD, Frei F: **Utilização de redes neurais artificiais para a determinação do número de refeições diárias de um restaurante universitário.** *Rev Nutr* 2011, **24**:735–742.
15. Guresen E, Kayakutlu G, Daim TU: **Using artificial neural network models in stock market index prediction.** *Expert Syst Appl* 2011, **38**:10389–10397.
16. Haykin S: *Neural Networks: A Comprehensive Foundation.* 2nd edition. NJ, USA: Prentice Hall PTR, Upper Saddle River; 1998.
17. Beale MH, Hagan MT, Demuth HB: *Neural Network Toolbox User's Guide.* [http://www.mathworks.com.au/help/pdf_doc/nnet/nnet_ug.pdf].
18. Abramoff MD, Magalhães PJ, Ram SJ: **Image processing with image.** *J Biophotonics Intern* 2004, **11**:36–42.
19. Yao X, Liu Y: **Towards designing artificial neural networks by evolution.** *Appl Math Comput* 1998, **91**:83–90.
20. Møller MF: **A scaled conjugate gradient algorithm for fast supervised learning.** *Neural Netw* 1993, **6**:525–533.

20

Meta-analysis of factors affecting milk component yields in dairy cattle

Junsung Lee[1], Jakyeom Seo[1], Se Young Lee[2], Kwang Seok Ki[3] and Seongwon Seo[1]*

Abstract

The objectives of this study were thus to identify most significant factors that determine milk component yield (MCY) using a meta-analysis and, if possible, to develop equations to predict MCY using variables that can be easily measured in the field. A literature database was constructed based on the research articles published in the *Journal of Dairy Science* from Oct., 2007 till May, 2010. The database consisted of a total of 442 observed means for MCY from 118 studies. The candidate factors that determine MCY were those which can be routinely measured in the field (e.g. DMI, BW, dietary forage content, chemical composition of diets). Using a simple linear regression, the best equations for predicting milk fat yield(MFY) and milk protein yield (MPY) were MFY = 0.351 (±0.068) + 0.038 (±0.003) DMI (R^2 = 0.27), and MPY = 0.552 (±0.071) + 0.031 (±0.002) DMI - 0.004 (±0.001) FpDM (%, forage as a percentage of dietary DM) (R^2 = 0.38), respectively. The best equation for predicting milk fat content (%) explained only 12% of variations in milk fat content, and none of a single variable can explain more than 5% of variations in milk protein content. We concluded that among the tested variables, DMI was the only significant factor that affects MFY and both DMI and FpDM significantly affect MPY. However, predictability of linear equations was relatively low. Further studies are needed to identify other variables that can predict milk component yield more accurately.

Keywords: Meta analysis, Milk fat, Milk protein, Dry matter intake, Dairy cattle

Background

The manipulation of milk composition has been of interest to improve the nutritional value of milk and to increase the efficiency in manufacturing and processing of raw milk for dairy products (Jenkins and McGuire, [1]). In addition, the decision of milk price depends on the amounts of milk component yield (MCY) in most countries including Korea, performing of quantitative analysis of factors that influence on MCY is crucially important in dairy industry.

Many studies have tried to identify such factors affecting milk yield (MY) and MCY for past decades. DMI, dietary energy and protein or individual carbohydrate (CHO) and protein fractions might be important factors in controlling MY and milk protein yield (MPY) in dairy cows ((Sutton, [2], Grummer, [3], DePeters and Cant, [4], Palmquist et al., [5], Hristov et al., [6,7], Jenkins and McGuire, [1], Huhtanen and Hristov, [8]). Smoler et al.

[9] suggested individual CHO fractions might be better predictors of MPY than total CHO. Hristov et al. [10] showed a moderate linear relationship (R^2 = 0.47) between DMI and MY using data set published in *Journal of Dairy Science* (Volumes 1 through 82). According to NRC [11], dietary CP was not correlated (P > 0.25) with milk protein percent, but was correlated weakly (r = 0.14; P < 0.01) with MPY. Huhtanen and Hristov [8] reported that metabolizable protein (MP) intake was better predictor of MPY compared with CP intake.

In addition, there have been attempts to develop equations to predict MY and MPY of dairy cows (NRC, [11], Hristov et al., [6]) through a meta-analysis approach. They, however, included RDP and RUP in the equation, which cannot be easily measured in the field. NRC [11] also presented equations to predict MY and MPY with DMI and CP contents in diets; however, the predictability was insufficient and study effect was not accounted for in developing equations. The objectives of this study were thus to identify most significant factors that determine MCY using meta-analysis based on recent studies conducted from last decade and, if possible, to derive

---
* Correspondence: swseo@cnu.kr
[1]Department of Animal Biosystem Sciences, College of Agriculture and Life Science, Chungnam National University, Daejeon 305-764, Republic of Korea
Full list of author information is available at the end of the article

equations to predict MCY with variables that can be easily measured in the field.

Results and discussions

Animal parameters, nutrient composition of diet, milk yield and composition were listed in Table 1. The variations in each variables used for developing equation in this study was large enough to represent a wide range of data.

The best equation for predicting milk fat content (%) explained only 12% of the variations in milk fat content, and none of a single variable could explain more than 5% of the variations in milk protein content. According to the review by Jenkins and McGuire [1], the most sensitive component of milk to dietary manipulation was fat content, which could be changed over a range of 3 percentage units. Milk protein was more responsive to diet (over a 0.5 percentage unit range) than lactose, but less responsive than fat. Sutton [2] reported that milk fat

concentration was affected by the amount of roughage, the forage-to-concentrate ratio, the carbohydrate composition of concentrate mix, lipids, intake, and meal frequency. A reduction in the dietary forage-to-concentrate ratio usually decreases milk fat content although the degree of response varies (Sutton, [2]). Milk fat content was fairly stable until the proportion of forage in the diet on a DM basis falls to about 50%, but with further reductions in the proportion of forage, a decrease in milk fat content occurs (Thomas and Martin, [12]). Smith et al. [13] indicated that the response in milk fat content to dietary supplementation of lipid was highly variable by the amount, physical form, and fatty acid composition of lipid. Sporndly [14] observed no significant correlation between protein content of milk and protein concentration of the diet (r = 0.06), while milk protein yield and dietary protein level were correlated (r = 0.37). Jenkins and McGuire [1] reported that reducing the proportion of

Table 1 Descriptive statistics for the database used for developing equations in this study

	N	Mean	S.D.	Median	Max	Min
Animal inputs						
BW, kg	442	638.21	53.41	641.00	778.00	443.00
DIM, day	358	115.23	56.12	116.50	277.00	1.00
DMI, kg/d	442	22.53	3.51	22.95	32.80	7.00
DMI,% BW	442	3.53	0.50	3.60	5.04	1.26
Forage DMI, kg	442	11.40	2.49	11.40	19.90	3.47
Forage,%DM	442	51.06	10.66	50.00	100.00	19.70
Nutrient composition						
DM,% AF	303	54.90	9.58	53.90	88.40	27.73
CP,% DM	441	17.35	1.89	17.30	25.98	7.10
EE,% DM	203	4.35	1.30	4.35	8.40	1.50
Ash,% DM	189	7.80	2.08	7.50	18.40	3.42
NFC,% DM	183	38.52	5.72	39.30	51.40	24.60
NDF,% DM	438	33.19	4.81	32.70	51.80	19.50
ADF,% DM	340	20.73	3.89	20.20	37.40	11.20
RDP,% DM	116	10.64	1.17	10.60	13.17	6.89
RUP,% DM	103	7.36	5.20	6.30	33.70	4.40
Starch,% DM	174	23.82	5.77	23.30	39.20	9.50
NEL, Mcal/kg	249	1.62	0.08	1.61	1.80	1.40
Milk yield and composition						
Milk yield, kg/d	442	34.53	6.76	35.55	50.90	15.70
3.5% FCM, kg	442	34.68	6.74	35.45	52.11	16.18
4.0% FCM, kg	442	32.08	6.23	32.79	48.20	14.97
Fat,%	442	3.57	0.51	3.53	5.93	2.16
Fat yield, kg	442	1.22	0.26	1.24	1.96	0.55
Protein,%	442	3.14	0.37	3.09	6.87	2.53
Protein yield, kg	442	1.06	0.22	1.09	1.73	0.47
MUN, mg/dl	242	13.63	4.34	13.01	35.80	6.77

forage in the diet increased both protein content and yield. Milk protein content could increase by 0.4 percentage units or more when forage proportion in the diet reduced to 10% or less of the diet DM. In addition, they indicated that low transfer efficiency (25 to 30%) of dietary protein to milk was a major factor accounting for the inability of diet to markedly alter milk protein content. Therefore, large variations or low responses to dietary manipulation resulted in low predictability for milk fat and protein content with dietary factors.

A further analysis was done to develop equations with variables that can be easily measured in the field for predicting MPY and MFY. DMI alone explained 27% of variations in MFY (Figure 1) and 35% of variations in MPY (Figure 2). There was negative correlation between MPY and FpDM (Figure 3), which was consistent with other report (DePeters and Cant, [4]). DePeters and Cant [4] indicated that the negative effect of forage content in the diet on MPY was related with reduction in energy density of the diet. Positive correlations between both the amount and the concentration of metabolizable energy and either milk protein content or protein yield were observed (Sporndly, [14]). However, an increase in energy intake by adding supplementary fat in the diet normally resulted in reduced milk protein content (Emery, [15]).

Using a random coefficient model with study as a random effect, we obtained 0.311 (±0.072) + 0.041 (±0.003) DMI (n = 442, −2 Res log likelihood = − 499) and 0.497 (±0.084) + 0.032 (±0.002) DMI - 0.003 (±0.001) FpDM (n = 442, −2 Res log likelihood = −721) for predicting MFY and MPY, respectively (Table 2). Unlike the model presented in NRC [11], CP content in the diet or CP intake was not significant variable to predict MPY. Using a simple linear regression, the best equations for predicting MFY and MPY were MFY = 0.351 (±0.068) + 0.038 (±0.003) DMI (R^2 = 0.27, root mean square error (RMSE) = 0.22),

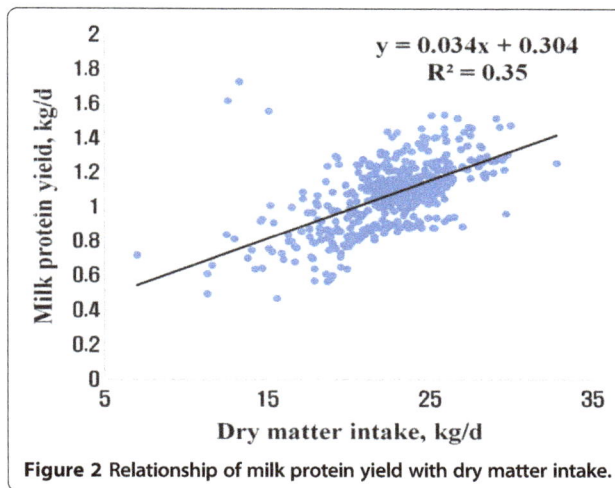

Figure 2 Relationship of milk protein yield with dry matter intake.

and MPY = 0.552 (±0.071) + 0.031 (±0.002) DMI - 0.004 (±0.001) FpDM (R^2 = 0.38, RMSE = 0.16), respectively.

The relationship of MCY with diet composition variables and intake of specific nutrients in dairy cows was also reported by previous studies (Sutton, [2], Grummer, [3], DePeters and Cant, [4], Palmquist et al., [5], Hristov et al., [6,7], Jenkins and McGuire, [1], Huhtanen and Hristov, [8]).The supplementation of lipids at up to 6 to 8% in the diet DM generally increases milk yields but the response in milk fat content varies widely, so that milk fat yield usually remains unchanged or increases (Macleod and Wood, [16], Van der Honing et al., [17]). Rook et al. [18] reported that DMI, dietary NDF concentration, and digestibility of dietary OM were important in predicting MPY. Sporndly [14] observed crude protein intake was correlated positively with milk protein content (r =0.25) and protein yield (r = 0.69). In addition

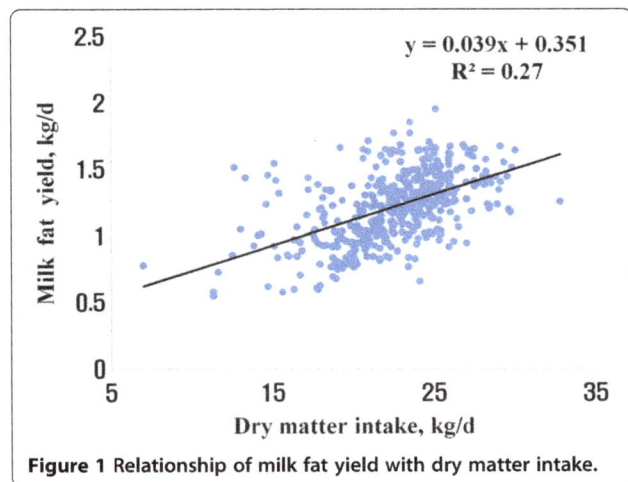

Figure 1 Relationship of milk fat yield with dry matter intake.

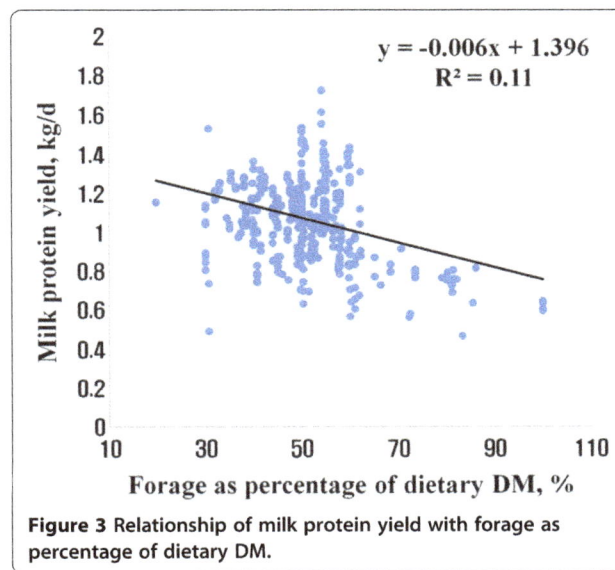

Figure 3 Relationship of milk protein yield with forage as percentage of dietary DM.

Table 2 Estimates, standard errors and significance for the random components of the final candidate models for milk component yield (kg/d; n = 442)

Item*	Parameter	Estimate	SE	P > |z|
MFY	Intercept	0.311	0.072	<0.0001
	DMI	0.041	0.003	0.0014
MPY	Intercept	0.497	0.084	<0.0001
	DMI	0.032	0.002	<0.0001
	FpDM	−0.003	0.001	0.0101

*MFY; milk fat yield, MPY; milk protein yield.

to dietary CP, estimated MP (NRC, [11]), intakes of RDP or RUP were significant predictors for MPY in dairy cows (Smoler et al., [9], NRC, [11], Hristov et al., [7]). Hristov et al. [7] showed the negative correlation of MPY with fermentable NDF and protein fraction B1 intakes. Diets for lactating dairy cows were formulated to be highly digestible and, in most situations, DMI was strongly related to intake of total digestible nutrients or energy (net energy for lactation or metabolizable energy) intake (Hristov et al., [19]). In Huhtanen and Hristov [8] the best prediction models for MPY were based on total digestible nutrients (TDN), CP intake and CP degradability. Hristov et al. [6] also suggested MY and MPY in dairy cows can be better predicted based on intake of individual nutrients than DMI alone.

Conclusions

Because genetic traits as well as production performance of dairy cows have been improved, a new development of equations for predicting MCY on the basis of animal parameters and/or feed characteristics was needed. Therefore, the recent experimental data obtained from 10 years before was used to re-evaluate major factors affecting MCY through meta-analysis in this study. We concluded that MFY and MPY can be predicted by DMI and FpDM which can easily be measured in field. However, predictability was relatively low. There was no variable or a combination of variables that were routinely measured and can be used to predict milk composition. More researches are needed to identify other variables that can predict milk component yield and milk composition more accurately.

Methods
Database construction

The database consisted of a total of 442 treatment means from 118 studies published in the *Journal of Dairy Science* from Oct., 2007 till May, 2010 (volumes 90 through 93). The detailed descriptive statistics of the database was described in Table 1. The average (±SD) body weight (BW) of the cows involved in this study was 638.2 ± 53.41 kg with a minimum of 443 kg and a maximum of 778 kg. The average DMI were 22.53 ± 3.51 kg/d with a minimum of 7.0 and a maximum of 32.8 kg/d. In nutrient composition of diets, average DM, CP, EE, NDF, RDP and RUP were 54.9 ± 9.58%, 17.4 ± 1.89%, 4.4 ± 1.30%, 33.2 ± 4.81%, 10.6 ± 1.17% and 7.4 ± 5.20%, respectively. Average MY was 34.53 ± 6.76 kg/d with 3.57 ± 0.51% of average milk fat content and 3.14 ± 0.37% of average milk protein content. Average MFY and MPY were 1.22 ± 0.26 and 1.06 ± 0.22 kg, respectively.

Model development and statistical analyses

The independent variables used for explaining the variations in MCY (i.e. MPY, MFY, milk protein content in milk, and milk fat content in milk) were DMI (kg/d), BW (kg), DMIpBW (%, dry matter intake as a percentage of BW), FpDM (%, forage as a percentage of dietary DM), NDF (% DM), CP (% DM), starch (% DM), ADF (% DM), CP intake (kg/d), NDF intake (kg/d), starch intake (kg/d), and forage intake (kg/d). Among these variables, four to five predictive variables that sufficiently explained the variations in each dependent variable were selected using step-wise regression.

The regression equations were developed in two phases. In the first phase, a random coefficients model was used using the MIXED procedure of SAS (SAS Institute, Inc., Cary, U.S.A) with studies as a random variable to identify independent variables that were statistically significant for each equation.

$$y = X\beta + Zu + e$$

where y is the vector of observed MCY of which size is N; X the n × p matrix of $x_{i,j}$; β the fixed effects parameter vector of which size is p; Z the designed N × (s × p) matrix that was blocked diagonally corresponding to each study (n_i × p) to account for the random effect of each study; e the unknown vector of independent, identically and normally distributed random errors with mean 0 and variance σ^2; N the total number of observations; s the number of studies; n_i the number of observations that ith study; p the number of parameters, which is one (intercept) plus the number of variables used in the equation; and u and e are the normally distributed.

Among the acceptable regression models that had a linear combination of significant fixed effect variables, a model that had the lowest value of −2 restricted log likelihood, Akaike information criterion (AIC), the corrected Akaike information criterion (AICc) and Schwarz's Bayesian criterion (SBC) was selected. The lowest value of those criteria above indicates a better model considering the number of observations, the number of parameters, and the maximum likelihood estimates. In the second phase, the parameters of the variables in the best model for each dependent variable, identified in the first phase, were estimated by fitting the prediction equations to a multiple regression model using GLM procedure of SAS (SAS Institute, Inc., Cary, U.S.A) [20].

Competing interests
The authors declare that they have no competing interests.

Authors' contributions
SS and JSL participated in the design of the study and performed the statistical analysis and the questionnaire. KSK reviewed the database construction. SYL reviewed the discussion section. JS helped to draft the manuscript. All authors read and approved the final manuscript.

Acknowledgement
This study was financially supported by research fund of Chungnam National University.

Author details
[1]Department of Animal Biosystem Sciences, College of Agriculture and Life Science, Chungnam National University, Daejeon 305-764, Republic of Korea. [2]Department of Animal Husbandry, Cheonan Yonam College, Cheonan 330-802, Korea. [3]National Institute of Animal Science, RDA, Cheonan 330-801, Korea.

References
1. Jenkins TC, McGuire MA: **Major advances in nutrition: Impact on milk composition.** *J Dairy Sci* 2006, **89:**1302–1310.
2. Sutton JD: **Altering milk-composition by feeding.** *J Dairy Sci* 1989, **72:**2801–2814.
3. Grummer RR: **Effect of feed on the composition of milk-fat.** *J Dairy Sci* 1991, **74:**3244–3257.
4. DePeters EJ, Cant JP: **Nutritional factors influencing the nitrogen composition of bovine milk: A review.** *J Dairy Sci* 1992, **75:**2043–2070.
5. Palmquist DL, Beaulieu AD, Barbano DM: **Feed and animal factors influencing milk-fat composition.** *J Dairy Sci* 1993, **76:**1753–1771.
6. Hristov AN, Price WJ, Shafii B: **A meta-analysis examining the relationship among dietary factors, dry matter intake, and milk and milk protein yield in dairy cows.** *J Dairy Sci* 2004, **87:**2184–2196.
7. Hristov AN, Price WJ, Shafii B: **A meta-analysis on the relationship between intake of nutrients and body weight with milk volume and milk protein yield in dairy cows.** *J Dairy Sci* 2005, **88:**2860–2869.
8. Huhtanen P, Hristov AN: **A meta-analysis of the effects of dietary protein concentration and degradability on milk protein yield and milk N efficiency in dairy cows.** *J Dairy Sci* 2009, **92:**3222–3232.
9. Smoler E, Rook AJ, Sutton JD, Beever DE: **Prediction of milk protein concentration from elements of the metabolizable protein system.** *J Dairy Sci* 1998, **81:**1619–1623.
10. Hristov AN, Hristova KA, Price WJ: **Relationship between dry matter intake, body weight, and milk yield in dairy cows: A summary of published data.** *J Dairy Sci* 2000, **83**(Suppl.1):260.
11. NRC: *Nutrient Requirements of Dairy Cattle.* 7th edition. Washington, DC, USA: National Academy Press; 2001.
12. SAS, Institute Inc: *User's guide: Statistics, Version 9th ed.* Cary, NC: SAS Institute, Inc.; 2002.
13. Thomas PC, Martin PA: **The influence of nutrient balance on milk yield and composition.** In *Nutrition and Lactation in the Dairy Cow.* Edited by. London, England: Butterworths; 1988.
14. Smith NE, Dunkley WL, Franke AA: **Effects of feeding protected tallow to dairy cows in early lactation.** *J Dairy Sci* 1978, **61:**747–756.
15. Sporndly E: **Effects of diet on milk composition and yield of dairy cows with specical emphasis on milk protein content.** *Swed J Agric Res* 1989, **19:**99–106.
16. Emery RS: **Feeding for increased milk protein.** *J Dairy Sci* 1978, **61:**825–828.
17. Macleod GK, Wood AS: **Influence of amount and degree of saturation of dietary fat on yield and quality of milk.** *J Dairy Sci* 1972, **55:**439–445.
18. Van der Honing Y, Weiman BJ, Steg A, Van Donselaar B: **The effect of fat supplementation of concentrates on digestion and utilization of energy by productive dairy cows.** *Neth J Agric Sci* 1981, **29:**79–85.
19. Rook AJ, Sutton JD, France J: **Prediction of the yields of milk constituents in dairy-cows offered silage ad-libitum and concentrates at a flat rate.** *Anim Prod* 1992, **54:**313–322.
20. Hristov AN, Price WJ, Shafii B: *An Overview of Dietary Factors Influencing Dry Matter Intake and Milk and Milk Protein Yields In Dairy Cows*, Proceedings of Pacific Northwest Animal Nutrition Conference. 2002:147–165.

21

Lesser known indigenous vegetables as potential natural egg colourant in laying chickens

Bolu Steven Abiodun*, Aderibigbe Simeon Adedeji and Elegbeleye Abiodun

Abstract

Background: A six-week study involving two hundred and fifty (250) Harco Black layer birds at point of lay was conducted to investigate the effects of potential natural colorant on performance and Egg quality traits. The birds were assigned to five (5) dietary treatments, each containing supplements either of control, Baobab Leaf (BL), Waterleaf (WL), Red Pepper (RP), Canthaxanthin (CTX) at 40 g/kg feed and 50 mg/kg feed of natural and commercial colorants, respectively.

Results: Performance records shows that there was no significant ($p > 0.05$) difference in feed intake across the supplements of Red pepper, Water leaf, Canthaxanthin and control diet, however, birds fed Baobab leaf treatment had a significantly lower ($p < 0.05$) feed intake value (94.07 g) when compared with other treatments. Body weight gain and Hen Day Production were not significant influenced ($p > 0.05$) by the dietary treatments, although laying hens fed Baobab leaf supplement had lowest mean HDP of 48.80%, while birds fed Red pepper and Water leaf supplement had an average value of 52.79%. There was no significant effect ($p > 0.05$) of colorants on egg external traits, compared with the control; birds fed Canthaxanthin treatment had higher mean egg weight (51.79 g), egg length (4.55 g), egg breadth (3.29 g); Red pepper treatment had highest mean shell thickness (0.29 g), however these differences were not significant ($p > 0.05$). Yolk height, Albumen height, Yolk index, and Haugh unit were not significantly affected ($p > 0.05$) across treatments. Yolk width was lowest ($p < 0.05$) in Baobab leaf treatment (2.54 cm); Red pepper, Water leaf and Canthaxanthin (2.89 cm, 2.62 cm and 2.89 cm respectively) were not significantly ($p > 0.05$) different from the control (2.73 cm). Yolk colour score was significantly highest ($p < 0.05$) in Red pepper treatment (7.50); Water leaf, Baobab leaf and Canthaxanthin ranged between 2.25- 3.31 on the DSM yolk colour fan, Control treatment had the lowest yolk colour score ($p < 0.05$) of 1.31.

Conclusion: The study showed Red pepper as a worthy alternative to commercial yolk colorant. Water leaf and baobab are not good substitutes for canthaxanthin as a yolk colourant.

Keyword: Canthaxanthin, Egg yolk colorant, Waterleaf, Red pepper, Baobab leaf

Background

Eggs are used in various food industries in the manufacture of confectionery cosmetics, production of vaccine as reported by [1]. Egg yolk colour is a major concern to consumers as it affects their purchasing behavior [2]. The colour of the egg yolk is considered to be one of the important factors for egg consumption. Consumers select eggs based on the egg yolk color and other egg qualities. However, laying birds cannot produce coloring pigments that is known to enhance yolk coloration to meet the demand of consumer for the desired colored

egg yolk and egg quality and thus, depends directly on dietary supply for egg yolk coloration [3]. Indeed, the perception of the intensity of the yolk colour depends directly on the quantity of feed consumed, the transfer efficacy and chemical composition of the carotenoid source [4]. The efficiency of pigment source depends on the digestibility, transfer, metabolism, and deposition of carotenoids in target tissue and upon their colour hue [5]. The most effective carotenoids in poultry diets are the synthetic forms which have been manufactured to ensure high transfer efficiency and colouring capacity (apoester: 50-60%; canthaxanthin 30-50%) [6]. The main vegetable sources of carotenoids are corn, corn gluten, lucerne, lucerne concentrates and flower (marigolds,

* Correspondence: biodunbolu@gmail.com
Department of Animal Production, University of Ilorin, Ilorin, Nigeria

tagetes) and plant (paprika) extracts. Synthetic oxycarotenoids which corresponded to natural carotenoids (canthaxanthin, citranaxathin) have been chosen in the laying hen rations for their colouring effectiveness, (which is two to three times more efficient as a yolk colorant than carotenoids of vegetable origin), and for their high stability (due to encapsulation against oxidation and degradation) [7]. These products are, however, banned in organic production, which rely mainly on yellower plant sources, explaining the paler yolk colouration found in this system of poultry production. Canthaxanthin has also been reported as a potential skin and eye irritant, and it is an expensive source of yolk colorant [8]. In practice, satisfactory colour of table eggs can be obtained with small amounts of yellow xanthophylls (15–25 mg/kg) combined with 1–2 mg/kg red carotenoids.

The attractive red color of peppers, yellow color of ripe pawpaw and the green pigment of vegetables (*Talinium triangulare, Telfairia occidentalis, Adansonia digitata, etc.*) is due to their various carotenoid pigment. These carotenoids include; capsanthin, capsorubin and cryptocapsin [8]. Maize (lutein), Red pepper (capsanthin and capsorubin), Fluted pumpkin, Pawpaw (cryptoxanthin), Sweet potato leaf and Marigold (lutein) contain colorants that can be used as feed supplements in diet of laying birds and are known to improve egg quality and yolk pigmentation.

Vegetable sources provide mainly yellow carotenoids. Paprika oleoresin is prepared from the dried fruit of *red pepper* by a similar process to *tagetes* involving dehydration, solvent extraction, saponification and stabilization are useful carotenoids for egg yolk coloration. Its effectiveness ratios relative to canthaxanthin varies from 3 to 5 for yolk deposition yield and 1 to 4 for pigmentation ability. Extract from algae containing Asthaxanthin (3,3'-dihydroxycanthaxanthin) has been used in combination with yellow xanthophylls for egg yolk coloration, but the rate of deposition of carotenoids in the egg yolk is lower than that of canthaxanthin [9].

This study focused on the use of lesser known indigenous vegetables as natural pigmenting plants; Red pepper (*Capsicum annuum*), Water leaf (*Talinum triangulare*) and Baobab leaf (*Adansonia digitata*), with the aim of improving egg quality traits through egg yolk coloration.

Method

Experimental animals and management

The experiment was carried out at the Teaching and Research farm, Faculty of Agriculture, University of Ilorin. Two hundred and Fifty (250) twenty-week old Black Harco layers were used for the experiment that lasted six weeks. The laying hens were housed in a 2-tiers battery cage system. Diet was formulated to meet NRC [10] requirements for energy and protein of laying birds.

Each natural plant colorants (Red Pepper, Baobab Leaf and Waterleaf) and a commercial yolk colorant (Canthaxanthin) were incorporated into the diet at separate rates of 40 g/kg and 50 mg/kg of feed, respectively. Other routine management practices such as vaccination, medication and proper hygiene recommended by animal science regulations in Nigeria were complied with.

The leaves of *Adansonia digitata* (Baobab) (BL), *Capsicum annuum* (Red pepper)(RP) and *Talinium triangulare* (Water leaf) (WL) were collected, destalked and washed. The leaves were air dried for four days to reduce the moisture content and achieve a constant dry weight while retaining their colours. After drying, the leaves were pulverized into fine powder using an electric blender (Moulinex Philips). A known commercial egg yolk colorant (Canthaxanthin, CTX) was procured from a commercial feedmill in Ilorin, Nigeria. Experimental Birds were randomly assigned to five treatments using the Completely Randomized design (CRD) comprising fifty (50) birds per treatments. Each treatment was replicated in five units of ten [10]. The treatments were allocated as such; Diet A containing *Capsicum annuum* as colorant at 40 g/kg of feed, Diet B containing *Talinium triangulare* as colorant at 40 g/kg of feed, Diet C containing *Adansonia digitata* as colorant at 40 g/kg of feed, Diet D containing Canthaxanthin at 50 mg/kg of feed, and Diet E (Control) diet with no yolk colorant (Table 1).

Data collection

Data were collected for performance and egg quality assessment. Average Feed Intake (AFI) was recorded daily, while Body weight gain (BWG) was recorded weekly as follows:

Table 1 Composition of experimental diet

Feed ingredients	Percentage %
White maize	45.0
Corn bran	10.0
Brewer's dried grain	10.0
Wheat offal	4.50
Fish meal (72%)	1.44
Soyabean meal	20.0
Bone meal	0.26
Oyster shell	8.00
Vitamin/mineral premix	0.25
Lysine	0.10
Methionine	0.15
Salt	0.30
Total	100

Nutrient composition according to (8).

Table 2 The relative effects of the potential natural colorant on Bird's performance

Parameter	Capsicum annuum	Talinium triangulare	Adansonia digitata	Canthaxanthin	Control	±SEM
BWG (g)	950 ± 0.06^{ab}	110 ± 0.03^{b}	470 ± 0.26^{a}	630 ± 0.15^{ab}	750 ± 0.45^{ab}	0.03
AFI (g/bird/d)	101.33 ± 3.32^{b}	103.16 ± 3.13^{b}	94.07 ± 1.37^{a}	103.56 ± 1.39^{b}	104.36 ± 2.24^{b}	0.54
HDP	52.97 ± 17.99	52.97 ± 19.09	48.80 ± 16.22	58.34 ± 18.91	58.60 ± 20.08	4.14

a, b, ab, Means across same row carrying different superscripts are significant (p < 0.05).

$$AFI = F_1 - F_0$$

F_1 = left over feed at the end of the day (g)
F_0 = feed offered in the morning (g)

$$BWG = W_1 - W_0$$

W_1 = weight change at the end of the week
W_0 = weight at the beginning of the week

Hen-Day Production (HDP) was calculated using the formula:

$$HDP = \frac{Number\ of\ eggs\ laid \times 100}{Number\ of\ days \times number\ of\ hens}$$

At the end of each week, fifty eggs per treatment were selected for analysis. Parameters measured include; Egg Weight (EW), Egg Length (EL), Egg Breadth (EB), Yolk Height (YH), Yolk Width (YW), Albumen Height (AH), Shell Thickness (ST), Yolk Colour (YC), Yolk Index (YI) and Haugh Unit (HU).

EW was collected by cleaning the eggs to remove impurities and weighing them using a sensitive electronic scale (Metler). ST was determined using a pair of micrometer screw gauge calibrated in millimetres. The accuracy of ST was ensured by measuring shell sample as one egg at the broad end, middle portion, and the narrow end all referred to as the thin, medium and thick, respectively. The average of these three parts measured was taken as the ST. YI was determined by relating the ratio of YH in millimetres to the YW measured in millimetres. The YH and YW were measured using a Spherometer and Venier calliper, respectively. YW was also taken as the maximum cross sectional diameter of the yolk which is the width at maximum point, usually across the centre of the yolk. The YC of collected eggs were determined using DSM yolk colour fan.

Albumen was separated from the yolk and carefully placed on a flat surface. The AH was measured using a Spherometer calibrated in millimetres. HU was measured relating the albumen height and the egg weight from the formula:

$$HU = 100 \times \log(h - 1.7w^{0.37} + 7.6)$$

H.U = Haugh unit
H = Observed albumen height.
W = Observed weight of egg in grams.

Statistical analysis
Data obtained from the experiment were subjected to analysis of variance for completely randomized experimental design using the SPSS package, according to [11]. Significant means were separated using the Duncan's Multiple Range Test [12].

Results and discussion
Effects of potential natural colorant on bird's performance
Birds differed significantly (P < 0.05) in AFI across the various treatments (Table 2). Diet supplemented with BL recorded a significantly lower (P < 0.05) mean intake value (94.07 g) when compared with other treatments. This could be due to the anti-nutritional agent present in baobab leaf. It has been reported [13] that high tannin content of BL prevent its use as a potential feed ingredient in poultry diets. However, BL has been used as a source of pigments in diets for laying birds where yolk colour was observed to have increased at 1% or 2% BL, with no adverse effects on feed intake, egg production and quality. Canthaxanthin has been reported to have no influence on AFI [6].

Table 3 Effects of potential natural colorant on egg external quality traits

Parameter	Capsicum annuum	Talinium triangulare	Adansonia digitata	Canthaxanthin	Control	±SEM
Egg weight	48.27 ± 1.20	47.56 ± 6.68	46.87 ± 5.20	51.79 ± 2.10	49.84 ± 2.34	0.93
Egg length	4.35 ± 0.06	4.24 ± 0.60	4.20 ± 0.46	4.55 ± 0.16	4.37 ± 0.13	0.08
Egg breadth	3.23 ± 0.09	3.08 ± 0.51	3.02 ± 0.39	3.29 ± 0.06	3.16 ± 0.30	0.07
Egg index	1.35 ± 0.48	1.38 ± 0.63	1.39 ± 0.03	1.38 ± 0.06	1.38 ± 0.05	0.01
Shell thickness	0.29 ± 0.18	0.27 ± 0.18	0.27 ± 0.44	0.28 ± 0.02	0.27 ± 0.03	0.02

Means across same row carrying different superscripts are significant (p < 0.05).

Table 4 Effects of natural colorant on internal egg quality traits

Parameter	Capsicum annuum	Talinum triangulare	Adansonia digitata	Canthaxanthin	Control	±SEM
Yolk height (mm)	14.83 ± 0.05	13.30 ± 2.07	13.75 ± 1.58	14.53 ± 0.52	14.62 ± 0.42	0.27
Yolk index	5.15 ± 0.84	4.81 ± 0.76	5.19 ± 0.86	5.04 ± 0.11	5.37 ± 0.14	0.12
Yolk width (cm)	2.89 ± 0.06	2.62 ± 0.37	2.54 ± 0.25	2.89 ± 0.12	2.73 ± 0.64	0.05
Albumen height (mm)	5.49 ± 0.21	5.06 ± 1.04	5.41 ± 1.19	5.78 ± 0.43	5.54 ± 0.30	0.17
Haugh unit	77.41 ± 1.68	76.08 ± 2.22	78.21 ± 4.03	77.54 ± 2.58	76.93 ± 2.25	0.60
Yolk color	$7.50 ± 1.47^c$	$3.31 ± 0.83^b$	$3.06 ± 0.13^b$	$2.25 ± 0.20^{ab}$	$1.31 ± 0.13^a$	0.17

a, b, ab, c Means across same row carrying different superscripts are significant (p < 0.05).

HDP increased comparatively over the period of experiment. Laying hens fed with the control diet had the highest mean value (58.60%) while the diet supplemented with BL had the lowest value (48.8%). RP supplemented diet did not influence (P > 0.05) the performance parameters measured viz: Body weight gain, feed intake, Hen-day production. The observation on HDP disagreed with the report of [6] that CTX increased egg production. CTX has been reported to have no effects on HDP [14].

There was no dietary effect of treatment (P > 0.05) on egg external quality traits (Table 3). Canthaxanthin and other dietary supplements had no significant effects on egg external quality traits such as; EW, EB, EL, EI and ST; this is consistent with the previous findings [15]. This study showed a shell thickness range of 0.27 mm-0.29 mm which is within normal range for poultry. The Egg ST is an important trait for hatchability and handling [16].

The various dietary treatments had no significant effect (P > 0.05) on egg internal qualities such as: AH, YH, YI and HU. This is also consistent with earlier reports [10].

There was however, a significant effect (P < 0.05) of dietary treatment on YC (Table 4). Laying hens fed Red pepper supplemented diet had significantly higher (P < 0.05) YC (colour index of 7.50) compared with the other treatments. However, diets supplemented with WL, BL (Adansonia digitata) and CTX had similar (P > 0.05) values for YC. CTX was reported (6) to enhance egg yolk pigmentation. The observed low YC score for CTX in this study may be related to reduced potency owing to post-importation storage. The control diet had significantly lowest YC score which further buttress the fact that yolk pigmentation is influenced mostly by diets given to the birds [17]. In this study, RP is the best natural source of potential egg colourant, this is consistent with the report of [18]. That canthaxanthin had similar score with other lesser known indigenous vegetables shows that these vegetables can be harnessed into layer diets for egg yolk colour pigmentation. BL and WL are potential good sources of xanthophylls, known to enhance yolk pigmentation. Natural xanthophylls are well absorbed by hen intestinal cells [19] and it is transferred into the yolk after being released into the circulatory system [20].

Conclusion

This study shows that lesser known indigenous vegetables are potential natural egg yolk colorants. RP supplement had superior performance in this regard over canthaxanthin (imported commercial egg yolk colourant). The period of importation and post-importation storage may have reduced the potency of CTX. The use of natural colorant such as lesser known indigenous vegetables is of utmost importance to improve yolk pigmentation and performance of birds with no adverse effect as compared to artificial colourants which could be of environmental and public health concern. Powdered RP supplementation could be used as an alternative for synthetic commercial yolk colorant which is quite expensive and potentially unhealthy. The use of yellow maize based diet with powdered WL and BL supplement can also help improve yolk color and performance.

Competing interests
The authors declare that they have no competing interests.

Authors' contributions
SA Bolu is the supervisor of the project from design to the end of the study. SA Aderibigbe is a Masters student while Elegbeleye is an undergraduate student under SA bolu. Both were involved in the collection of data and preliminary write up. All authors helped draft the manuscript, read and approved the final manuscript.

References
1. Oluyemi JA, Roberts FA: Poultry production in warm wet climates. London: Macmillan Press Ltd; 2007:PP 118–PP 120.
2. Fletcher DL: Broiler breast colour variation, pH, and texture. Poult Sci 1999, 78:1323–1327.
3. Soto-Salanova MF: Natural pigments: practical experiences. In Recent Advances in Animal Nutrition. Edited by Garnworty PC, Wiseman J. Nottingham, UK: Nothingham University Press; 2003.
4. Nys Y: Dietary carotenoids and egg yolk coloration. Archiv für Geflugelkunde 2000, 64:45–54.
5. Hamilton PB: The use of high-performance liquid chromatography for studying pigmentation. Bri Poult Sci 1992, 71:718–724.
6. Cho JH, Zhang ZF, Kim IH: Effects of Canthaxanthin on egg production, egg quality and Egg yolk color in laying hens. J Agric Sci 2014, 5(1):269–274.
7. Deepa N, Kaur C, George B, Singh B, Kapoor HC: Antioxidant constituents in some sweet pepper (Capsicum anuum L.) genotypes during maturity. LWT 2007, 40:121–128.
8. EFSA: Scientific opinion on the safety and efficacy of canthaxanthin as a feed additive for poultry and for ornamental birds and ornamental fish.

EFSA J 2014, **12**(1):3527. [24 pp.]. doi:10.2903/j.efsa.2014.3527. http://www.efsa.europa.eu/en/efsajournal/pub/3527.htm.

9. Marusich WL, Bauernfeind JC: **Oxycarotenoids in poultry feeds.** In *Carotenoids as Colorants and Vitamin A Precursors.* Edited by Bauerfeind JC. New York: Academic Press; 1981:320–462.

10. NRC: *Nutrient Requirements of Poultry,* 9th Revised edition. Washington, DC: National Academic Press; 1994.

11. Steel RGD, Torrie JH: *Principle and Procedure of Statistics.* 2nd edition. London: McGraw- Hill; 1982.

12. Duncan DB: **Multiple range and F- test.** *Biometrics* 1955, **11**:1–42.

13. Butswat IS, Nelson FN, Oyawole EO, Akande FO: **Utilization of baobab leaf meal for egg yolk pigmentation in layers.** *Ind J Animal Sci* 1997, **67**(1):82–83.

14. Kanda L, Koh-en Y, Tsutomu K, Keiko S: **Enhancement of yolk colour in raw and boiled egg yolk with lutein from marigold flower meal and marigold flower extract.** *J Poultry Sci* 2001, **48**:25–32.

15. Hasin BM, Feudaus AJM, Islam MA, Uddin MJ, Islam MS: **Marigold and orange skin as egg yolk color promoting agents.** *Int J Poult Sci* 2006, **5**:979–987.

16. Khan M, Khatun M, Kibria A: **Study the quality of eggs of different genotypes of chicken under scavenging system at Bangladesh.** *Pak J Biol Sci* 2004, **7**(12):2163–2166.

17. Colin GS, George B, Ensminger ME: *Poultry Science.* 4th edition. New Jersey: Pearson Prentice Hall, Upper Saddle River; 2004.

18. Niu Z, Fu J, Gao Y, Liu F: **Influence of paprika extract supplement on egg quality of laying hens fed wheat-based diet.** *Int J Poultry Sci* 2008, **7**:887–889. http://dx.doi.org/10.3923/ijps.2008.887.889.

19. Gouveia L, Veloso V, Reis A, Fernandes H, Novais J, Empis J: *Chlorella vularis* **used to colouregg yolk.** *J Sci Food Agr* 1996, **70**:167–172. http://dx.doi.org/10.1002/(SICI)1097-0010(199602)70:2<167::AID-JSFA472>3.0.CO;2.

20. Selma UA, Miah G, Tareq KMA, Maki T, Tsujii H: **Effects of dietary** *Rhodobacter capsulatus* **on egg-yolk cholesterol and laying hen performance.** *Poultry Sci* 2002, **86**:714–719.

22

Estimation of effective population size using single-nucleotide polymorphism (SNP) data in Jeju horse

Kyoung-Tag Do[1†], Joon-Ho Lee[2†], Hak-Kyo Lee[2], Jun Kim[3] and Kyung-Do Park[2*]

Abstract

This study was conducted to estimate the effective population size using SNPs data of 240 Jeju horses that had raced at the Jeju racing park. Of the total 61,746 genotyped autosomal SNPs, 17,320 (28.1%) SNPs (missing genotype rate of >10%, minor allele frequency of <0.05 and Hardy–Weinberg equilibrium test P-value of $<10^{-6}$) were excluded after quality control processes. SNPs on the X and Y chromosomes and genotyped individuals with missing genotype rate over 10% were also excluded, and finally, 44,426 (71.9%) SNPs were selected and used for the analysis. The measures of the LD, square of correlation coefficient (r^2) between SNP pairs, were calculated for each allele and the effective population size was determined based on r^2 measures. The polymorphism information contents (PIC) and expected heterozygosity (HE) were 0.27 and 0.34, respectively. In LD, the most rapid decline was observed over the first 1 Mb. But r^2 decreased more slowly with increasing distance and was constant after 2 Mb of distance and the decline was almost linear with log-transformed distance. The average r^2 between adjacent SNP pairs ranged from 0.20 to 0.31 in each chromosome and whole average was 0.26, while the whole average r^2 between all SNP pairs was 0.02. We observed an initial pattern of decreasing N_e and estimated values were closer to 41 at 1 ~ 5 generations ago. The effective population size (41 heads) estimated in this study seems to be large considering Jeju horse's population size (about 2,000 heads), but it should be interpreted with caution because of the technical limitations of the methods and sample size.

Keywords: Jeju horse, Linkage disequilibrium (LD), Effective population size

Background

According to the literature, horses began to be raised in Jeju Island before the Goryo Dynasty. However, historically in 1276 Mongolian Yuan Dynasty of China established a horse ranch in Jeju Island and 160 Mongolian horses were introduced to produce warhorse. Through adaptation to the harsh environment of the Jeju Island and long term isolation, Jeju horses have developed their own conformation. They have several coat colours and body size is smaller than that of Mongolian horse. Since 1960s due to the industrialization and the development of agricultural machines and means of transportation, demand for horses decreased. In 1986, dozens of Jeju

horses with pedigree registry were designated as a natural monument (No.347) because of their historical importance. In May, 2000, Livestock Promotion Agency was designated as Jeju horse registration agency and Jeju horse registration started. Currently, about 2,000 heads of Jeju horses are being raised at local ranches. Domestic animals are well suited for genetic studies, since they enable comparisons of populations exposed to different selection criteria and environmental challenges [1,2]. Jeju horses are very valuable animals to preserve historically and economically and it is very important to investigate unique genetic characteristics of Jeju horses [3,4]. Jeju horses have been isolated for more than 700 years and it is estimated that their homozygocity of genotype increased by inbreeding and genetic drift. The increase of recessive homozygosity caused inbreeding and decreased growth and reproductive performance [5,6]. Especially,

* Correspondence: doobalo@hknu.ac.kr
†Equal contributors
[2]The Animal Genomics and Breeding Center, Hankyong National University, Anseong 456-749, Republic of Korea
Full list of author information is available at the end of the article

average withers height of Jeju horse, approximately 122 cm, is shorter than that of Mongolian (140 cm).

As the rapid development of microarray technology, high density whole genome SNPs (SNP chip) became a strong tool for the researches of quantitative and population genetics. Recently, these genome-wide SNPs were commonly used for estimation of historical effective population size in livestock [7-13] and human [14,15]. Closely-linked loci give information on population sizes over historical periods of time, while loosely-linked loci estimate population sizes in the immediate past [16-18]. Using high density SNPs, LD of many SNP pairs which have either close linkage or loose linkage by the distance between SNPs can be measured and used for estimation of historical effective population size.

This experiment was conducted to investigate the LD in population level and to estimate the effective population size for systemic preservation using genomic information of Juju horses.

Material and methods
Single-nucleotide polymorphism (SNP) data
DNA samples were obtained from 240 Jeju horses (race-horses) that were randomly chosen and had raced at the Jeju racing park and they were genotyped for the initial genome-wide scan using Equine SNP70 BeadChips (Geneseek, Lincoln, NE). Genomic DNA was isolated from nasal area according to the procedure of Performagene™-LIVE-STOCK PG-AC1 Reagent Package (DNA Genotek INC, Canada). The quantity and quality of the genomic DNA was evaluated using 0.8% Agarose gel electrophosis and Nanodrop ND-100 electrophotometer. Genotyping was

performed using the InfiniumHD iselect Custom BC Neo-gen_Equine_Community_Array (Illumina, USA), which contained 65,157 SNPs across the whole genome. Geno-mestudio softwareV.2011.1.9.4 (Illumina, USA) was used to call the genotypes from the samples. The chip includes 65,157 SNPs that are uniformly distributed on the 31 equine autosomes, X and Y chromosomes from the Equ-Cab2 SNP database of the horse genome (Figure 1). We excluded the SNPs with a missing genotype rate of over 10%, minor allele frequency (MAF) of less than 0.05, and Hardy–Weinberg equilibrium (HWE) test P-value of less than 10^{-6} as a quality control procedure [13]. SNPs on the X and Y chromosomes and genotyped individuals with missing genotype rate over 10% were also excluded, remaining 44,426 autosomal SNPs from 218 heads for further analysis.

Linkage disequilibrium (LD)
The measures of the LD were square of correlation co-efficient (r^2) between SNP pairs and calculated for each allele at locus A with each allele at locus B [7,19].

$$r^2 = \frac{D^2}{P_A P_a P_B P_b} \tag{1}$$

Where $D = P_{AB} - P_A P_B$ and P_A, P_a, P_B and P_b are the frequencies of alleles A, a, B and b, respectively.

Effective population size
The effective population size was determined based on r^2 measures. Because LD breaks down more rapidly over

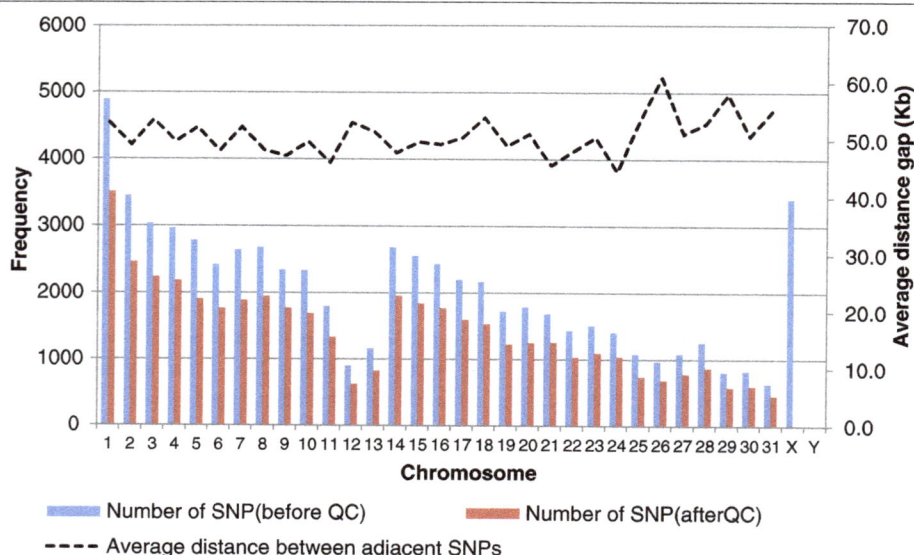

Figure 1 Number of SNPs and average distance between adjacent SNPs per chromosome after quality control processes.

generations for loci further apart, LD at large distances reflects N_e at recent generations.

$$E(r^2) = \frac{1}{1 + 4N_e c} \quad (2)$$

Where, N_e is effective population size and c is the recombination distance (in Morgans) between the SNPs. Equation (2) can be rearranged as follows [17,20-22]:

$$2_e = \frac{\left[(r_c^2)^{-1} - 1\right]}{2c} \quad (3)$$

Where, N_e is the effective population size t generations ago, c is the distance between markers in Morgans, r_c^2 is the mean value of r^2 for markers c Morgans apart, and c = $(2 t)^{-1}$. Megabase to centimorgan conversion rate was applied for generation grouping based on the result of Corbin et al. [21]. The estimation of LD measure and

Table 1 Simple statistics for single-nucleotide polymorphism (SNP) data by chromosome

| Chromosome | No. of SNPs | Mean | | | | No. of SNP pairs | Mean |
		Distance[1]	MAF[2]	HE[3]	r-square[4]		r-square[5]
1	3,509	53.0	0.24	0.33	0.25	6,154,786	0.02
2	2,463	49.1	0.25	0.34	0.29	3,031,953	0.02
3	2,234	53.5	0.25	0.34	0.27	2,494,261	0.02
4	2,185	49.7	0.24	0.33	0.28	2,386,020	0.02
5	1,909	52.2	0.25	0.34	0.26	1,821,186	0.02
6	1,762	48.1	0.24	0.33	0.25	1,551,441	0.02
7	1,887	52.2	0.24	0.33	0.28	1,779,441	0.02
8	1,946	48.2	0.25	0.34	0.27	1,892,485	0.03
9	1,767	47.2	0.26	0.34	0.29	1,560,261	0.03
10	1,688	49.7	0.24	0.33	0.27	1,423,828	0.03
11	1,332	46.0	0.24	0.34	0.27	886,446	0.03
12	622	53.0	0.25	0.34	0.21	193,131	0.03
13	826	51.4	0.24	0.33	0.20	340,725	0.03
14	1,954	47.7	0.24	0.34	0.28	1,908,081	0.03
15	1,841	49.6	0.24	0.33	0.26	1,693,720	0.02
16	1,775	49.2	0.24	0.33	0.25	1,574,425	0.02
17	1,598	50.5	0.25	0.33	0.31	1,276,003	0.03
18	1,532	53.8	0.24	0.33	0.27	1,172,746	0.02
19	1,225	48.9	0.24	0.33	0.26	749,700	0.03
20	1,252	51.0	0.25	0.34	0.26	783,126	0.02
21	1,257	45.6	0.24	0.33	0.25	789,396	0.03
22	1,036	48.1	0.24	0.34	0.24	536,130	0.02
23	1,097	50.4	0.25	0.34	0.28	601,156	0.03
24	1,039	44.5	0.25	0.34	0.25	539,241	0.03
25	737	53.2	0.24	0.33	0.25	271,216	0.03
26	684	61.0	0.24	0.33	0.22	233,586	0.03
27	778	51.1	0.25	0.34	0.25	302,253	0.03
28	867	52.9	0.24	0.33	0.27	375,411	0.03
29	579	58.0	0.24	0.33	0.23	167,331	0.03
30	593	50.7	0.24	0.33	0.23	175,528	0.03
31	452	55.1	0.25	0.33	0.20	101,926.	0.03
Overall	44,426	50.4	0.24	0.34	0.26	38,766,939	0.02

[1]Kilo base pairs (Kb) between adjacent SNPs, [2]minor allele frequency, [3]Expected heterozygosity, [4]between adjacent SNP pairs, [5]between all SNP pairs.

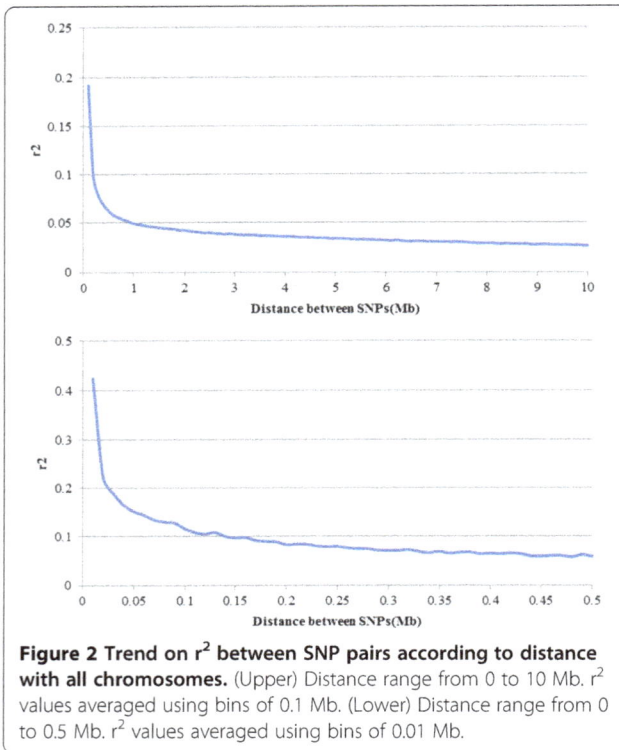

Figure 2 Trend on r² between SNP pairs according to distance with all chromosomes. (Upper) Distance range from 0 to 10 Mb. r² values averaged using bins of 0.1 Mb. (Lower) Distance range from 0 to 0.5 Mb. r² values averaged using bins of 0.01 Mb.

effective population size was used programs that we developed by GNU Fortran.

Results and discussion

Single-nucleotide polymorphism (SNP) data

Of the total 61,746 genotyped autosomal SNPs, 17,320 (28.1%) SNPs were excluded after quality control

processes (missing genotype rate of >10%, minor allele frequency of <0.05 and Hardy–Weinberg equilibrium test P-value of <10^{-6}) and finally, 44,426 (71.9%) SNPs were selected and used for the analysis. The minor allele frequencies (MAF) in each chromosome followed a uniform distribution and averaged to be 0.24 and the average χ^2 value (p-value) of Hardy-Weinberg disequilibrium (HWE) test, polymorphism information contents (PIC) and expected heterozygosity (HE) were 1.32 (0.25), 0.27 and 0.34, respectively. The number of SNPs per autosome ranged from 452 to 3,509 and average distance between adjacent SNPs was 50.4 kb (Table 1), and their relationships are shown in Figure 1. The frequency of adjacent SNP pairs which are aparted between 10 Mb (Mega base pairs = 1,000,000 bp) and 100 Mb was 27,289 (61.4%), and that of adjacent SNP pairs less than 10 Mb was 14,764 (24.9%).

Linkage disequilibrium (LD)

The results of this study provide an overview of LD in the Jeju Horse using a high density SNP panel. Linkage disequilibrium decreased with increasing distance between SNP pairs (Figure 2) and the most rapid decline was observed over the first 1 Mb. But r² decreased more slowly with increasing distance and was constant after 2 Mb of distance and the decline in LD was almost linear with log-transformed distance [21]. The average r² between adjacent SNP pairs ranged from 0.20 to 0.31 in each chromosome and whole average was 0.26, while the whole average r² between all SNP pairs was 0.02 (Table 1).

According to reports [21,23] in a sample of 817 and 24 Thoroughbred horses, LD in r² decreased from 0.6 to

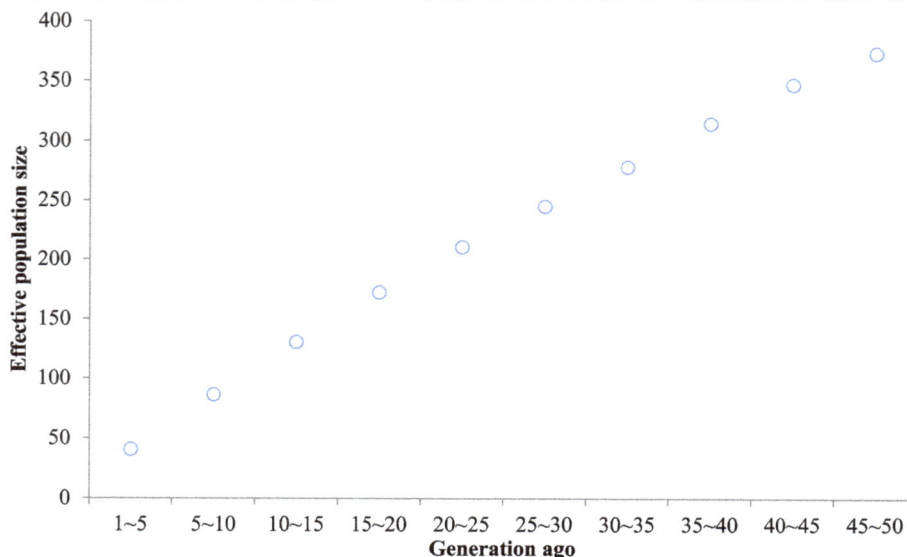

Figure 3 Effective population size (Nₑ) plotted against generations in the past, truncated at 50 generations.

0.2 when the distance between markers increased to 0.5 Mb. The pattern of decline of LD with distance in our population was similar (Figure 2), but the LD observed was lower (0.49 ~ 0.07) when compared with other reports [21,23].

Validation work by Corbin et al. [21] on their Thoroughbred (817 head) data suggests that our sample size of 218 heads is more accurate to obtain an unbiased result of LD in our population. On the other hand, the pattern and magnitude of decline of LD with distance at less than 10 Mb were almost similar and linkage disequilibrium declined more slowly in Jeju horse population than in Thoroughbred populations [21].

Effective population size

We observed an initial pattern of decreasing N_e and estimated values were closer to 41 at 1 ~ 5 generations ago (Figure 3). This result is in agreement with the previous approach [17] by calculating historical N_e, assuming linear population growth. The observed pattern showed a decrease in N_e upto around 1 ~ 5 generations. Corbin et al. [21] reported the effective population size (N_e) was estimated to be 100 heads at 20 generations in Thoroughbreds and Cunningham et al. [24] calculated the effective number of studbook founders of the Thoroughbred to be 28.2 from pedigree analyses.

The 41 heads (N_e) estimated in this study seems to be large considering Jeju horse's population size. Currently, there is about 2,000 Jeju horses in Jeju Island and it may be difficult to interpret inflated N_e. There may be a few speculations, such as an immigration event, a hybridization event or any combination of these. Therefore, it is useful to consider our observation in the context of what is known about the demographic history of Jeju horses. In 1986, 150 Jeju horses with pedigree registry were designated as a natural monument (No.347). In October, 1990, Jeju horse racing park was open and Jeju horse racing started and the names of various horses raised in Jeju was unified to Jeju horse. As the sales of Jeju horse racing park increased, the demand for Jeju horse increased and since the horses raised at ranches were selected as basic registered horses and included to Jeju horse management system, bloods of other breeds might be introduced.

On the other hand, since intensive selection for racing performance of Throughbred has been conducted for long period, the effective population size of Throughbred can be relatively small. However, for Jeju horse, fundamental effective population size can be larger than that of Throughbred since almost no selection has been conducted for Jeju horses. The effective population size (41 heads) estimated at 1 ~ 5 generations should be interpreted with caution because of the technical limitations of the methods and sample size.

Conclusions

Jeju horses are very valuable animals to preserve historically and economically and it is very important to investigate unique genetic characteristics of Jeju horses for the stable maintenance. Also, we should make efforts to prevent inbreeding coefficient increase and to increase effective population size through the reduction of generation interval.

Competing interests
The authors declare that they have no competing interests.

Authors' contributions
K-TD and H-K wrote the manuscript of this paper and the manuscript was revised by K-DP. JK collected DNA samples from the Jeju horses and SNP chip data, and J-HL performed statistical analysis. All authors read and approved the final manuscript.

Acknowledgments
This work was supported by the National Research Foundation of Korea Grant funded by the Korean Government (NRF-2013R1A1A2012586). We are grateful to Jeju horse breeder's Association for helping us.

Author details
[1]Department of Equine Sciences, Sorabol College, Gyeongju 780-711, Republic of Korea. [2]The Animal Genomics and Breeding Center, Hankyong National University, Anseong 456-749, Republic of Korea. [3]Provincial Livestock Promotion, Jeju 690-802, Republic of Korea.

References
1. Bergman IM, Rosengren KJ, Edman K, Edfors I: European wild boars and domestic pigs display different polymorphic patterns in the Toll-like receptor (TLR) 1, TLR2, and TLR6 genes. Immunogenetics 2010, 62:49–58.
2. Kawahara-Miki R, Tsuda K, Shiwa Y, Arai-Kichise Y, Matsumoto T, Kanesaki Y, Oda S, Ebihara S, Yajima S, Yoshikawa H, Kono T: Whole-genome resequencing shows numerous genes with nonsynonymous SNPs in the Japanese native cattle Kuchinoshima-Ushi. BMC Genomics 2011, 12:103.
3. Kim KI, Yang YH, Lee SS, Park C, Ma R, Bouzat JL, Lewin HA: Phylogenetic relationships of Cheju horses to other horse breeds as determined by mtDNA Dloop sequence polymorphism. Anim Genet 1999, 30(2):102–108.
4. Yang YH, Kim KI, Cothran EG, Flannery AR: Genetic diversity of Cheju horses (equus caballus) determined by using mitochondrial DNA D-loop polymorphism. Biochem Genet 2002, 40:175–186.
5. Weigel KA: Controlling inbreeding in modern breeding programs. J Dairy Sci 2001, 84:E177–E184.
6. Wall E, Brotherstone S, Kearney JF, Woolliams JA, Coffey MP: Impact of nonadditive genetic effects in the estimation of breeding values for fertility and correlated traits. J Dairy Sci 2005, 88:376–385.
7. Alam M, Han KI, Lee DH, Ha JH, Kim JJ: Estimation of effective population size in the sapsaree: a Korean native dog (canis familiaris). Asian-Australas J Anim Sci 2012, 25(8):1063–1072.
8. Garcia-Gamez E, Sahana G, Gutierrez-Gil B, Arranz JJ: Linkage disequilibrium and inbreeding estimation in Spanish Churra sheep. BMC Genet 2012, 13:43–54.
9. Kim ES, Kirkpatric BW: Linkage disequilibrium in the North American Holstein population. Anim Genet 2009, 40:279–288.
10. Lee YS, Lee JW, Kim HB: Estimating effective population size of thoroughbred horses using linkage disequilibrium and theta (4Nμ) value. Livest Sci 2014, 168:32–37.
11. Qanbari S, Pimentel ECG, Tetens J, Thaller G, Lichtner P, Sharifi AR, Simianer H: The pattern of linkage disequilibrium in German Holstein cattle. Anim Genet 2010, 41:346–356.
12. Qanbari S, Hansen M, Weigend S, Preisinger R, Simianer H: Linkage disequilibrium reveals different demographic history in egg laying chickens. BMC Genet 2010, 11:103–113.

13. Uimari P, Tapop M: **Extent of linkage disequilibrium and effective population size in Finnish Landrace and Finnish Yorkshire pig breeds.** *J Anim Sci* 2010, **89**:609–614.

14. Park LY: **Effective population size of current human population.** *Genet Res (Camb)* 2011, **93**:105–114.

15. Tenesa A, Navarro P, Hayes BJ: **Recent human effective population size estimated from linkage disequilibrium.** *Genome Res* 2007, **17**:520–526.

16. Hill WG: **Estimation of effective population size from data on linkage disequilibrium in finite populations.** *Genet Res* 1981, **38**:209–216.

17. Hayes BJ, Visscher PM, McPartlan HC, Goddard ME: **Novel multilocus measure of linkage disequilibrium to estimate past effective population size.** *Genome Res* 2003, **13**:635–643.

18. Sved JA, Cameron EC, Gilchrist CA: **Estimating effective population size from linkage disequilibrium between unlinked loci: theory and application to fruit Fly outbreak populations.** *Genome Res* 2003, **13**:635–643.

19. Hill WG, Robertson A: **Linkage disequilibrium in finite populations.** *Theor Appl Genet* 1968, **38**:226–231.

20. De Roos A, Hayes BJ, Spelman R, Goddard ME: **Linkage disequilibrium and persistence of phase in Holstein-Friesian, Jersey and Angus cattle.** *Genetics* 2008, **179**:1503–1512.

21. Corbin L, Blott S, Swinburne J, Vaudin M, Bishop S, Woolliams J: **Linkage disequilibrium and historical effective population size in the Thoroughbred horse.** *Anim Genet* 2010, **41**:8–15.

22. Shin DH, Cho KH, Park KD, Lee HJ, Kim HB: **Accurate estimation of effective population size in the Korean dairy cattle based on linkage disequilibrium corrected by genomic relationship matrix.** *Asian-Australas J Anim Sci* 2013, **26**:1672–1679.

23. Wade CM, Giulotto E, Sigurdsson S, Zoli M, Gnerre S, Imsland F, Lear TL, Adelson DL, Bailey E, Bellone RR, Blöcker H, Distl O, Edgar RC, Garber M, Leeb T, Mauceli E, MacLeod JN, Penedo MC, Raison JM, Sharpe T, Vogel J, Andersson L, Antczak DF, Biagi T, Binns MM, Chowdhary BP, Coleman SJ, Della Valle G, Fryc S, Guérin G, *et al*: **Genome sequence, comparative analysis, and population genetics of the domestic horse.** *Science* 2009, **326**:865–867.

24. Cunningham EP, Dooley JJ, Splan RK, Bradley DG: **Microsatellite diversity, pedigree relatedness and the contributions of founder lineages to thoroughbred horses.** *Anim Genet* 2001, **32**:360–364.

23

Effects of repeated tuberculin skin testings on immune responses in experimental monkeys

Fangui Min[1,2], Jing Wang[1,2], Wen Yuan[1,2], Huiwen Kuang[1,2] and Weibo Zhao[1,2*]

Abstract

Though many alternative methods to tuberculin skin testing (TST) have been established and evaluated in recent years, sensitivities and specificities of most methods could not meet the requirements of golden standards. In this study, we sought to identify whether repeated TSTs could affect the immune responses in experimental monkeys. Nine natural tuberculosis (TB) monkeys receiving repeated TSTs biweekly were used to demonstrate the effect on TST responsiveness. Two healthy monkeys were administrated with repeated TSTs to analyze the immune response profiling. Intrapalpebral reactions in TB infections gradually weakened or presented intermittent positive reactions. The leukocyte counts, cytokine responses, and antibody responses to all antigens except Old tuberculin (OT) and MPT64L showed no specific changes for TB in healthy monkeys. Positive antibody responses to OT and MPT64L emerged during the first half experimental period, which may cause by their cross-reactivity with mycobacterial species. Results showed that repeated TSTs had no significant effects on immune responses in healthy monkeys but a progressive reduction in TST responsiveness in TB infections.

Keywords: Tuberculin skin testing, Tuberculosis, Monkeys, Immune responses

Background

Primate tuberculosis (TB) is a zoonotic disease caused primarily by the bacterial pathogen *Mycobacterium tuberculosis* or, less commonly, *M. bovis*. Though most nonhuman primate species are susceptible to TB infection, Old World monkeys are considered more susceptible than New World monkeys [1,2]. The commonly used rhesus (*Macaca mulatta*) and cynomolgous (*Macaca fasicularis*) monkeys in researches belong to Old World monkeys. So the TB surveillance in experimental monkeys becomes a severe challenge.

Diagnosis of primate TB is difficult for the nonspecific clinical features and difficulty in interpreting chest radiographs. Additionally, the clinical features and chest radiographs do not give conclusive evidence for the disease. The golden standard for human TB diagnosis based on bacterial culture of clinical specimens is often not possible in experimental monkeys due to difficulty in collection of sputum specimens. Despite significant improvement in technology and the availability of *M. tuberculosis* genomic sequence data has facilitated the development of newer diagnostic techniques of TB, most of them are still under evaluation or remain limitations [3]. The palpebral tuberculin skin test (TST) has been the mainstay of TB diagnosis in nonhuman primates and the only approved method for TB screening of animals in primary import quarantine. Since the TST is an in vivo test detecting the delayed-type hypersensitivity to tuberculin antigens, the administration of mycobacterial derived antigens may influence the immune response and subsequent immunodiagnostic tests. These were demonstrated in cattle naturally infected with *M. bovis* [4], the repeat short-interval skin-testing could affect not only the skin-test responsiveness but also gamma interferon test.

In present study, we performed a longitudinal investigation to make clear whether repeated TSTs could affect TST responsiveness in TB infections, leukocyte counts, serum cytokines and antibody responses in health monkeys. Results have demonstrated that the repeated TST shows no effect on immune response in TB-negative monkeys but a progressive reduction in TST responsiveness in TB infections.

* Correspondence: 183775444@qq.com
[1]Guangdong Laboratory Animals Monitoring Institute, Guangzhou 510663, PR China
[2]Guangdong Provincial Key Laboratory of Laboratory Animals, Guangzhou 510663, PR China

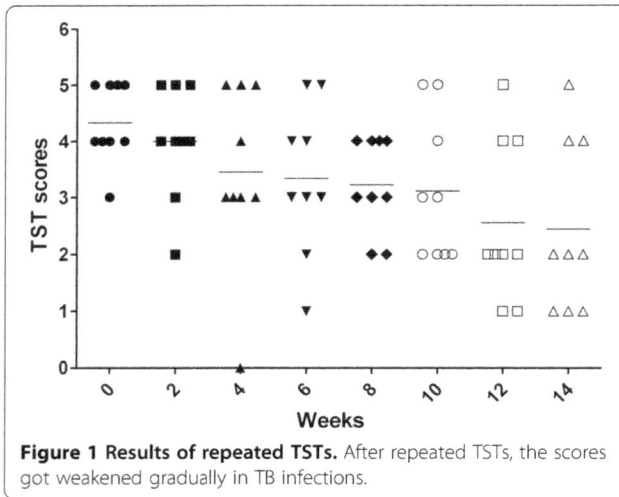

Figure 1 Results of repeated TSTs. After repeated TSTs, the scores got weakened gradually in TB infections.

Methods

Animals

Nine TB-positive rhesus monkeys with at least one time of TST-positive reaction by routine quarantines and two healthy monkeys (06-1885R and 06-1891R) aged 3 4 years were used in this study. The healthy monkeys were tested negative for monkey B virus, simian immunodeficiency virus (SIV) and simian T-cell leukemia virus 1 (STLV-1) by ELISA and simian retrovirus (SRV) by immunofluorescence. Animal work was conducted in a negative air pressure facility.

Animal use protocols were reviewed and approved by the Institutional Animal Care and Use Committee of Guangdong Laboratory Animal Monitoring Institute in accordance with the *Guide for the Care and Use of Laboratory Animals* [5].

Study design

At the interval of 2 weeks, all monkeys were anesthetized intramuscularly with ketamine in combination with xylazine for administration of TSTs and blood samples collection.

TST procedures

Two kinds of tuberculin, Old tuberculin (OT) and purified protein derivative (PPD), were used in TSTs. OT was from *Synbiotics Corp.* and PPD was from *Harbin Pharmaceutical Group Bio-vaccine Co.* In this study, intradermal palpebral skin testing was performed using 0.1 mL of OT in right palpebra and 0.1 mL of PPD in left palpebra respectively biweekly. TST results were graded at 24, 48, and 72 h using the standard 1 to 5 scoring system [6], and in this study, palpebral reactions of grade 0 ~ 2 were considered negative, those of grade 2 ~ 3 were suspect, and those graded ≥3 were considered positive.

Total leukocyte and leukocyte population counts

For total leukocyte and leukocyte population counts, about 0.5 ml of heparinized blood samples were assayed on automated hematology analyzer (Sysmex XT-2000iv, Japan).

Serum antibody detection

Serum antibody responses to 10 *M. tuberculosis* purified recombinant antigens (Ag85b, CFP10, ESAT-6, CFP10-ESAT-6, 38 kDa, 14 kDa, MPT64L, 16 kDa, TB16.3, and U1 protein), PPD, and OT were detected by ELISA as previous report [7].

Serum cytokine detection

Serum cytokines (TNF-α, IL-8, IL-12/23p40, IFN-γ, IL-2, IL-4, IL-6, and IL-10) were measured by ELISA as

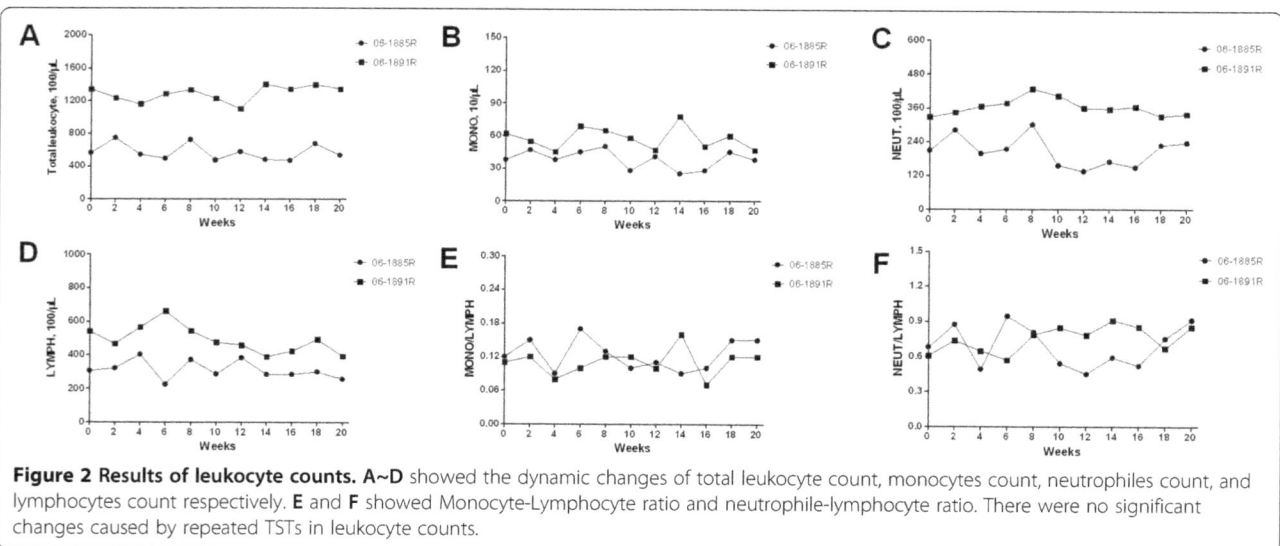

Figure 2 Results of leukocyte counts. A~D showed the dynamic changes of total leukocyte count, monocytes count, neutrophiles count, and lymphocytes count respectively. **E** and **F** showed Monocyte-Lymphocyte ratio and neutrophile-lymphocyte ratio. There were no significant changes caused by repeated TSTs in leukocyte counts.

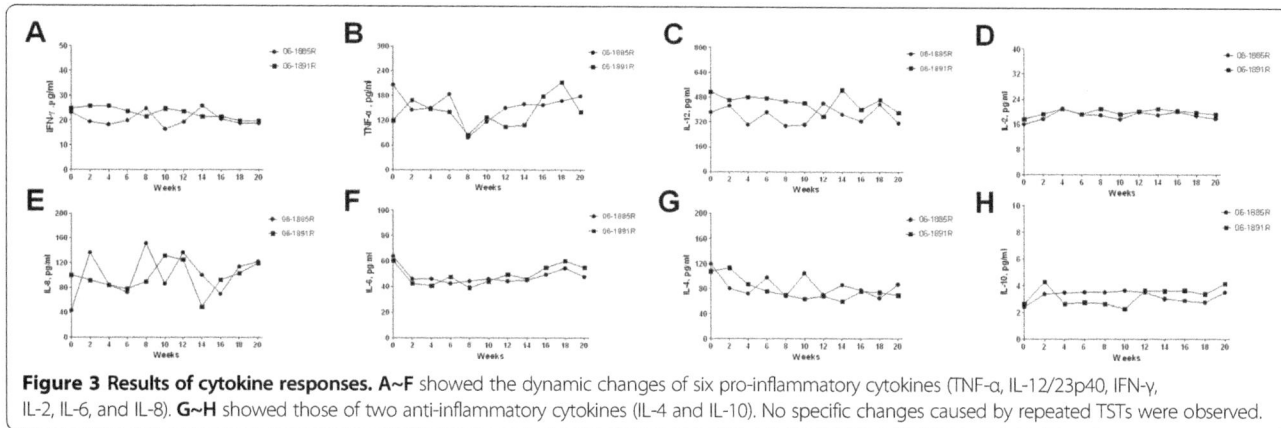

Figure 3 Results of cytokine responses. A~F showed the dynamic changes of six pro-inflammatory cytokines (TNF-α, IL-12/23p40, IFN-γ, IL-2, IL-6, and IL-8). **G~H** showed those of two anti-inflammatory cytokines (IL-4 and IL-10). No specific changes caused by repeated TSTs were observed.

previous report [8]. The kits for IFN-γ, IL-2, IL-4, IL-6, and IL-12/23p40 assays were purchased from Mabtech Inc. (Ohio, USA) and for IL-8, IL-10, TNF-α from Bender Medsystems, Inc. (California, USA).

Results

Repeated TSTs showed intermittent positive reaction in TB infections

Among nine TB-positive monkeys, eight monkeys were strongly positive (graded 4 ~ 5) at week 0 and one monkey were scored 3. Afterward, intrapalpebral reactions weakened gradually, one monkey (scored 3 at week 0) gave intermittent positive results. At week 12 (the seventh TST) only three monkeys showed positive intrapalpebral reactions (Figure 1). There were no intrapalpebral reactions for the two healthy monkeys.

Changes in total leukocyte and leukocyte populations

There were no significant changes in total leukocyte count and leukocyte population counts including monocytes, neutrophiles, and lymphocytes during the TSTs performed period (Figure 2A~D). The values of monkey 06-1891R were higher than those of monkey 06-1885R. Monocyte-Lymphocyte ratio and neutrophile-lymphocyte ratio, calculated from the numbers of monocytes, neutrophiles, and lymphocytes, showed no obvious elevations (Figure 2E~F).

Changes in serum cytokines responses

In this study, six serum pro-inflammatory cytokines and two serum anti-inflammatory cytokines were detected. They were TNF-α, IL-12/23p40, IFN-γ, IL-2, IL-6, IL-8, IL-4 and IL-10. Animals kept baseline all the time for most cytokines. Though there were multipeak kinetics under low

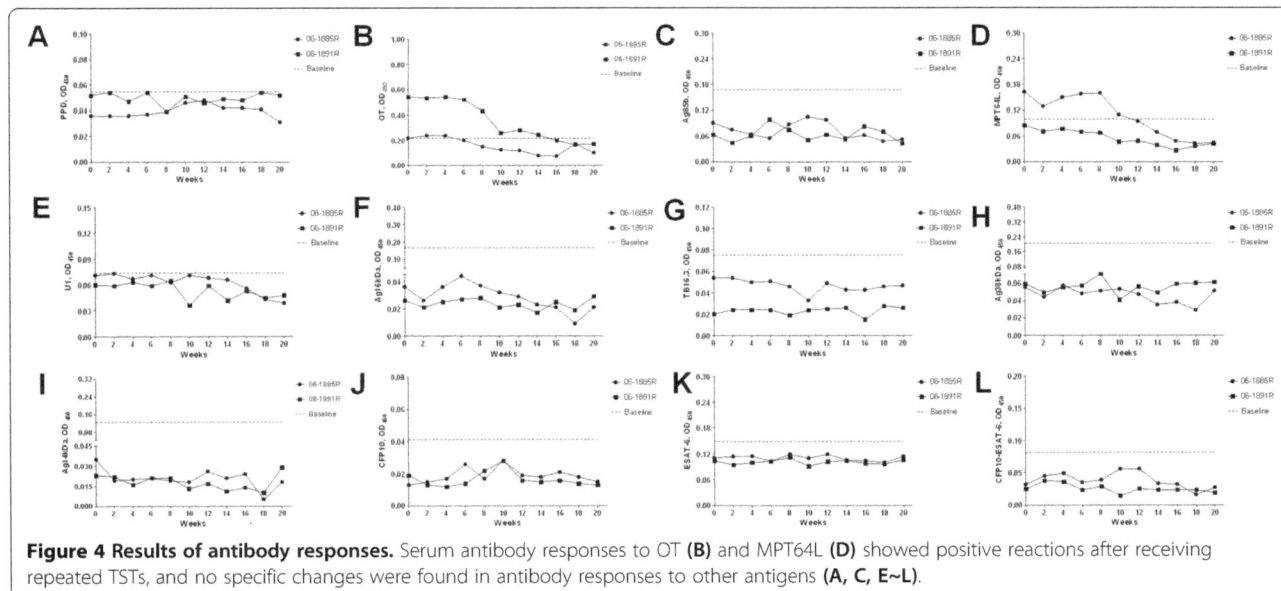

Figure 4 Results of antibody responses. Serum antibody responses to OT **(B)** and MPT64L **(D)** showed positive reactions after receiving repeated TSTs, and no specific changes were found in antibody responses to other antigens **(A, C, E~L)**.

levels in some cytokines, no specific changes for TB were observed during TSTs administration course (Figure 3).

Changes in serum antibody responses

Sequential serum samples were tested against selected antigens. Data obtained for all animals in every antigen are present in Figure 4. The baselines were calculated by average plus 3SD according to our previous data [7]. No positive antibody reactions against those antigens except OT and MPT64L were found during TST-treated period. The antibody responses characterizations of OT and MPT64L were similar to each other, moving from positive to negative reactions during the experiment period.

Discussion

Although the TST in nonhuman primates has many limitations to sensitivity and specificity, it is still kept as the mainstay of tuberculosis screening and antemortem diagnosis. In this study, we tried to make clear whether repeated TSTs in rhesus monkeys affect immune responses, including TST-reaction, total leukocyte and leukocyte populations, serum cytokines responses and antibody responses.

Repeated TSTs in natural TB-infections showed gradually weakened intrapalpebral reactions and intermittent positive results. Our study once again demonstrated the chief limitations of TST in documents. As we previously suggested [8], keeping long enough intervals of TSTs is important in animals routine quarantine to draw optimal results, such as quarter quarantine and annual quarantine.

Primary or secondary TB infections were always accompanied by elevated total leukocyte count and leukocyte population counts and a peak in monocyte-lymphocyte ratio and neutrophile-lymphocyte ratio [8,9]. In our study, no significant changes caused by repeated TSTs in healthy monkeys were observed in total leukocyte count and leukocyte population counts. For TB infection, serum cytokines were always induced or suppressed. There were also no specific changes for TB infection in serum cytokines of healthy monkeys. Whether repeated TSTs would affect TB serodiagnostic efficacy still remained open. Now the answer could be obtained from this study. Among 12 selected antigens, only the antibody responses to OT and MPT64L presented periodic positive reactions in healthy monkeys, which may be caused by their non-specificity and cross-reactivity with mycobacterial species [10,11]. Our results showed that repeated TSTs in healthy monkeys could not induce antibody responses and affect TB seradiagnostics.

Conclusions

In conclusion, repeated TSTs may induce immune tolerance for TST reaction in TB infections, but no effects in leukocyte counts, serum cytokines, and antibody responses in healthy monkeys. Results indicate that the combinatorial use of TST and TB-specific antibody tests may become the overall effectiveness of TB surveillance programs for nonhuman primates.

Competing interests
The authors declare that they have no competing interests.

Authors contributions
FM and WZ conceived and designed the study. FM, JW, WY and HK performed the experiments. FM wrote the paper. All authors read and approved the final manuscript.

Acknowledgements
This work was supported by grant 2013B020307005 of Guangdong Provincial Science & Technology Project and Grant [2012] 224 41 of Guangzhou Science & Technology Project.

References
1. Centers for Disease Control and Prevention: Tuberculosis in imported nonhuman primates: United States, June 1990 May 1993. *MMWR Morb Mortal Wkly Rep* 1993, **42**:572 576.
2. McMurray DN: A nonhuman primate model for preclinical testing of new tuberculosis vaccines. *Clin Infect Dis* 2000, **30**(Suppl 3):S210 S212.
3. Lerche NW, Yee JL, Capuano SV, Flynn JL: New approaches to tuberculosis surveillance in nonhuman primates. *ILAR J* 2008, **49**(2):170 178.
4. Ryan TJ, Buddle BM, de Lisle GW: An evaluation of the gamma interferon test for detecting Bovine tuberculosis in cattle 8 to 28 days after tuberculin skin testing. *Res Vet Sci* 2000, **69**(1):57 61.
5. National Research Council: *Guide for the Care and Use of Laboratory Animals.* 8th edition. Washington, DC: National Academy Press; 2010.
6. Capuano SV 3rd, Croix DA, Pawar S, Zinovik A, Myers A, Lin PL, Bissel S, Fuhrman C, Klein E, Flynn JL: Experimental Mycobacterium tuberculosis infection of cynomolgus macaques closely resembles the various manifestations of human. *M Tuberculosis Infection Infect Immun* 2003, **71**(10):5831 5844.
7. Min FG, Zhang Y, Huang R, Li WD, Wu Y, Pan JC, Zhao WB, Liu XM: Serum antibody responses to 10 Mycobacterium tuberculosis proteins, purified protein derivative, and old tuberculin in natural and experimental tuberculosis in rhesus monkeys. *Clin Vaccine Immunol* 2011, **18**(12):2154 2160.
8. Min FG, Zhang Y, Pan JC, Wang J, Yuan W: *Mycobacterium tuberculosis* infection in rhesus monkeys (*Macaca mulatta*) and evaluation of ESAT-6 and CFP10 as immunodiagnostic antigens. *Exp Anim* 2013, **62**(4):281 293.
9. Smithburn KC, Sabin FR, Hummel LE: 1937. **Haematological studies in experimental tuberculosis: variations in the blood cells of rabbits inoculated with cultures differing in virulence.** *American Review of Tuberculosis XXXVI.* 673 691. Article. 19 Images.
10. Harboe M: Antigen of PPD, old tuberculin and autoclaved mycobacterium bovis BCG studied by Immunoelectrophoresis. *Am Rev Respir Dis* 1981, **124**:80 87.
11. Weldingh K, Rosenkrands I, Okkels LM, Doherty TM, Andersen P: **Assessing the serodiagnostic potential of 35 *Mycobacterium tuberculosis* proteins and identification of four novel serological antigens.** *J Clin Microbiol* 2005, **43**(1):57 65.

24

Evaluation of different milking practices for optimum production performance in Sahiwal cows

Naveed Aslam[1], Muhammad Abdullah[1], Muhammad Fiaz[2], Jalees Ahmad Bhatti[1], Zeeshan Muhammad Iqbal[1], Nasrullah Bangulzai[3], Chang Weon Choi[4] and Ik Hwan Jo[4*]

Abstract

The production performance of multiparous lactating Sahiwal cows (n = 24) was evaluated according to both milking frequency and method. Selected animals were randomly divided into four groups containing six animals each under a completely randomized design. Cows in groups A & B were milked by the hand milking method three times per day, respectively. Similarly, cows in groups C & D were milked by the machine milking method two and three times per day, respectively. All animals were maintained under uniform feeding and management conditions. Dry matter intake was high in animal groups milked three times per day, and it remained unchanged between the hand and machine milking methods. Milk yield was higher (P < 0.05) in cows milked three times compared to those milked twice per day, and it did not differ between hand and machine milking methods. Milk fat percentage was higher (P < 0.05) in cows milked twice per day compared to those milked three times using both machine and hand milking methods. The percentage of total solids showed a similar pattern as the fat percentage. However, percentages of protein, lactose, and non-fat solids in milk were not significantly different (P > 0.05) among the treatment groups. Collectively, the results show that milking three times per day instead of twice at 8-hour intervals can enhance milk yield in Sahiwal cows using both hand and machine milking methods.

Keywords: Sahiwal cows, Machine milking, Milking frequency, Milk yield, Composition

Background

Demand for milk in Pakistan is rising due to an increasing human population and higher inclination towards consumption of milk and milk products. Milk is the largest commodity amongst dairy products, making Pakistan the 4th largest milk-producing country in worldwide. A total of 49.5 billion kg of milk was produced during the 2012-13 year [1]. Major sources of milk in Pakistan are cattle and buffaloes, as the country is blessed with suitable tropical breeds of dairy cattle. Sahiwal is one such renowned tropical dairy cattle breed due to its excellent heat and tick resistance [2,3]. As such, the performance of this breed in a tropical environment is also better than other cattle breeds due its better resistance to tick-borne diseases [4].

Cow milking is considered to be one of the most laborious and time-consuming activities at livestock farms. Additionally, considerable cost goes into this major farm operation. As a result, machine milking was invented to improve labor efficiency due to growing costs [5]. Machine milking has been shown to have the potential to increase milk production by up to 12%, reduce labor by up to 18%, and improve dairy cow welfare.

Increasing milking frequency from twice to three times per day may correspondingly increase milk yield from 6 to 25% during complete lactation as well as improve udder health by reducing the somatic cell count (SCC) [6-8]. The chance of clinical mastitis has also been reported to be lower in cows milked three times or more than in those milked only twice per day [9]. Milk yield can be increased by machine milking, which has beneficial effects on udder health [10,11]. Moreover, milking

* Correspondence: greunld@daegu.ac.kr
[4]Department of Animal Resources, International Research Center for Eradication of Poverty and Hunger, Daegu University, Gyeongsan, South Korea
Full list of author information is available at the end of the article

three times per day has been shown to increase milk yield by up to 14% compared to milking only twice per day [12]. The literature is limited regarding the effects of milking frequency as well as machine milking on milk yield, composition, and udder health of Pakistani cattle. Therefore, this study investigated the optimal milking frequency and milking method for improved production performance in Sahiwal cows.

Methods

Precinct of study

The study was conducted at the Livestock Experiment Station Jahangirabad, Khanewal. This station is situated in south central irrigated Punjab, Pakistan. Its geographical coordinates are 30° 18' 0" North, 71° 56' 0" East. Jahangirabad town in Khanewal District is well known as the homeland of Sahiwal cows, a renowned tropical dairy cattle breed in Pakistan [2,3,13]).

Experimental animals and treatment groups

Twenty-four lactating Sahiwal cows in their 3rd to 4th parity after parturition with similar average bodyweight, conditions, and production performance were separated into groups A, B, C, & D. Cow milking was carried out by the hand and machine milking methods twice or three times a day, as shown in Table 1. The animals were kept for a period of 90 days, excluding a 15-day adjustment period. The cows were also properly tagged and treated for internal and external parasites. All animals were kept in separate pens under identical conditions.

Feeding of animals

Animals were given standard diet: daily green fodder (Maize) @ 40 kg per cow, wheat straw @ 4 kg per cow, and Anmol wanda (concentrate consisting of CP 16 and 68% TDN) @ 1.0 kg per 2.5 liters of milk produced by each cow. Composition of Anmol Wanda is given in Table 2. Clean premises were made available to the experimental cows, and animals had *ad limitum* access to fresh clean water.

Milking of cows

Cows in groups A & B were subjected to the hand milking method, whereas cows in groups C & D were milked using a portable milking machine. However, groups A &

Table 2 Ingredient and nutrient composition

S. No.	Ingredient (s)	Inclusion level (%)
1	Cotton seed Cake	18
2	Maize gluten 30	15
3	Canola meal	10
4	Maize grain broken	14
5	Wheat Bran	25
6	Molasses	16
7	Minerals mixture*	2
Total		100
CP (%)		16
TDN (%)		68

*100 Kg minerals mixture included DCP 70.81 kg, NaCl 18.91 kg, MgSO$_4$ 8.64 kg, FeSO$_4$ 0.89 kg, MnSO$_4$ 0.49 kg, ZnSO$_4$ 0.22 kg, CuSO$_4$ 0.03 kg, KI 8.77 g, CoCl$_2$ 0.89 g, and NaSiO$_3$ 1.50 g.

C were milked twice per day at 12-hour intervals at 3:00 and 15:00 hours, whereas groups B & D were milked three times per day at 8-hour intervals at 2:00, 10:00, and 18:00 hours. Prior to milking, the udder of each cow was thoroughly washed with moderate warm water & dried properly. The milking machine was thoroughly washed with acid and alkali solutions in lukewarm water.

Data recording and parameters

Feed intake of each experimental animal was recorded daily. Milk yields of all cows were recorded daily using a calibrated spring balance. To screen for mastitis, a few milk were tested prior to milking. Milk samples were collected from all cows every 2 weeks for milk composition analysis. The following parameters were studied to determine the response to different treatments.

1. Dry matter intake (kg)
2. Milk yield (kg)
3. Milk protein%
4. Milk fat%
5. Lactose%
6. Total solids%

Laboratory analysis

Analyses were performed in the Food & Nutrition and Quality Operation Labs at the University of Veterinary

Table 1 Layout of experiment

Treatment groups	Replicates	Milking method	Frequency of daily milking
A	6	Manual	Twice
B	6	Manual	Thrice
C	6	Machine	Twice
D	6	Machine	Thrice

Table 3 Effect of milking frequency & method on dry matter intake and milk yield in Sahiwal cows

Parameters	Hand milking twice (A)	Hand milking thrice (B)	Machine milking twice (C)	Machine milking thrice (D)
Dry matter intake (Kg/day)	7.50 ± 0.18^b	9.30 ± 0.11^a	7.90 ± 0.15^b	9.50 ± 0.15^a
Milk yield (Kg)	9.08 ± 0.15^b	11.50 ± 0.18^a	9.25 ± 0.21^b	11.75 ± 0.28^a

Different superscripts in same row differ significantly (P < 0.05).

and Animal Sciences (UVAS), Lahore. Dry matter content of feed samples was determined according to the procedures of [14]). Milk composition analysis was carried out using a Lactoscan-S Milk Analyzer (50 W, Milkotronic Ltd., Bulgaria) in WTO-Quality Operation Labs at the UVAS, Lahore for the following milk constituents: milk fat, non-fat solids, milk protein, lactose, and total solids.

Statistical analysis

Data were analyzed using ANOVA techniques [15] under a Completely Randomized Design (CRD) using SAS 9.1.3 portable software. Differences among treatment means were tested by DMRt. The mathematical model was:

$$Yij = \mu + \tau i + \varepsilon ij$$

where

Yij = each observation on i^{th} treatment due to j^{th} animal.

μ = overall mean.

τi = effect of ith treatment ($\Sigma \tau i = 0$ and i = 1, 2, 3, 4).

εij = random error associated with i^{th} treatment and j^{th} animal with the restriction that variance σ_2 and mean zero.

Results and discussion

Dry Matter Intake

Dry matter intake (DMI) in cows is shown in Table 3. Cows milked three times daily showed higher DMI compared to those milked twice per day. However, DMI was not significantly different between hand and machine milking methods. The higher DMI consumption of cows milked three times per day can be attributed to higher milk yield, which increases nutrient demands. In support

of the current results, [16] previously studied the effects of milking frequency on production efficiency in Holstein cows and found that milking three times daily increased milk yield and DMI. In another study, [17] asserted that DMI increases (p < 0.05) with greater milking frequency. The current results are also supported by previous works ([16,18,19]) in which increased DMI was shown to be accompanied by higher milk yield due to greater milking frequency. Milk yield is an indicator of feeding behavior as increased milk yield is positively correlated with DMI [20].

Milk yield

Average daily milk yields are presented in Table 3. Milk yield was higher (P < 0.05) in cows milked three times daily compared to cows milked twice. Milk production between the machine and hand milking treatment groups was not significantly different (P > 0.05). However, higher milking frequency significantly (P < 0.05) enhanced milk yield.

Our observation of an association between enhanced milking frequency and greater milk yield has been substantiated by the literature. In a previous study, [7] reported that milk yield increased by up to 10.4% in cows milked three times per day compared to those milked twice per day. Similarly, [21] showed that milking three times daily increased milk yield in cows by up to 18%. In another study, [22] showed that the yield of energy corrected milk was higher in cows milked three times per day compared to those milked twice per day.

Similarly, [23] affirmed that milking three times per day rather than twice increased milk yield in crossbred (different blood ratios among Friesian, Hariana, Brown Swiss, and Jersey breeds) cows, but at the cost of their body condition. The higher milk yield in cows milked three times per day might be due to the fact that milk

Table 4 Effects of milking frequency & method on milk composition in Sahiwal cows

Parameters	Hand milking twice (A)	Hand milking thrice (B)	Machine milking twice (C)	Machine milking thrice (D)
Protein%	3.35 ± 0.06	3.25 ± 0.06	3.35 ± 0.06	3.22 ± 0.08
Fat%	3.68 ± 0.04^a	3.45 ± 0.03^b	3.70 ± 0.04^a	3.50 ± 0.04^b
Lactose%	4.82 ± 0.06	4.87 ± 0.02	4.90 ± 0.04	4.92 ± 0.02
Total Solids%	12.7 ± 0.04^a	12.4 ± 0.04^b	12.7 ± 0.04^a	12.4 ± 0.04^b

Different superscripts in same row differ significantly (P < 0.05).

secretion is a continuous process resulting in gradual elevation of internal udder pressure. Thus, more frequent milking might reduce internal udder pressure and consequently stimulate milk-secreting cells to operate at full capacity for a longer time.

Milk composition

The impacts of milking frequency and method on milk composition in Sahiwal cows are shown in Table 4. Milk fat percentage was higher (P < 0.05) in cows milked twice compared to those milked three times per day. However, milking method did not influence fat percentage in milk. Similarly, the percentage of total solids was higher in milk from cows milked twice per day compared to those milked three times.

However, the percentages of protein and lactose in milk were not significantly different (P > 0.05) among cows in the different treatment groups. Milk fat content has been shown to be affected by milking frequency [7,24,25]). Similarly, cows milked three times per day produce milk with a lower fat content compared to those milked once a day during early lactation [26]. The negative effect of frequent milking on fat content can be attributed to increased air exposure due to frequent milking, enzymatic activity of fatty acid syntethase, and increased production of short-chain fatty acids [7]. Another factor might be the shortened time for fat synthesis in the case of 8-hour intervals. The high percentage of total solids in milk from cows milked twice daily might be due to a high fat percentage.

Conclusion

It can be concluded that milking three times instead of only twice per day at 8-hour intervals can enhance milk yield in Sahiwal cows using both hand and machine milking methods.

Competing interests
The authors declare that they have no competing interests.

Authors' contributions
NA, MA, MF, CWC and IHJ participated in the design of the study and the performance of the entire experiment. ZMI made the laboratory work and data interpretation. NB and JAB performed the statistical analysis and preparing the manuscript. All authors helped to draft the manuscript, and they read and approved the final manuscript.

Acknowledgements
This work was supported by the Directorate of Research Centre for Conservation of Sahiwal Cattle, Livestock & Dairy Development Department, Government of Punjab, Pakistan as well as the Cooperative Research Program for Agriculture Science & Technology Development (PJ009289042014), Rural Development Administration, Republic of Korea.

Author details
[1]Department of Livestock Production, Faculty of Animal Production and Technology, University of Veterinary and Animal Sciences, Lahore, Pakistan. [2]Department of Livestock Production and Management, Pir Mehr Ali Shah Arid Agriculture University Murree Road Shamsabad, Rawalpindi, Pakistan. [3]Department of Livestock Management, Lasbela University of Agriculture, Water and Marine Sciences Balochistan, Karachi, Pakistan. [4]Department of Animal Resources, International Research Center for Eradication of Poverty and Hunger, Daegu University, Gyeongsan, South Korea.

References
1. GOP (Government of Pakistan): Economic Survey of Pakistan. Islamabad, Pakistan: Government of Pakistan, ministry of finance, Published by Economic Adviser Wing 2012-13.
2. Khan MS, Rehman Z, Khan MA, Ahmad S: Genetic resources and diversity in Pakistani cattle. Pakistan Vet J 2008, 28:95–102.
3. Rehman Z, Khan MS, Bhatti SA, Iqbal J, Iqbal A: Factors affecting first lactation performance of Sahiwal cattle in Pakistan. Arch Anim Breed 2008, 51. In press.
4. Glass EJ, Preston PM, Springbett A, Craigmile S, Kirvar E, Wilkie G, Brown CGD: Bos taurus and Bos indicus (Sahiwal) calves respond differently to infection with Theileria annulata and produce markedly different levels of acute phase proteins. Int J Parasit 2005, 35:337–347.
5. Lind O, Ipema AH, de Koning C, Mottram TT, Hermann HJ: Automatic Milking. Brussels, Belgium: International Dairy Federation; 2000. Bulletin 348/2000, 1–14.
6. Erdman RA, Varner M: Fixed yield responses to increased milking frequency. J Dairy Sci 1995, 78(5):1199–1203.
7. Klei LR, Lynch JM, Barbano DM, Oltenacu PA, Lednor AJ, Bandler DK: Influence of milking three times a day on milk quality. J Dairy Sci 1997, 80:427–436.
8. Vorst YVD, Hogeveen H: Automatic Milking Systems and Milk Quality in the Netherlands. In Proc. Int. Symp. Robotic Milking. Edited by Hogeveen H, Meijering A. Lelystad, the Netherlands: Wageningen Pers; 2000:73–82.
9. Hillerton JE: The Effects of Milking Frequency on Mastitis. In Proceedings of the British Mastitis Conference, Stoneleigh, UK.; 1991:61–69.
10. Hillerton JE, Winter A: The Effects of Frequent Milking on Udder Physiology and Health. In Prospects for Automatic Milking. Edited by Ipema AH, Lippus AC, Metz JHM, Rossing W. Wageningen, the Netherlands: Pudoc Scientific Publishers; 1992:201–212.
11. Paape MJ, Capuco AV, Lefcourt A, Burvenich C, Miller RH: Physiological Response of Dairy Cows to Milking, Proc. Int. Symp. Prospects for Auto. Milking. Wageningen, The Netherlands: Pudoc Scientific Publishers; 1992.
12. Ipema AH, Benders E: Production, duration of machine-milking and teat quality of dairy cows milked 2, 3 or 4 times daily with variable intervals. In Proc. Int. Symp. Prospects For Automatic Milking. Edited by Ipema AH, Lippus AC, Metz JHM, Rossing W. Pudoc Scientific Publishers, Wageningen, The Netherlands: Pudoc Scientific Publishers; 1992:244–252.
13. Fiaz M, Abdullah M, Pasha TN, Jabbar MA, Babar ME, Bhatti JA, Nasir M: Evaluating varying dietary energy levels for optimum growth and early puberty in Sahiwal heifers. Pakistan J Zool 2012, 44:625–634.
14. AOAC (Association of Official Analytical Chemists): Official Methods of Analysis. 17th edition. Arlington, Virginia, USA: Association of Analytical Chemists; 2000.
15. Steel RGD, Torrie JH, Dickie AD: Principles and Procedures of Statics a Biometric Approach. 3rd edition. Tronto: McGraw-Hill Publishing Company; 1997.
16. Barnes MA, Pearson RE, Lukes-Wilson AJ: Effects of milking frequency and selection for milk yield on productive efficiency of Holstein Cows. J Dairy Sci 1990, 73(6):1603–1611.
17. Williams TJ, Osinowo OA, Smith OF, James IJ, Ikeobi CON, Onagbesan OM, Shittu OO, Solola FT: Effects of milking frequency on milk yield, dry matter intake and efficiency of feed utilization in wad goats. Archivos de Zootecnia 2012, 61:457–465. No. 235.
18. Bar-Pelled U, Maltz E, Bruckental I, Folman Y, Kali Y, Gacitua H, Lehrer AR, Knight CH, Robinson B, Voet H, Tagari H: Relationship between frequent milking or suckling in early lactation and milk production of high producing dairy cows. J Dairy Sci 1995, 78(12):2726–2736.
19. Royle C, Gamsworth PC, McArthur AJ, Mepham TB: Effect of Frequent Milking on Heart Rate and Other Physiological Variables in Dairy Cows, In: Int. Symp. Prospects for Automatic Milking, Hrdoc. Wageningen. The Netherlands; 1992:237.
20. Dado RG, Allen MS: Variation in relationships among feeding, chewing and drinking variables for lactating dairy cows. J Dairy Sci 1994, 77:132.

21. Stelwagen K: **Effect of milking frequency on mammary functioning and shape of the lactation curve.** *J Dairy Sci* 2001, **84**(supple):E204–E211.

22. Osterman S, Bertilsson J: **Extended calving interval in combination with milking two or three times per day: effects on milk production and milk composition.** *Livest Prod Sci* 2003, **82**(2–3):139–149.

23. Sirohi AS, Pandey HN, Singla M: **Effects of milking frequency on feed intake, body weight and haemato-biochemical changes in crossbred cows.** *J Appl Anim Res* 2012, **40**(1):63–68.

24. O'Brien B, Ryan G, Meaney WJ, McDonagh D, Kelly A: **Effect of frequency of milking on yield, composition and processing quality of milk.** *J Dairy Res* 2002, **69**(3):367–374.

25. Friggens NC, Rasmussen MD: **Milk quality assessment in automatic milking systems: accounting for the effects of variable interval between milkings on milk composition system.** *Livest Prod Sci* 2001, **73**:45–54.

26. Patton J, Kenny DA, Mee JF, O'Mara FP, Wathes DC, Cook M, Murphy JJ: **Effect of milking frequency and diet on milk production, energy balance, and reproduction in dairy cows.** *J Dairy Sci* 2006, **89**(5):1478–1487.

Carcass and body organ characteristics of broilers supplemented with dietary sodium and sodium salts under a phase feeding system

Mirza Muhammad Haroon Mushtaq[*], Rana Parvin and Jihyuk Kim

Abstract

The effect of sodium and sodium salts on carcass and body organ characteristics of broilers under a four phase feeding program were investigated. A basal diet (0.08% dNa with NaCl) was formulated and one of two sources of dNa ($NaHCO_3$ and Na_2SO_4) were supplemented to obtain four different percentages of dNa (0.17, 0.26, 0.35, and 0.44%) for each treatment. There was a linear decrease in dressing percentage (DP) with source × level interaction ($p \leq 0.001$), while there was a linear increase in breast yield and thigh yield with increasing dNa supplementation ($p \leq 0.001$). Chicks fed 0.35% $NaHCO_3$ and 0.44% dNa Na_2SO_4 supplemental salts had lower abdominal fat ($p \leq 0.04$). Chicks that received increasing levels of dNa (from 0.17 to 0.44%) showed increasing gizzard weight ($p \leq 0.002$) and decreasing spleen weight ($p \leq 0.02$). When both salts were supplemented at 0.26% dNa, the chicks showed their lowest bursa weight ($p \leq 0.001$). Consequently, chicks at higher dNa showed an increase in breast and thigh meat yield, and increasing capacity of their digestive organ. The higher levels of dNa should be tested with other cations and anions to fully understand acid base homoeostasis.

Keywords: Sodium, Salt, Carcass and body organ characteristics, Phase feeding program, Broiler

Introduction

Sodium (Na^+), the principal cation of extracellular fluid, is involved in numerous functions including the regulation of extracellular fluid volume, acid base balance, cell membrane potential, nerve function, and the absorption of glucose and amino acids ([1]; Leeson and Summers, [2]). Dietary Na (dNa) and chloride (Cl) are inexpensive in terms of meeting dietary requirements, as Pakistan has huge reserves of sodium chloride (NaCl) [3].

Mongin [1] described the effect and interrelationship of Na^+, K^+, and Cl^- in an equation for dietary electrolyte balance (DEB = Na + K-Cl, mEq/kg diet). Rondon et al. [4] and Borges et al. [5] reported different DEB values with different concentrations of Na^+, K^+, or Cl^-. Recently, Mushtaq et al. [6,7] tested 3 levels of dNa and dCl adjusted with dK and reported that different ions acted differently to yield a similar DEB of 250 mEq/kg. The addition of various salts changes osmotic balance by contributing electrolytes [5,8-10]. Most researchers consider Na-bicarbonate ($NaHCO_3$) as the best supplemental salt for Na^+ and HCO_3^- [11-16]. Sodium sulphate (Na_2SO_4) induced severe blood acidosis when it was compared with other sulphate sources in growth trials; hence it was suggested that the acidic properties of sulphates are directly linked to supplemental salts [16,17].

A significant effect of temperature on abdominal fat was found for treatments (5% KCl or 5% $NaHCO_3$ in water) under normal but not under cyclic environmental conditions [18]. Sharma and Gangwar [8] reported a decrease in the concentration of Na^+ and K^+ in the breast and thigh muscles of broilers from 4 to 8 wks old under high temperature (32°C). The author, moreover, observed that breast muscles had significantly lower Na^+ and higher K^+ concentrations than thigh muscles. Pourreza and Edriss [19] reared straight-run broilers at 20°C or 30°C and noticed that the high temperature decreased slaughter, carcass, and abdominal fat weight and increased dressing percentage. Johnson and Karunajeewa [11], Karunajeewa et al. [20], and

* Correspondence: haroonuaf@gmail.com
Poultry Science Division, National Institute of Animal Science, RDA, 114, Sinbang 1-gil, Seonghwan-eup, Seobuk-gu, Cheonan-si, Chungcheongnam-do 331–801, Republic of Korea

Borges et al. [21] observed no effect on carcass and abdominal fat yield under different DEB treatments under normal environmental condition.

The development of housing systems and genetics has necessitated a look into the changing nutrient requirements of current broilers. The present study evaluated the effect of dietary sodium supplementation with the applicability of DEB using different sodium salts on carcass and body organ characteristics of a modern broiler strain fed under a four phase feeding program.

Materials and methods

All experimental birds and procedures were maintained in compliance with laws proposed by the Advanced Studies and Research Board, University of Veterinary and Animal Sciences, Lahore, Pakistan.

Birds' husbandry

A total of 1280 day-old straight-run Hubbard broiler chicks (Hubbard × Hubbard) were given one of eight dietary treatments replicated four times; under this system each replicate contained 40 birds (without considering the sex). Each replicate pen was equipped with a separate overhead, transparent, and volume-graduated 20 L water bottle linked to a nipple drinker line. Water bottles were cleaned and filled with fresh water on a daily basis. One flat bottom round feeder was provided for each experimental pen. Birds were housed in environment control systems where variation in temperature and relative humidity were recorded and maintained according to the production manual [22]. Continuous light was provided 24 h for the first 3 days and then a 23 L:1D light pattern was adopted for the rest of the experimental period. Fresh sawdust (7.5 cm deep) was used as litter material over a concrete floor. For the first 3 days, the house temperature was maintained at 32°C and thereafter reduced by 0.5°C per day until 24°C was attained at d 19.

Birds were vaccinated against Newcastle Disease (ND) plus Infectious Bronchitis viruses at 4 d, Infectious Bursal Disease virus at 8 d and again at 14 d; Hydropericardium Syndrome virus at 18 d, and ND-Lasota strain at d 22.

Dietary plan and experimental design

A basal diet with dNa, K and Cl at 0.08, 0.71, and 0.20%, respectively, with a DEB value of 160 mEq/kg (Table 1 and Table 2) was formulated. For this purpose, a large batch of basal diet was prepared for each phase and then experimental diets were prepared such that four levels of dNa (0.17, 0.26, 0.35, and 0.44%) were supplemented to this basal diet with either commercially available feed-grade sodium bicarbonate ($NaHCO_3$) or sodium sulphate (Na_2SO_4). The levels of dNa corresponded to DEB values of 200, 240, 280, and 320 mEq/kg, respectively. The

Table 1 Ingredient composition of basal diets fed four levels of sodium with two sources of sodium salts at different stages of growth in broilers[1]

Ingredients (%)	Pre-starter (1 – 10 d)	Starter (11 – 20 d)	Grower (21 – 33 d)	Finisher (34 – 42 d)
Corn	46.93	47.54	67.46	67.94
Broken rice	13.46	16.16	0.07	–
Soybean meal	27.63	29.11	26.93	24.47
Canola Meal	6.39	1.24	–	–
Oil[2]	1.62	2.21	1.88	3.94
DCP	2.16	2.02	1.89	1.60
Limestone	1.04	0.94	1.05	1.10
L-Lysine HCl	0.23	0.22	0.20	–
L-Lysine sulphate	–	–	–	0.34
NaCl	0.15	0.16	0.16	0.16
KCl	0.04	0.03	0.05	0.14
DL-methionine	0.20	0.20	0.17	0.17
L-threonine	0.05	0.07	0.03	0.04
Premix[3]	0.10	0.10	0.10	0.10

[1]All diets were supplemented with four levels of either $NaHCO_3$ (0.33, 0.66, 0.99, or 1.32%) or Na_2SO_4 (0.28, 0.56, 0.84, or 1.12%) to make final Na concentrations of 0.17, 0.26, 0.35, or 0.44%, respectively. The basal diet contained 0.08% Na.
[2]Residual bakery oil.
[3]Provides per kg of finished diet: vitamin A, 12 mg; vitamin D_3, 7 mg; vitamin E, 100 mg; vitamin K_3 (50% as MNB), 3 mg; vitamin B_1 (98%), 3 mg; vitamin B_2 (800,000 mg), 12 mg; vitamin B_3 (niacin; 99%), 600 mg; vitamin B_6 (98%), 4 mg; vitamin B_9 (folic acid; 95%), 2 mg; vitamin B_{12} (0.10%), 20 mg; biotin (0.10%), 5 mg; Ca-Pentothenate (98%), 12 mg; cholin (70% as choline sodium), 1 g; MnO (60%), 169 mg; $FeSO_4$ (21%), 200 mg; $ZnSO_4$ (36%), 150 mg; $CuSO_4$ (25%), 40 mg; Se (sodium selenite 0.40%), 100 mg; KI (68%), 2 mg; salinomycin, 60 mg; zinc bacitracin (as Albac 10%), 50 mg.

experimental period was divided into four phases: pre-starter (1 to 10 d), starter (11 to 20 d), grower (21 to 33 d), and finisher (34 to 42 d) which met or exceeded the nutrient specifications recommended by the Hubbard management guide ([22]; Table 2) except for amino acid composition.

All ingredients were assayed for their proximate composition [23] prior to diet formulation and actual values were used in the formulation. The Na^+ and K^+ contents of each diet were analyzed by flame photometer [23] and Cl^- content was analyzed by titration with $AgNO_3$ [24]. Prior to starting the experiment, the Na^+, K^+, and Cl^- contents of the final diet were verified. The ME of each ingredient was calculated by the appropriate regression equation suggested by the NRC [25]. The amino acid composition of each ingredient was calculated using AminoDat™ 3.0 Platinum (Degussa AG, Germany) based on the DM and CP contents of each ingredient [6]. The amino acid composition of each diet met or exceeded the ideal amino acid ratio suggested by Han and Baker [26]. The experiment, offering mash diets, lasted until 42 d of age.

Table 2 Nutrient composition of basal diets for different phases of birds fed four levels of sodium with two sources of sodium salts[1]

Nutrients	Pre-starter	Starter	Grower	Finisher
	(1 – 10 d)	(11 – 20 d)	(21 – 33 d)	(34 – 42 d)
ME (kcal/kg)	2900	3000	3000	3147
Crude Protein (%)	21.00	20.00	19.00	18.00
Calcium (%)	1.00	0.90	0.90	0.85
Available Phos. (%)	0.45	0.42	0.40	0.35
Sodium (%)	0.08	0.08	0.08	0.08
Potassium (%)	0.71	0.71	0.71	0.71
Chloride (%)	0.20	0.20	0.20	0.20
DEB[2] (mEq/kg)	160	160	160	160
Dig Lys (%)	1.10	1.05	0.97	0.93
Dig Met/Dig Lys	0.45	0.45	0.44	0.44
Dig Met + Dig Cys/Dig Lys (%)	0.72	0.71	0.72	0.72
Dig Thr/Dig Lys	0.66	0.67	0.66	0.66
Dig Try/Dig Lys	0.18	0.18	0.18	0.17

[1]$NaHCO_3$ and Na_2SO_4.
[2]Dietary Electrolyte Balance (mEq/kg) = (% Na × 10,000 / 23) + (% K × 10,000 / 39.1) − (% Cl × 10,000 / 35.5).

Growth response

Feed intake (FI; g/bird), BW gain (BWG; g/bird) and feed-to-gain ratio (FG; g:g) were recorded at the end of the experiment. The feed was withheld for 6 h before weighing the birds to ensure the emptying of the digestive tract of the bird. Mortality was recorded on a daily basis and dead bird was weighed prior to removal to correct FG.

Carcass and organ characteristics

At the end of 42 d, two birds were randomly selected from each replicate and subjected to carcass and body organ evaluation. The feed was withheld for 6 h before slaughter to ensure emptying of the digestive tract. Carcass and body organ responses were evaluated in terms of dressing percentage, breast, thigh, abdominal fat, gizzard, proventriculus, heart, liver, kidney, spleen, pancreas, bursa, gallbladder, intestine, and lung weights, and for the shank and intestine lengths (Table 3 and Table 4). The carcass and body organ weights were taken on a fresh basis. Dressing percentage (DP) was calculated by dividing dressed weight, without viscera, by live weight and multiplied by 100. The weight of the abdominal fat pad was expressed as DP (without visceral weight), while the weights of other body organs (gizzard, proventriculus, heart, liver, kidney, spleen, pancreas, bursa, gall bladder, lungs, and intestine) were taken as a percentage of dressing weight (with visceral weight). Intestinal length was measured (in centimeters) from the start of the duodenal loop to the ileocaecal junction [27].

Table 3 Effect of dietary sodium and sodium salts on growth performance of broilers on day

Item	BW gain (g/bird)	Feed intake (g/bird)	Feed:gain (g:g)
Dietary Na (%)			
0.17	1,893	3,564	1.95
0.26	1,959	3,634	1.80
0.35	1,883	3,605	1.89
0.44	1,864	3,622	1.95
SEM	30.5	45.9	0.076
Salts			
NaHCO$_3$	1,884	3,604	1.88
Na$_2$SO$_4$	1,916	3,608	1.91
SEM	21.6	32.4	0.053
Na × Salts			
0.17 × NaHCO$_3$	1,941	3,586	1.82
0.26 × NaHCO$_3$	1,941	3,631	1.83
0.35 × NaHCO$_3$	1,829	3,649	1.91
0.44 × NaHCO$_3$	1,824	3,551	1.98
0.17 × Na$_2$SO$_4$	1,844	3,542	2.08
0.26 × Na$_2$SO$_4$	1,977	3,637	1.76
0.35 × Na$_2$SO$_4$	1,936	3,560	1.88
0.44 × Na$_2$SO$_4$	1,905	3,694	1.91
SEM	43.1	64.9	0.107
ANOVA		Probability	
Na	NS	NS	NS
Na$_L$	NS	NS	NS
Na$_Q$	NS	NS	NS
Salt	NS	NS	NS
Salt × Na	0.036	NS	NS

NS - Non-significant.
Na_L and Na_Q are linear and quadratic terms for Na, respectively.

Water characteristics

Water alters carcass responses because of its concentrations of electrolytes (Na^+, K^+ and Cl^-); therefore, supplied water was analyzed for these electrolytes. Water characteristics were also recorded twice (morning and noon) daily to check pH by pH meter (LT-Lutron pH-207 Taiwan) and dissolved oxygen by DO meter (DO; YSI 55 Incorporated, Yellow Springs, Ohio, 4387, USA). Moreover, temperature, electrical conductivity (EC), total dissolved solids (TDS), and salinity were recorded by the Combo meter (H M Digital, Inc. CA 90230; Table 5). These observations were randomly recorded from different replicates.

Statistical analyses

The experiment was executed under a completely randomized design with factorial arrangement using four

Table 4 Effect of dietary sodium and sodium salts on carcass and body organ responses of broilers

Item	Dressing weight[1]		Breast weight[2]		Thigh weight[2]		Intestinal weight[3]		Abdominal fat[1]	
Dietary Na (%)										
0.17	56.12		31.85		45.26		58.2		3.01	
0.26	54.18		32.98		46.28		54.3		3.12	
0.35	52.94		33.79		47.34		54.9		2.44	
0.44	52.07		34.31		48.30		56.1		2.61	
SEM		0.423		0.347		0.627		1.15		0.24
Salts										
NaHCO₃	53.95		33.16		46.28		56.3		2.74	
Na₂SO₄	53.70		33.30		47.31		55.4		2.84	
SEM		0.299		0.246		0.443		0.81		0.17
Na × Salts										
0.17 × NaHCO₃	55.68		32.11		45.29		58.8		2.96	
0.26 × NaHCO₃	54.85		32.56		45.05		56.2		2.93	
0.35 × NaHCO₃	53.38		33.57		46.79		58.7		1.93	
0.44 × NaHCO₃	51.90		34.40		48.01		51.6		3.12	
0.17 × Na₂SO₄	56.56		31.60		45.24		57.7		3.06	
0.26 × Na₂SO₄	53.50		33.40		47.51		52.4		3.31	
0.35 × Na₂SO₄	52.51		34.01		47.90		51.1		2.94	
0.44 × Na₂SO₄	52.24		34.22		48.58		60.7		2.11	
SEM		0.599		0.492		0.886		1.62		0.35
ANOVA			——————— Probability ———————							
Na	≤0.001		≤0.001		0.008		NS		NS	
Na$_L$	≤0.001		≤0.001		≤0.001		NS		NS	
Na$_Q$	NS		NS		NS		0.03		NS	
Salt	NS		NS		NS		NS		NS	
Salt × Na	NS		NS		NS		≤0.001		0.04	

NS - Non-significant.
[1] % of live weight (without visceral organs).
[2] % of dressed weight (with organs weights).
[3] measured in grams.
Na$_L$ and Na$_Q$ are linear and quadratic terms for Na, respectively.

levels of dNa from two salt sources. The experimental pen was an experimental unit. The data obtained at the end of the experiment were subjected to ANOVA using GLM of Minitab 15.1 (Minitab Inc., State College PA). A statistical significance of 0.05 was used unless stated otherwise.

Results and discussion

Water was evaluated on a daily basis for quality parameters including temperature, pH, EC, TDS, DO, and salinity (Table 5). At the start of the experiment, water was also analyzed for its sodium absorption ratio (25.6) and residual sodium carbonate (9.02). As the concentration of various minerals (cations plus anions) and values of other quality parameters in drinking water could alter the electrolyte concentration of digesta ([12,28]; b) their concentration in water was evaluated. The water electrolyte

concentration was too low to impact carcass and organ yields. Water pH values (7.17–7.49) were within the range (6.0–8.5) considered optimal for broiler performance [28-30]. Previous reports [31,32] showed retarded growth up to a pH level of 6.3. The water TDS level ranged from 1000–3000 ppm was considered satisfactory for broilers by Chiba [33], however the analyzed values i.e. 1060-1284 did not appear to disturb the present experiment.

All growth responses were unaffected by dietary treatments, except for sodium level and salt interaction on BW gain (BWG; p ≤ 0.036; Table 3). The supplementation of Na₂SO₄ at 0.26% showed higher BWG at day 42. However this change in BWG was not sufficient to positively influence feed:gain (FG).

Carcass and intestinal responses were affected by supplementation of dNa from NaHCO₃ and Na₂SO₄ (Table 4). A highly linear drop in DP (p ≤ 0.001) was observed with

Table 5 Drinking water properties during the experimental period

Phase	Item	Salinity	TDS[1]	EC[2]	Temperature	pH	DO[3]
Phase 1	Max	1.30	1284	1.39	29.7	7.49	5.40
	Min	1.20	1198	1.27	26.2	7.31	3.90
	Average	1.25	1251	1.32	27.3	7.40	4.65
Phase 2	Max	1.20	1111	1.21	26.4	7.35	5.07
	Min	1.20	1060	1.06	25.8	7.17	3.70
	Average	1.20	1088	1.12	26.0	7.27	4.39
Phase 3	Max	1.30	1187	1.24	25.2	7.40	5.60
	Min	1.20	1108	1.08	24.3	7.21	3.70
	Average	1.22	1144	1.15	24.9	7.33	4.65
Phase 4	Max	1.20	1194	1.21	24.8	7.33	5.07
	Min	1.20	1110	1.07	24.0	7.23	3.70
	Average	1.20	1141	1.11	24.2	7.29	4.39

[1]TDS - Total Dissolved Solids.
[2]EC - Electric Conductivity.
[3]DO - Dissolved Oxygen.
Salinity, TDS, EC, temperature, and DO were measured as parts per thousand (ppt), parts per million (PPM), millisemen/centimeter (mS/cm), centigrade (°C) and milligram/litre (mg/L), respectively.

increasing supplementation of dNa. Mushtaq et al. [7] observed no difference in DP with increasing dNa from 0.20 to 0.30%. Their difference in results might be due to the lower levels of dNa studied. Salt sources or interaction effects in the present study were not beneficial on DP *per se*. Breast (p ≤ 0.001) and thigh (p ≤ 0.001) meat yield increased with increasing supplementation of dNa (Table 4). This contradiction in results might be due the measurement of breast and thigh meat as a percent of dressing weight with organs. These findings were not in line with the results of Mushtaq et al. [7] who observed reduced breast and leg meat by increasing dNa from 0.20 to 0.30%. This difference could be due to heat stress conditions in their experiment as more nutrients, consumed to maintain acid base balance, may not be converted to meat.

The abdominal fat pad was affected with source × level interaction (Table 6). The abdominal fat pad was lowest at 0.35% dNa in diets supplemented with NaHCO₃. A similar response of abdominal fat to a high level of dNa (0.30%) was observed by Mushtaq et al. [7]. In the present study, increasing dNa under normal physiological conditions did not disturb basal metabolism and energy was mainly utilized for meat production and not wasted as abdominal fat.

The interaction (level × source) effect was found to change intestinal weight (p ≤ 0.001); the lowest weight (51.6 vs. 51.1 gm) was recorded at 0.44% (NaHCO₃) and 0.35% dNa (Na₂SO₄). Intestinal weight reflects the gut's capacity to absorb nutrients, which reflects better health; therefore, higher levels of dNa showed a lower carcass yield.

Table 6 Effect of dietary sodium and sodium salts on body organ weights of broilers at the end of the experiment

Item	Gizzard	Kidney	Spleen	Bursa
	——— % of dressed weight[1] ———			
Dietary Na (%)				
0.17	2.41	0.31	0.13	0.23
0.26	2.57	0.36	0.08	0.15
0.35	2.98	0.37	0.07	0.21
0.44	2.99	0.37	0.08	0.25
SEM	*0.15*	*0.02*	*0.02*	*0.02*
Salts				
NaHCO₃	2.85	0.42	0.09	0.21
Na₂SO₄	2.62	0.28	0.09	0.22
SEM	*0.10*	*0.02*	*0.01*	*0.01*
Na × Salts				
0.17 × NaHCO₃	2.72	0.42	0.11	0.16
0.26 × NaHCO₃	2.65	0.47	0.10	0.18
0.35 × NaHCO₃	2.98	0.41	0.08	0.22
0.44 × NaHCO₃	3.05	0.39	0.07	0.26
0.17 × Na₂SO₄	2.11	0.21	0.15	0.29
0.26 × Na₂SO₄	2.48	0.24	0.61	0.13
0.35 × Na₂SO₄	2.98	0.33	0.06	0.20
0.44 × Na₂SO₄	2.93	0.35	0.08	0.24
SEM	*0.21*	*0.04*	*0.02*	*0.02*
ANOVA	——————— Probability ———————			
Na	0.01	NS	NS	0.001
Na$_L$	0.002	NS	0.02	NS
Na$_Q$	NS	NS	NS	0.01
Salt	NS	≤0.001	NS	NS
Salt × Na	NS	0.03	NS	0.001

NS - Non-significant.
[1]% of dressed weight (with organ weights).
Na$_L$ and Na$_Q$ are linear and quadratic terms for Na, respectively.

Organ weights of the proventriculus, heart, liver, pancreas, gall bladder, lungs, and intestinal and shank lengths were measured and found to be non-significant (data not shown). In contrast, weights (% of dressed weight) of gizzard, kidney, spleen and bursa significantly changed (Table 4). Gizzard weight was increased linearly with increasing levels of dNa from 0.17 to 0.44% (P ≤ 0.002; main effect). The increased weight of the gizzard reflects the increasing digestive or metabolic capacity of birds. Kidney weight was almost double the lowest level of dNa in the case of NaHCO₃ when compared with the lowest level of Na₂SO₄ (p ≤ 0.03; interaction effect). This suggests the bicarbonate buffer system mainly determines blood acid–base balance for optimal production performance and functions under regulatory control of the kidneys [14].

Kidney weight is indicates broiler dietary nutrient insufficiencies [34] or the presence of anti-nutritional factors [35,36]. In the present study, the acid base imbalance might cause the higher kidney weight. A linear increase in dNa decreased spleen weight ($p \leq 0.02$). The interaction (source × level) effect influenced the bursa weight ($p \leq 0.001$). The low weight of the spleen was observed at 0.35% dNa. The low weight of the bursa was observed at 0.26% dNa for both salt sources. As the spleen and bursa are associated with immune function (as lymphoid organs) this may explain the poor DP at these levels.

Conclusions

Birds showed increased breast and thigh meat yield, and increasing capacity of the gizzard at higher levels of dietary sodium. In contrast, a reduced dressing percentage with increasing supplementation of dietary sodium was unclear. Lower levels of dietary sodium were sufficient for supporting immune organs (bursa and spleen). Therefore a verification of requirements by changing other electrolytes (K^+ and Cl^-), keeping a constant DEB level, and changing salt sources is suggested.

Competing interests

The authors declare that they have no competing interests.

Authors' contributions

RP and JK participated in the design of the study and did the farm visits and clinical investigations. MMHM performed the statistical analysis. All authors helped to draft the manuscript, and all authors read and approved the final manuscript.

Acknowledgements

This experiment was conducted and sponsored by the Research Model Broiler Unit at the Poultry Research and Training Center (University of Veterinary and Animal Sciences, Pakistan). All data collection of carcass and body characteristics was facilitated by students of BS (Hons) Poultry Science, Batch (2007–2011). The contributions of Zulfiqar Ali (Ex-Nutritional Consultant, Big Feeds (Pvt) Ltd, Lahore, Pakistan) in experimental diet formulation, preparation, and analyses are highly appreciated. This study was supported by Cooperative Research Program for Agriculture Science and Technology Development (Project No. PJ009422), Rural Development Administration, Republic of Korea. Higher Education Commission, Pakistanalso supported study financially.

References

1. Mongin P: *Electrolytes in nutrition. A review of basic principles and practical applications in poultry and swine.* Orlando, FL: Proceedings of 3rd Annul International Minerals Conference; 1980:1.
2. Leeson S, Summers JD: *Scott's Nutrition of the Chicken.* 4th edition. Guelph, Ontario, Canada: University Books; 2001.
3. Fletcher T: **Natural resources of Pakistan.** 2011. http://insider.pk/national/natural-resources-of-pakistan/.
4. Rondon EOO, Murakami AE, Furlan AC, Garcia J: Exigenciasnutricopmaos de sodio e cloro e estimativa do melhor balance electrolitico da racao para frangos de corte na fasepre–inicial (1–7 dias de idade). *Rev Bras Zoot* 2000, **29:**1162–1166.
5. Borges SAAV, Fischer da Silva ADA, Meira T, Moura AM, Ostrensky A: **Electrolyte balance in broiler growing diets.** *Intl J Poult Sci* 2004, **3:**623–628.
6. Mushtaq T, Sarwar M, Nawaz H, Mirza MA, Ahmad T: **Effect and interactions of dietary sodium and chloride on broiler starter performance (hatching to twenty-eight days of age) under subtropical summer conditions.** *Poult Sci* 2005, **84:**1716–1722.
7. Mushtaq T, Mirza MA, Athar M, Hooge DM, Ahmad T, Ahmad G, Mushtaq MMH, Noreen U: **Dietary sodium and chloride for twenty-nine to forty-two-day-old broiler chickens at constant electrolyte balance under subtropical summer conditions.** *J Appl Poult Res* 2007, **16:**161–170.
8. Sharma ML, Gangwar PC: **Electrolyte changes in the blood plasma of broilers as influenced by cooling during summer.** *Int'l J Biometeorol* 1987, **31**(3):211–216.
9. Smith MO, Teeter RG: **Practical application of potassium chloride and fasting during naturally occurring summer heat stress.** *Poult Sci* 1988, **67**(Suppl. 1):36. Abstr.
10. Borges SA, Fischer da Silva AV, Majorka A, Hooge DM, Cummings KR: **Physiological responses of broiler chickens to heat stress and dietary electrolyte balance (sodium plus potassium minus chloride, milliequivalents per kilogram).** *Poult Sci* 2004, **83:**1551–1558.
11. Johnson RJ, Karunajeewa H: **The effects of dietary minerals and electrolytes on the growth and physiology of the young chick.** *J Nutr* 1985, **115:**1680–1690.
12. Teeter RG, Smith MO, Owens FN, Arp SC, Sangiah S, Breazile JE: **Chronic heat stress and respiratory alkalosis, Occurrence and treatment in broiler chicks.** *Poult Sci* 1985, **64:**1060–1064.
13. Gorman I, Balnave D: **Effects of dietary mineral supplementation on the performance and mineral excretions of broilers at high ambient temperatures.** *Br Poult Sci* 1994, **35:**563–572.
14. Hooge DM: *Practicalities of using dietary sodium and potassium supplements to improve poultry performance. Proceedings of Arkansas Nutrition Conference.* 2003:19.
15. Ahmad T, Sarwar M, Mahr-un-Nisa, Ahsan-ul-Haq, Zia-ul-Hasan: **Influence of varying sources of dietary electrolytes on the performance of broilers reared in a high temperature environment.** *Anim Feed Sci Technol* 2005, **120:**277–298.
16. Ahmad T, Sarwar M: **Dietary electrolyte balance, Implications in heat stressed broilers.** *Worlds Poult Sci J* 2006, **62:**638–653.
17. Ruiz-Lopez B, Austic RE: **The effects of selected minerals on acid–base balance of growing chicks.** *Poult Sci* 1993, **72:**1054–1062.
18. Whiting TS, Andrews LD, Stamps L: **Effects of sodium bicarbonate and potassium chloride drinking water supplementation. 1. Performance and exterior carcass quality of broilers grown under thermoneutral or cyclic heat-stress conditions.** *Poult Sci* 1991, **70:**53–9.
19. Pourreza J, Edriss MA: **The effects of high vs. normal temperatures on the physical characteristics of the broilers carcass.** *J Agric Sci Technol* 1992, **1:**35–41.
20. Karunajeewa H, Barr DA, Fox M: **The effects of high vs. normal temperatures on the growth performance of broiler chickens.** *Br Poult Sci* 1986, **27:**601–612.
21. Borges SA, Ariki J, Martins JCL, De-Moraes VMB: **Potassium chloride supplementation in heat stressed broilers.** *Rev Bras Zool* 1999, **28:**313–319.
22. Hubbard: *Broiler Management Guide.* Duluth, GA: Hubbard LLC; 2004.
23. AOAC: *Official Methods of Analysis.* 18th edition. Gaithersburg, MD: Association of Official Analytical Chemists; 2005.
24. Lacroix RL, Keeney DR, Welsh LM: **Potentiometric titration of chloride in plant tissue extracts using the chloride ion electrode.** *Commun Soil Sci Plant Anal* 1970, **1:**1–6.
25. NRC: *Nutrient Requirements of Poultry.* 9 Revisedth edition. Washington, DC: National Academic Press; 1994.
26. Han Y, Baker DH: **Digestible lysine requirement of male and female broiler chicks during the period three to six weeks post-hatching.** *Poult Sci* 1994, **73:**1739–45.
27. Suresh BN, Reddy BSV, Manjunatha PBH, Jaishankar N: **Carcass Characteristics of Broilers Fed Sugarcane Press Residue with Biotechnological Agents.** *Int J Poult Sci* 2009, **8**(7):671–676.
28. Borges SA, Fischer da Silva AV, Ariki J, Hooge DM, Cummings KR: **Dietary electrolyte balance for broiler chickens under moderately high ambient temperatures and relative humidities.** *Poult Sci* 2003, **82:**301–308.
29. Socha MT, Ensley SM, Tomlinson DJ, Johnson B: *Variability of water consumption and potential impact on animal performance. Proceedings of California Animal Nutrition Conference;* 2002:81–91.
30. Borges SA, Fischer da Silva AV, Ariki J, Hooge DM, Cummings KR: **Dietary electrolyte balance for broiler chickens exposed to thermoneutral or heat¯stress environments.** *Poult Sci* 2003, **82:**482–435.

31. Good B: Water quality affects poultry production. *Poult Dig* 1985, **44**(517):108–109.

32. Grizzle JT, Armbrust MB, Saxton A: Water quality I, The effect of water nitrate and pH on broiler growth performance. *J Appl Poult Res* 1996, **5**:330–336.

33. Water and Electrolytes (& Iodine), Sec. 4. In *Animal Nutrition Handbook*. 2 revth edition. Edited by Chiba LI. Auburn University Press, ALabama; 2009.

34. Carew L, Mcmurtry J, Alster F: Effects of lysine deficiencies on plasma levels of thyroid hormones, insulin-like growth factors I and II, liver and body weights, and feed intake in growing chickens. *Poult Sci* 2005, **84**:1045–1050.

35. Bailey CA, Stipanovic RD, Ziehr MS, Haq AU, Sattar M, Kubena LF, Kim HL, Vieira RD: Cottonseed with a high (+) to (−)-gossypol enantiomer ratio favorable to broiler production. *J Agric Food Chem* 2000, **48**:5692–5695.

36. Farran MT, Halaby WS, Barbour GW, Uwayjan MG, Sleiman FT, Ashkarian VM: Effects of feeding ervil (Viciaervilia) seeds soaked in water or acetic acid on performance and internal organ size of broilers and production and egg quality of laying hens. *Poult Sci* 2005, **84**:1723–1728.

26

Effects of citrus pulp, fish by-product and *Bacillus subtilis* fermentation biomass on growth performance, nutrient digestibility, and fecal microflora of weanling pigs

Hyun Suk Noh[1], Santosh Laxman Ingale[1], Su Hyup Lee[1], Kwang Hyun Kim[1], Ill Kyong Kwon[2], Young Hwa Kim[3] and Byung Jo Chae[1*]

Abstract

An experiment was conducted to investigate the effects of dietary supplementation with citrus pulp, fish by-product, and *Bacillus subtilis* fermentation biomass on the growth performance, apparent total tract digestibility (ATTD) of nutrients, and fecal microflora of weanling pigs. A total of 180 weaned piglets (Landrace × Yorkshire × Duroc) were randomly allotted to three treatments on the basis of body weight (BW). There were six replicate pens in each treatment with 10 piglets per pen. Dietary treatments were corn-soybean meal-based basal diet supplemented with 0 (control), 2.5, and 5.0% citrus pulp, fish by-product, and *B. subtilis* fermentation biomass. The isocaloric and isoproteineous experimental diets were fed in mash form in two phases (d 0 ~ 14, phase I and d 15 ~ 28, phase II). Dietary treatments had significant linear effects on gain to feed ratio (G:F) in all periods, whereas significant linear effects on ATTD of dry matter (DM), gross energy (GE), and ash were only observed in phase I. Piglets fed diet supplemented with 5.0% citrus pulp, fish by-product, and *B. subtilis* fermentation biomass showed greater ($p < 0.05$) G:F (phase I, phase II, and overall) as well as ATTD of DM, GE, and ash (phase I) than pigs fed control diet. Dietary treatments also had significant linear effects on total anaerobic bacteria populations by d 14 and 28. In addition, piglets fed diet supplemented with 5.0% citrus pulp, fish by-product and *B. subtilis* fermentation biomass showed greater ($p < 0.05$) fecal total anaerobic bacteria populations (d 14 and 28) than pigs fed control diet. Dietary treatments had no significant effects (linear or quadratic) on average daily gain (ADG), average dial feed intake (ADFI; phase I, phase II, and overall), or fecal populations of *Bifidobacterium* spp., *Clostridium* spp., and coliforms (d 14 and 28). These results indicate that dietary supplementation with 5.0% citrus pulp, fish by-product, and *B. subtilis* fermentation biomass has the potential to improve the feed efficiency, nutrient digestibility, and fecal microflora of weanling pigs.

Keywords: Citrus pulp, *Bacillus subtilis*, Performance, Fecal microflora, Weanling pigs

Background

Worldwide production of citrus fruits approaches 90 million tons per year [1]. Most of these fruits are squeezed to juice and by-products, including peels, segment membranes, and other parts, which are considered as citrus juice waste or pulp [2]. Due to their high processing cost, most citrus juice industry by-products are dumped into the ocean, leading to environmental pollution [3]. Recent approaches advocating the use of citrus juice waste have focused on reduction of its moisture contents and its use as an animal feed [4]. Dried citrus pulp contains relatively large amounts of pectins [5] and soluble carbohydrates [6]. Further, several health-promoting bioactive compounds such as limonene, hesperidin, naringin, quercetin, and bioflavonoids have been identified in citrus pulp [3,7-9]. Contreras Esquivel et al. [10] and Sen et al. [11] reported that citrus juice industry by-products have the necessary characteristics required for substrates of probiotic growth during fermentation.

* Correspondence: bjchae@kangwon.ac.kr
[1]Department of Animal Resources Science, College of Animal Life Sciences, Kangwon National University, Chuncheon 200-701, South Korea
Full list of author information is available at the end of the article

Among probiotic microbes, *Bacillus* spp. are well known for their ability to produce pectinase using pectin in citrus peel as the sole carbon source [12,13]. Previous have reported that citrus juice waste can be used as a substrate for the growth of *B. subtilis*, and the resulting fermentation biomass has potential for improving the performance, intestinal morphology, and cecal microflora of broilers and weanling pigs [11,14]. Therefore, the objective of the present study was to investigate the effects of dietary supplementation with citrus pulp, fish by-product, and *B. subtilis* fermentation biomass on growth performance, apparent total tract digestibility (ATTD) of nutrients, and fecal microflora in weanling of pigs.

Methods

The protocol for the present experiment was approved by the Institutional Animal Care and Use Committee of Kangwon National University, Chuncheon, Republic of Korea.

Preparation of fermentation biomass

The *B. subtilis* used in the present study was isolated and characterized by Yoo et al. [3] and maintained in the laboratory at −80°C as stock culture. Culture broth medium consisting of 6% corn steep liquor, 4% molasses, 0.3% yeast extract, 0.5% KH_2PO_4, and 0.25% K_2HPO_4 in distilled water was prepared and autoclaved before being used. Stock culture (2 mL) was then added to 2 L of autoclaved culture broth and incubated at 37°C at pH 7.0 for 48 h.

The *B. subtilis* grown on culture broth medium was used as a starter to produce fermentation biomass. Dried citrus pulp and fish by-product with 30% moisture was used as the sole substrate. The substrate was inoculated with 2 L of starter per 10 kg of substrate and fermented for 7 d at 32°C and pH 7.0. After 7 d, the fermentation biomass was dried in a forced-air drying oven at 40°C for 72 h.

Animals and experimental design

A total of 180 weaned piglets (Landrace × Yorkshire × Duroc) were randomly allotted to three treatments on the basis of initial body weight (BW). There were six replicate pens in each treatment with 10 piglets per pen. All piglets were clinically healthy at the start of the trial and originated from 20 sows in their third parity. Dietary treatments were corn-soybean meal-based basal diet supplemented with 0 (control), 2.5, and 5.0% citrus pulp, fish by-product, and *B. subtilis* fermentation biomass. The citrus pulp, fish by-product and *B. subtilis* fermentation biomass were added to the weanling pig diets by replacing equal volumes of fish meal. The isocaloric and isoproteineous experimental diets were fed in mash form

in two phases (d 0 ~ 14, phase I and d 15 ~ 28, phase II). Diets for phase I were formulated to contain 3,400 kcal/kg metabolizable energy (ME), 21.0% crude protein (CP), and 1.6% lysine (Table 1), whereas diets for phase II were formulated to contain 3,360 kcal/kg ME, 20.0% CP, and 1.4% lysine (Table 2). All diets met or exceeded the nutrient requirements recommended by NRC [15].

The experiment was conducted at the Kangwon National University farm facility. The piglets were housed in partially slatted concrete floor pens (2.8 × 5.0 m). The temperature in the barn was 30°C at the beginning of the experiment and was slowly reduced to 25°C on d 8, after which it was kept constant until the end of the

Table 1 Ingredient and chemical compositions of experimental diets (d 0 ~ 14; as-fed basis)

	Citrus pulp, fish by-product, and *Bacillus subtilis* fermentation biomass, %		
	0	2.5	5
Ingredients, %			
Corn	45.38	42.36	39.34
Whey powder	15.38	15.38	15.38
Deh-SBM	24.88	27.08	29.29
Soy oil	3.00	3.00	3.00
Fish meal	5.0	2.5	0.0
L-lysine (78%)	0.50	0.51	0.52
DL-Methionine (100%)	0.12	0.14	0.15
Choline-chloride (50%)	0.07	0.07	0.07
MCP	0.50	0.79	1.08
Limestone	0.67	0.73	0.78
Salt	0.20	0.20	0.20
Mineral premix[1]	0.30	0.30	0.30
Vitamin premix[2]	0.30	0.30	0.30
ZnO	0.34	0.34	0.34
Sucrose	3.36	3.80	4.25
Citrus pulp, fish by-product, and *Bacillus subtilis* fermentation biomass	0.0	2.50	5.00
Calculated chemical composition[3]			
ME, kcal/kg	3,400	3,400	3,400
CP, %	21.00	21.00	21.00
Ca, %	0.80	0.80	0.80
Av. P, %	0.40	0.40	0.40
Lys, %	1.60	1.60	1.60
Met + Cys, %	0.80	0.80	0.80

[1]Supplied per kilogram of diet: 45 mg Fe, 0.25 mg Co, 50 mg Cu, 15 mg Mn, 25 mg Zn, 0.35 mg I, 0.13 mg Se.
[2]Supplied per kilogram of diet: 16,000 IU vitamin A, 3,000 IU vitamin D3, 40 IU vitamin E, 5.0 mg vitamin K3, 5.0 mg vitamin B1, 20 mg vitamin B2, 4 mg vitamin B6, 0.08 mg vitamin B12, 40 mg pantothenic acid, 75 mg niacin, 0.15 mg biotin, 0.65 mg folic acid, 12 mg antioxidant.
[3]Based on NRC [15] values.

Table 2 Ingredient and chemical compositions of experimental diets (d 15 ~ 28; as-fed basis)

	Citrus pulp, fish by-product, and *Bacillus subtilis* fermentation biomass, %		
	0	2.5	5
Ingredients, %			
Corn	61.55	59.05	56.55
Whey powder	5.00	5.00	5.00
Deh-SBM	23.11	25.23	27.35
Soy oil	2.21	2.23	2.25
Fish meal (60%)	5.00	2.50	-
L-lysine (78%)	0.38	0.39	0.40
DL-Methionine (100%)	0.05	0.06	0.07
MCP	0.71	1.00	1.29
Limestone	0.75	0.80	0.85
Salt	0.30	0.30	0.30
Mineral premix[1]	0.30	0.30	0.30
Vitamin premix[2]	0.30	0.30	0.30
ZnO	0.34	0.34	0.34
Citrus pulp, fish by-product, and *Bacillus subtilis* fermentation biomass	0.0	2.50	5.00
Calculated chemical composition[3]			
ME, kcal/kg	3,360	3,360	3,360
CP, %	20.00	20.00	20.00
Ca, %	0.78	0.78	0.78
Av. P, %	0.37	0.37	0.37
Lys, %	1.40	1.40	1.40
Met + Cys, %	0.72	0.72	0.72

[1]Supplied per kilogram of diet: 45 mg Fe, 0.25 mg Co, 50 mg Cu, 15 mg Mn, 25 mg Zn, 0.35 mg I, 0.13 mg Se.
[2]Supplied per kilogram of diet: 16,000 IU vitamin A, 3,000 IU vitamin D_3, 40 IU vitamin E, 5.0 mg vitamin K_3, 5.0 mg vitamin B_1, 20 mg vitamin B_2, 4 mg vitamin B_6, 0.08 mg vitamin B_{12}, 40 mg pantothenic acid, 75 mg niacin, 0.15 mg biotin, 0.65 mg folic acid, 12 mg antioxidant.
[3]Based on NRC [15] values.

trial. The humidity ranged between 60 and 70%. Each pen was equipped with an infrared heating lamp, self-feeder, and low-pressure nipple drinker to allow *ad libitum* access to feed and water.

Experimental procedure and sampling

The pigs were individually weighed at the beginning of the trial as well as on d 14 and 28 of the experiment. Feed that was not consumed was weighed at the end of each phase, and consumption was calculated for phase I (d 0 to 14), phase II (d 15 to 28), and for the overall study period (d 0 to 28). As feed wastage was considered minimal, feed disappearance was determined to be a reliable estimate of feed consumption. Feed consumption was calculated at the end of each phase, and average

daily feed intake (ADFI) and gain to feed ratio (G:F) were calculated. Average daily gain (ADG) and ADFI were calculated by dividing total pen weight gain and total pen feed consumption by the number of days. The G:F for each pen was calculated by dividing the ADG by the ADFI. To evaluate the effects of dietary treatments on the apparent total tract digestibility (ATTD) of energy and nutrients, 2.5 g kg^{-1} of chromium (as an inert, indigestible indicator) was included in the diets from d 8 to 14 (phase I) or d 22 to 28 (phase II). Fecal grab samples (100 gm/d per pen) were collected from each pen during the last 3 d of each phase to determine the ATTD of nutrients. Fecal samples from each pen were pooled, dried in a forced air-drying oven at 60°C for 72 h, and ground in a Wiley laboratory mill (Thomas Model 4 Wiley® Mill, Thomas scientific, Swedesboro, NJ, USA) using a 1-mm screen. Additionally, fresh fecal samples were collected from two pigs from each pen on d 14 and 28 and then measured for fecal bacterial counts. The samples collected for microbial analysis were immediately placed on ice (2–3 h) and transported to the laboratory for further analysis on the same day.

Chemical and microbial analyses

Experimental diets and excreta samples were analyzed in triplicate for dry matter (DM, method 930.15; [16]), crude protein (CP, method 990.03; [16]), ash (method 942.05; [16]), calcium, and phosphorus (method 985.01; [16]). Gross energy was measured by a bomb calorimeter (Model 1261, Parr Instrument Co., Moline, IL), and chromium concentrations of experimental diets and excreta samples were determined with an automated spectrophotometer (Jasco V-650, Jasco Corp., Tokyo, Japan) according to the procedure of Fenton and Fenton [17]. The ATTD (%) of nutrients was calculated by the following formula:

$$ATTD\ (\%) = [1-(N_f \times C_d)/(N_d \times C_f)] \times 100$$

Where

N_f = nutrient concentration in feces (%)
N_d = nutrient concentration in diet (%)
C_f = chromium concentration in feces (%)
C_d = chromium concentration in diet (%)

The microbiological assay of excreta was carried out by the procedure suggested by Choi et al. [18]. The microbial groups enumerated were total anaerobic bacteria (TAB; plate count agar, Difco Laboratories, Detroit, MI, USA), *Bifidobacterium* spp. (MRS agar), coliforms (violet red bile agar, Difco Laboratories, Detroit, MI, USA), and *Clostridium* spp. (Tryptose sulphite cycloserine agar, Oxoid, Hampshire, UK). The anaerobic conditions during the TAB and *Clostridium* spp. assays were created by using the gas-pak anaerobic system (BBL, No. 260678, Difco, Detroit, MI, USA).

Statistical analysis

Data generated in the present study were subjected to statistical analysis using the GLM procedure of SAS (SAS Inst. Inc., Cary, NC) in a randomized complete block design. When significant difference were identified among treatment means, they were separated using Tukey's Honestly Significant Difference test. The linear and quadratic contrasts were used to compare effects of increasing levels of citrus pulp, fish by-product, and *B. subtilis* fermentation biomass (0. 2.5 and 5.0% of basal diet). The pen was used as the experimental unit for the analysis of all parameters. Probability values of ≤ 0.05 were considered significant.

Results

Growth performance

Dietary treatments had significant linear effects on gain to feed ratio (G:F) in all periods (Table 3). Moreover, piglets fed diet supplemented with 5.0% citrus pulp, fish by-product and *B. subtilis* fermentation biomass showed greater ($p < 0.05$) G:F than piglets fed control diet. The G:F of piglets fed diet supplemented with 2.5% citrus pulp and *B. subtilis* fermentation biomass was not different ($p > 0.05$) from that of piglets fed control or 5.0% citrus pulp, fish by-product, and *B. subtilis* fermentation

biomass. Dietary treatments had no effects (linear or quadratic; $p > 0.05$) on the ADG or ADFI of piglets in all periods.

Apparent total tract digestibility

Significant linear effects on ATTD of DM, GE, and ash were only observed in phase I (Table 4). In addition, in phase I, piglets fed diet supplemented with 5.0% citrus pulp and *B. subtilis* fermentation biomass showed greater ($p < 0.05$) ATTD of DM, GE, and ash than piglets fed control diet. However, ATTD of DM, GE, and ash of piglets fed diet supplemented with 2.5% citrus pulp, fish by-product, and *B. subtilis* fermentation biomass was not different ($p > 0.05$) than those of pigs fed control and 5.0% citrus pulp, fish by-product, and *B. subtilis* fermentation biomass. During phase II, dietary treatments had no effects (linear or quadratic; $p > 0.05$) on ATTD of nutrients.

Fecal microflora

Dietary treatments also had significant linear effects on total anaerobic bacteria populations by d 14 and 28 (Table 5). Moreover, piglets fed diet supplemented with 5.0% citrus pulp, fish byproduct, and *B. subtilis* fermentation biomass showed greater ($p < 0.05$) fecal total

Table 3 Effect of dietary inclusion of citrus pulp, fish by-product and *Bacillus subtilis* fermentation biomass on growth performance of weanling pigs[1]

Item	Citrus pulp, fish by-product, and *Bacillus subtilis* fermentation biomass, %			SEM[2]	P-values[3]	
	0	2.5	5.0		Linear	Quadratic
Phase I (d 0~14)						
ADG, g	441	460	474	14.86	0.145	0.871
ADFI, g	721	706	689	8.51	0.053	0.937
G:F, g/kg	613[b]	652[ab]	688[a]	19.96	0.018	0.960
Phase II (d 15~28)						
ADG, g	515	540	576	16.80	0.057	0.688
ADFI, g	1,087	1,129	1,123	13.36	0.163	0.686
G:F, g/kg	474[b]	478[ab]	512[a]	9.06	0.018	0.199
Overall (d 0~28)						
ADG, g	478	500	525	11.40	0.250	0.944
ADFI, g	904	917	906	13.74	0.058	0.747
G:F, g/kg	529[b]	545[ab]	579[a]	11.53	0.024	0.545

[ab]Values with different superscripts of the row significantly differ (p < 0.05).
[1]Data represent means based on 6 replicates per treatment.
[2]Standard error of means.
[3]Linear and quadratic effects of increasing citrus pulp and *Bacillus subtilis* fermentation biomass in the diet.

Table 4 Effect of dietary inclusion of citrus pulp, fish by-product, and *Bacillus subtilis* fermentation biomass on apparent total tract digestibility (%) of nutrients in weanling pigs[1]

Item	Citrus pulp, fish by-product, and *Bacillus subtilis* fermentation biomass, %			SEM[2]	P-values[3]	
	0	2.5	5.0		Linear	Quadratic
Phase I (d 8–14)						
DM	84.63[b]	84.99[ab]	85.49[a]	0.15	0.003	0.714
GE	83.56[b]	83.89[ab]	84.19[a]	0.12	0.005	0.890
CP	79.66	79.93	80.21	0.275	0.098	0.906
Ash	60.02[b]	60.76[ab]	61.72[a]	0.26	0.005	0.461
Ca	56.48	55.52	54.99	1.77	0.548	0.851
P	45.58	45.22	45.04	1.03	0.690	0.592
Phase II (d 22–28)						
DM	81.87	82.04	82.77	0.35	0.106	0.535
GE	82.14	82.67	83.34	0.47	0.101	0.909
CP	75.95	76.78	77.39	0.61	0.127	0.885
Ash	54.98	55.34	55.57	0.75	0.595	0.947
Ca	52.08	52.34	52.53	1.38	0.821	0.983
P	46.96	47.57	48.67	1.52	0.448	0.894

[ab]Values with different superscripts of the row significantly differ (p < 0.05).
[1]Data represent means based on 6 replicates per treatment.
[2]Standard error of means.
[3]Linear and quadratic effects of increasing citrus pulp and *Bacillus subtilis* fermentation biomass in the diet.

Table 5 Effect of dietary inclusion of citrus pulp, fish by-product, and *Bacillus subtilis* fermentation biomass on fecal microflora (Log$_{10}$cfu/g) of weanling pigs[1]

Item	Citrus pulp, fish by-product, and *Bacillus subtilis* fermentation biomass, %			SEM[2]	P-values[3]	
	0	2.5	5.0		Linear	Quadratic
Phase I (d 14)						
Total anaerobic bacteria	9.23^b	9.32^{ab}	9.50^a	0.05	0.020	0.446
Bifidobacterium spp.	8.73	8.78	8.79	0.05	0.434	0.706
Clostridium spp.	7.21	7.17	7.10	0.07	0.277	0.846
Coliforms	6.98	6.95	6.93	0.05	0.517	0.907
Phase II (d 28)						
Total anaerobic bacteria	9.38^b	9.46^{ab}	9.57^a	0.16	0.020	0.786
Bifidobacterium spp.	8.82	8.85	8.88	0.18	0.379	0.976
Clostridium spp.	7.30	7.26	7.24	0.22	0.400	0.887
Coliforms	7.16	7.11	7.08	0.23	0.362	0.814

^{ab}Values with different superscripts of the row significantly differ (p < 0.05).
[1]Data represent means based on 6 replicates per treatment.
[2]Standard error of means.
[3]Linear and quadratic effects of increasing citrus pulp and *Bacillus subtilis* fermentation biomass in the diet.

anaerobic bacteria populations than piglets fed control diet. However, total anaerobic bacteria populations of piglets fed diet supplemented with 2.5% citrus pulp, fish by-product, and *B. subtilis* fermentation biomass were not different (*p* > 0.05) than those of piglets fed control or 5.0% citrus pulp and *B. subtilis* fermentation biomass. Dietary treatments had no effects (linear or quadratic; *p* > 0.05) on populations of fecal *Bifidobacterium* spp., *Clostridium* spp., and coliform (d 14 and 28).

Discussion

Among several bacterial species used as probiotics, spore-forming *Bacillus* spp. are preferred due to the high resistance of their spores to harsh environments and long-term storability at ambient temperatures [19,20]. We have reported previously that *B. subtilis* fermentation biomass prepared by solid substrate fermentation with corn-soybean meal or citrus juice waste substrate has potential for improving growth performance, nutrient retention, and intestinal morphology as well as reducing harmful intestinal bacteria in broilers and weanling pigs [11,14,18,21]. In this study, dried citrus pulp and fish by-product was fermented with *B. subtilis*, and the resulting fermentation biomass was supplemented to weanling pig diets.

In the present study, dietary supplementation with 5.0% citrus pulp, fish by-product and *B. subtilis* fermentation biomass improved the feed efficiency of weanling pigs. These results are in agreement with data reported by Lee et al. [14], who observed improved feed efficiency in pigs fed diet supplemented with *B. subtilis* LS 1–2 grown on citrus juice waste. Similarly, a previous study reported that dietary inclusion of *B. subtilis* LS 1–2 and corn-soybean meal or citrus juice waste fermentation biomass can improve the feed conversion efficiency of

broilers [11,21]. In contrast, this study found that citrus pulp, fish by-product, and *B. subtilis* fermentation biomass had no effects on ADG and ADFI of weanling pigs. This discrepancy could be attributed to variations in the administration level of probiotic products, health status within herds, farm hygiene, diet composition, feed forms, and interactions with other dietary feed additives [22,23]. Other studies also reported improved growth performance in pigs fed diet supplemented with probiotic products containing *Bacillus* spp. [24-27]. The fermentation biomass in the present study was prepared by fermentation of citrus pulp with *B. subtilis*, and the resultant biomass included probiotic microbes as well as secondary metabolites produced during microbial fermentation, as previously described [28,29].

In this study, dietary supplementation with 5.0% citrus pulp, fish by-product, and *B. subtilis* fermentation biomass improved ATTD of DM, GE, and ash during phase I (d 0–14). Our results confirm the findings of Lee et al. [14], who evaluated the effects of *B. subtilis* fermentation biomass dietary supplementation on weanling pigs and observed improved ATTD of DM and GE during phase I (d 0–14). Similarly, it was reported that weanling pigs fed diet supplemented with probiotic products prepared using *B. subtilis* and corn-soybean meal as a substrate show improved coefficient of total tract digestibility of DM and GE [30]. In contrast to the results of Choi et al. [30], Lee et al. [14], and the present study, Chen et al. [26] and Wang et al. [27] observed no effects of *Bacillus*-based probiotic products on the ATTD of nutrients in growing and finishing pigs. Variations in nutrient digestibility in weanling, growing, and finishing pigs indicate that the age of pigs is a considerable factor affecting the efficacy of *B. subtilis* and its fermentation biomass.

In the present study, the ATTD of nutrients during phase II (d 15–28) was not affected by dietary treatments, possibly due to the presence of developed and stable intestinal microflora during phase II. Stavric and Kornegay [31] reported that probiotics are more effective in pigs during microflora development or when microflora stability is impaired.

It has been well established that probiotic products favorably affect the host animal by improving intestinal balance [19], by creating gut micro-ecological conditions that suppress harmful microorganisms, and by favoring beneficial microorganisms [14,18,32]. Positive effects of probiotic products containing *B. subtilis* [11,18,21,33] have been reported previously. In the present study, pigs fed diets supplemented with 5.0% citrus pulp, fish by-product, and *B. subtilis* fermentation biomass showed greater total anaerobic bacteria populations by d 14 and 28, whereas there were no effects on *Bifidobacterium* spp., *Clostridium* spp., and coliform populations. In contrast, Lee et al. [14] observed that weanling pigs fed diet supplemented with corn-soybean and *B. subtilis* fermentation biomass show no difference in total anaerobic bacterial populations, whereas cecal *Clostridium* spp. and coliform populations are significantly reduced. This discrepancy might be due to variations in the type and dose of probiotic product, method of probiotic fermentation, and health status of piglets. In this study, we used healthy piglets with no symptoms of diarrhea.

Conclusions

In conclusion, the results obtained in present study indicate that dietary supplementation with 5.0% citrus pulp, fish by-product, and *B. subtilis* fermentation biomass has the potential to improve feed efficiency, nutrient digestibility, fecal microflora, and cost savings in weanling pigs without affecting growth performance.

Competing interests
The authors declare that there are no conflicts of interest.

Authors' contributions
BJC and YHK designed experiment, HSN, LSH, KHK Carried out animal trial, HSN, SLI, LSH and IKK done lab analysis, SLI, LSH analyzed data, SLI, YHK and BJC written manuscript. All authors read and approved the final manuscript.

Acknowledgments
This work was carried out with the support of the Cooperative Research Program for Agriculture Science & Technology Development (Project No. 009410) Rural Development Administration, Republic of Korea. The authors sincerely acknowledge the technical facilities provided by Kangwon National University, Chuncheon, Republic of Korea.

Author details
[1]Department of Animal Resources Science, College of Animal Life Sciences, Kangwon National University, Chuncheon 200-701, South Korea.
[2]Department of Animal Products and Food Science, College of Animal Life Sciences, Kangwon National University, Chuncheon 200-701, South Korea.
[3]Department of Animal Resources Development, Swine Science Division, National Institute of Animal Science, RDA, Suwon, South Korea.

References
1. Marin FR, Soler-Rivas C, Benavente-Garcia O, Castillo J, Perez-Alvarez JA: By-products from different citrus processes as a source of customized functional fibers. *Food Chem* 2007, **100**:736–741.
2. Martinez Pascual J, Fernandez Carmona J: Citrus pulp in diets for fattening lambs. *Anim Feed Sci Technol* 1980, **5**:11–22.
3. Yoo JH, Lee JJ, Lee HB, Choi SW, Kim YB, Sumathy B, Kim EK: Production of an antimicrobial compound by *Bacillus subtilis* LS 1–2 using a citrus-processing byproduct. *Korean J Chem Eng* 2011, **28**:1400–1405.
4. Gladine C, Morand C, Rock E, Bauchart D, Durand D: Plant extracts rich in polyphenols (PERP) are efficient antioxidants to prevent lipoperoxidation in plasma lipids from animals fed n-3 PUFA supplemented diets. *Anim Feed Sci Technol* 2007, **136**:281–296.
5. Wilkin MR, Widmer WW, Grohmann K, Cameron RG: Hydrolysis of grapefruit peels waste with cellulase and pectinase enzymes. *Bioresour Technol* 2007, **98**:1596–1601.
6. Rihani N, Guessous F, Johnson WL: Nutritive value of dried citrus and beet pulps produced in Morocco. *J Anim Sci* 1986, **63**(Suppl. 1):428. Abstr.
7. Belshaw F: Citrus flour - a new fiber, nutrient source. *Food Prod Dev* 1978, **12**(7):36.
8. Braddock RJ: Utilization of citrus juice vesicle and peel fiber. *Food Technol* 1983, **37**:85–87.
9. Formica JV, Regelson W: Review of the biology of Quercetin and related bioflavonoids. *Food Chem Toxicol* 1995, **33**:1061–1080.
10. Contreras Esquivel JC, Hours RA, Voget CE, Mignone CF: *Aspergillus kawachii* produces an acidic pectin releasing enzyme activity. *J Biosci Bioeng* 1999, **88**:48–52.
11. Sen S, Ingale SL, Kim YW, Kim JS, Kim KH, Khong C, Lohakare JD, Kim EK, Kim HS, Kwon IK, Chae BJ: Effect of supplementation of *Bacillus subtilis* LS 1–2 grown on citrus-juice waste and corn-soybean meal substrate on growth performance, nutrient retention, caecal microbiology and small intestinal morphology of broilers. *Asian-Aust J Anim Sci* 2011, **24**:1120–1127.
12. Mahmood AU, Greenman J, Scragg AH: Orange and potato peel extracts: Analysis and use as *Bacillus* substrates for the production of extracellular enzymes in continuous culture. *Enzy Microb Technol* 1998, **22**:130–137.
13. Matsumoto T, Sugiura Y, Kondo A, Fukuda H: Efficient production of protopectinases by *Bacillus subtilis* using medium based on soybean flour. *Biochem Eng J* 2000, **6**:81–86.
14. Lee SH, Ingale SL, Kim JS, Kim KH, Lokhande A, Kim EK, Kwon IK, Kim YH, Chae BJ: Effects of dietary supplementation with *Bacillus subtilis* LS 1–2 fermentation biomass on growth performance, nutrient digestibility, cecal microflora and intestinal morphology of weanling pig. *Anim Feed Sci Technol* 2014, **188**:102–110.
15. National Research Council: *Nutrient Requirement of Swine*. 10th edition. Washington, DC: National Academy Press; 1998.
16. AOAC: *Official Methods of Analysis of the Association of Official Analytical Chemists International*. 18th edition. Gaithersburg, MD, USA: 2007.
17. Fenton TW, Fenton M: An improved method for chromic oxide determination in feed and feces. *Can J Anim Sci* 1979, **59**:631–634.
18. Choi JY, Shinde PL, Ingale SL, Kim JS, Kim YW, Kim KH, Kwon IK, Chae BJ: Evaluation of multi-microbe probiotics prepared by submerged liquid or solid substrate fermentation and antibiotics in weaning pigs. *Livest Sci* 2011, **138**:144–151.
19. Fuller R: Probiotics in man and animals. *J Appl Bacteriol* 1989, **66**:365–378.
20. Hong HA, le Duc H, Cutting SM: The use of bacterial spore formers as probiotics. *FEMS Microbiol Rev* 2005, **29**:813–835.
21. Sen S, Ingale SL, Kim YW, Kim JS, Kim KH, Lohakare JD, Kim EK, Kim HS, Ryu MH, Kwon IK, Chae BJ: Effect of supplementation of *Bacillus subtilis* LS 1–2 to broiler diet on growth performance, nutrient retention, caecal microbiology and small intestinal morphology. *Res Vet Sci* 2012, **93**:264–268.
22. Chesson A: Probiotics and other intestinal mediators. In *Principles of Pig Science*. Edited by Cole DJA, Wiseman J, Varley MA. Loughborough, U.K: Nottingham University Press; 1994:197–214.
23. Lessard M, Dupuis M, Gagnon N, Nadeau E, Matte JJ, Goulet J, Fairbrother JM: Administration of *Pediococcus acidilactici* or *Saccharomyces cerevisiae boulardii* modulates development of porcine mucosal immunity and

reduces intestinal bacterial translocation after *Escherichia coli* challenge. *J Anim Sci* 2009, **87**:922–934.

24. Gracia MI, Hansen S, Sanchez J, Medel P, Imasde Agropecuaria SL: **Efficacy of addition of B. licheniformis and B. subtilis in pig diets from weaning to slaughter.** *J Anim Sci* 2004, **82**(Suppl. 1):26.

25. Alexopoulos C, Georgoulakis IE, Tzivara A, Kritas SK, Siochu A, Kyriakis SC: **Field evaluation of the efficacy of a probiotic containing Bacillus licheniformis and Bacillus subtilis spores, on the health status and performance of sows and their litters.** *J Anim Physiol Anim Nutr* 2004, **88**:381–392.

26. Chen YJ, Min BJ, Cho JH, Kwon OS, Son KS, Kim HJ, Kim IH: **Effects of dietary Bacillus-based probiotic on growth performance, nutrients digestibility, blood characteristics and fecal noxious gas content in finishing pigs.** *Asian-Aust J Anim Sci* 2006, **19**:587–592.

27. Wang Y, Cho JH, Chen YJ, Yoo JS, Huang Y, Kim HJ, Kim IH: **The effect of probiotic BioPlus 2B® on growth performance, dry matter and nitrogen digestibility and slurry noxious gas emission in growing pigs.** *Livest Sci* 2009, **120**:35–42.

28. Ohno A, Ano T, Shoda M: **Effect of temperature on production of lipopeptide antibiotics, iturin A and surfactin by a dual producer, Bacillus subtilis RB14, in solid-state fermentation.** *J Ferment Bioeng* 1995, **80**:517–519.

29. Scholten RHJ, van der Peet-Schwering CMC, Verstegen MWA, den Hartog LA, Schrama JW, Vesseur PC: **Fermented co-products and fermented compound diets for pigs: a review.** *Anim Feed Sci Technol* 1999, **82**:1–19.

30. Choi JY, Kim JS, Ingale SL, Kim KH, Shinde PL, Kwon IK, Chae BJ: **Effect of potential multimicrobe probiotic product processed by high drying temperature and antibiotic on performance of weanling pigs.** *J Anim Sci* 2011, **89**:1795–1804.

31. Stavric S, Kornegay ET: **Microbial probiotic for pigs and poultry.** In *Biotechnology in Animal Feeds and Animal Feeding.* Edited by Wallace RJ, Chesson A. Weinheim: VCH Verlagsgesellschaft mbH; 1995:205–231.

32. Kim JS, Ingale SL, Kim YW, Kim KH, Sen S, Ryu MH, Lohakare JD, Kwon IK, Chae BJ: **Effect of supplementation of multi-microbe probiotic product on growth performance, apparent digestibility, cecal microbiota and small intestinal morphology of broilers.** *J Anim Physiol Anim Nutr* 2012, **96**:618–626.

33. Teo AY, Tan HM: **Evaluation of the performance and intestinal gut microflora of broilers fed on corn-soy diets supplemented with Bacillus subtilis PB6 (CloSTAT).** *J Appl Poult Res* 2007, **16**:296–303.

27

Effects of feed form and feed particle size with dietary L- threonine supplementation on performance, carcass characteristics and blood biochemical parameters of broiler chickens

Vahid Rezaeipour[*] and Sepideh Gazani

Abstract

An experiment was conducted to evaluate the effect of form and particle size of feed supplemented with L- threonine on growth performance, carcass characteristic and blood biochemical parameters of broiler chickens. The experimental design was a $2 \times 2 \times 2$ factorial arrangement of treatments evaluating two feed forms (pellet or mash), two feed particle sizes (fine or course), and two inclusion rates of dietary L-threonine (with or without) which adopted from 7 to 42 days of age. In this experiment, 360 a day old chicks in two sexes were assigned in each treatment and each experimental unit was included 15 chicks. Feed consumption and weight gain were measured weekly. At 35 days of age, blood samples were taken to analysis blood biochemical parameters. At the end of the experimental period, two birds were slaughtered in each treatment and carcass analysis was carried out. The results showed that the effect of feed form on body weight gain and feed intake in whole of experimental period was significant ($P < 0.05$). Broilers fed pelleted diets had more weight gain than the mash group. Growth performance parameters were not affected by feed particle size and dietary L-threonine supplementation in whole of experimental period ($P > 0.05$). The results of carcass analysis showed that liver and gizzard relative weights were influenced by feed form ($P < 0.05$). However, pancreas and liver relative weights were affected by feed particle size and dietary L-threonine supplementation, respectively ($P < 0.05$). Triglyceride and VLDL levels were affected by feed form and dietary L-threonine supplementation ($P < 0.05$). The effect of feed particle size on blood biochemical parameters was not significant ($P > 0.05$). In conclusion, the experimental results indicated that feed form increased feed consumption and weight gain in whole of experimental period (1 to 42 days of age) while feed particle size and dietary L-threonine had no effect on broiler performance.

Keywords: L-threonine, Feed form, Broiler, Performance

Background

Interest in the effects of feed form and particle size has increased in recent years in Iran. Therefore, some of the poultry feed industry members search for ways to optimize the utilization of feed and improve poultry production. Particle size reduction increases both the number of particles and the surface area per unit volume allowing greater access to digestive enzymes and easier mixing of the ingredients [1]. Recommendations regarding optimum particle size, however, have been contradictory, as the results from feeding trials are confounded by a number of factors including feed physical form, complexity of the diet, grain type, endosperm hardness, grinding method, pellet quality and particle size distribution [2].

Different types of feed form have been evolved in broiler production at the present time. Mash is a form of a complete feed that is finely ground and mixed so that birds cannot easily separate out ingredients. Pellet system of feeding is really a modification of the mash system. It consists of mechanically pressing the mash into hard dry pellets. The greatest advantage in using pellets is that there is little waste in feeding. It is suggested that the large-grain particle size influenced broiler performance to a greater extent when birds were fed mash diets than when fed pelleted or crumbled diets [3,4].

---

* Correspondence: vrezaeipour@gmail.com
Department of Animal Science, Qaemshahr Branch, Islamic Azad University,
PO Box 163, Qaemshahr, Iran

Threonine is the third most limiting amino acid, especially in a low crude protein diet. L-threonine is added to the diet of pigs and poultry in order to exactly match the dietary amino acid balance with the unique nutritional requirements of the animal [5]. Dozier et al. [6] indicated that broilers fed inadequate threonine had decreased live performance, but no effects were apparent on carcass fat. It is reported that supplementation of L-threonine as a source of dietary-threonine in combination with *Saccharomyces cerevisiae* improved growth performance and intestinal morphology traits in broilers [5]. Therefore, it is necessary to balance threonine in broiler diets by adding L-threonine supplementation or use of soybean meal and meat meal as most important ingredients which supply threonine in the chick diets.

Therefore, the objectives of this study were to investigate the effects of feed form and particle size with L-threonine supplementation on performance, carcass characteristics and blood biochemical parameters of broilers.

Methods

Birds and diets

Three hundred and sixty 1 day-old Ross 308 broiler chicks (Mixed sex) with initial body weight of 46 ± 2 g were obtained from a local hatchery (Chalak Jojeh, Qaemshahr, Iran). Birds were weighed and randomly assigned to 24 floor pens with 15 birds per pen so that the average weight per pen was similar. Average initial body weight of chicks was 46 ± 3 g. The chicks were raised on floor pens for 42d and had free access to feed and water during the whole period of experiment. The ambient temperature was maintained at 32°C for the first 3d and then gradually decreased until 24°C was reached by 21d. Experimental procedures were approved by Islamic Azad University, Qaemshahr branch.

Corn grain was ground in hammer mill to achieve a size of 3 or 6 mm for fine and course grades, respectively. Half of each formulated diet was fed in mash form and other half was cold pelleted with a pellet mill. Dietary L-threonine supplementation was added to each starter and grower diets. The ingredients profile and chemical composition of dietary treatments are shown in Table 1.

Experimental procedure and sampling

All birds were group weighed by pen after overnight fasting at 21 and 42d. Feed intake on group basis was measured at the same time intervals. Feed conversion ratio for each pen was calculated by dividing feed intake to body weight gain. Mortality was checked daily and weighed for adjusting feed conversion ratio.

Blood sampling of two birds per each pen was done at 35 d of age. Samples were left for three hours at room temperature to coagulate the blood and then Serum was isolated by centrifugation (3000 g, 15 min) to blood

Table 1 The ingredients and chemical composition of basal diets

Ingredients	Basal diets	
	Starter	Grower
	(g/kg)	
Corn grain	624.0	645.0
Soybean meal	322.0	292.0
Corn oil	15.0	27.0
DCP	19.0	18.0
Limestone	5.0	4.0
Common salt	2.5	2.1
Mineral premix[1]	5.0	5.0
Vitamin premix[1]	5.0	5.0
DL-Methionine	2.5	2.0
L-Lysine	1.8	1.3
L-Threonine	0.6	0.4
Sodium bicarbonate	1.2	2.0
Salinomycine	0.5	0.5
Calculated analysis (%)		
AME (kcal/kg)	3000	3100
Crude protein	19.5	18.00
Calcium	0.90	0.90
Phosphorous, available	0.45	0.45
Met + Cys	0.88	0.88
Lysine	1.15	1.12

[1]Provides per kg of diet: 9000 I.U. vitamin A; 2000 I.U. vitamin D3; 18 I.U. vitamin E; 2 mg menadion; 1.8 mg thiamine; 6.6 mg riboflavin; 30 mg niacin; 3 mg pyridoxine; 15 mg vitamin B12; 100 mg D-pantothenic acid; 1 mg folic acid; 0.1 mg biotin; 500 mg choline chloride; 100 mg antioxidant; 100 mg manganese; 84.7 mg zinc; 50 mg iron; 10 mg copper; 1 mg iodine; 0.2 mg selenium.

chemistry measurements. Serum Triglyceride, total cholesterol and glucose were analyzed using enzymatic colorimetric kits (Pars Azemon, Iran).

At the end of the experiment (on day 42), after overnight fasting , two birds from each pen (four birds per each treatment) with a body weight close to pen mean were selected and killed. Viscera were manually removed and carcass characteristics were determined. Then, empty weight of total digestive tract, caeca, gizzard, liver (without gallbladder) and pancreas were measured. All carcass data are presented based on percent of live weight of each bird.

Statistical analysis

The experiment was conducted using completely randomized design with factorial structure. Pen mean was an experimental unit. Effect of feed particle size (fine or course), feed form (pellet or mash), and dietary L-threonine (with or without) were statistically analyzed as $2 \times 2 \times 2$ factorial design by General Linear Models (GLM) procedure of SAS [7]. Probability values <0.05 were taken to indicate statistical significance using Duncan multiple range test.

Results

The results of growth performance of birds are shown in Table 2. The results showed that weight gain and feed intake were not affected by particle size and L-threonine supplementation in whole of experimental period (P >0.05). However, pelleted form and L-threonine supplementation increased body weight gain of broilers at 1–21 days of age (P <0.05). The birds fed with pellet form had a better weight gain than those fed with mash form in whole of experimental period. The results showed that feed consumption increased in birds fed pelleted diets at all phases of growth period. The results also revealed that the birds fed diets without L-threonine supplementation had better weight gain (1-21d) and feed intake (21-42d) than those fed with L-threonine supplementation. Effects of all dietary treatments on feed conversion ratio were not statistically significant (P >0.05).

The results of carcass parameters (Table 3) showed that except for liver and gizzard relative weights, none of carcass parameters affected by dietary treatments (P >0.05). The results of Table 3 indicated that liver relative weight was greater in birds fed pelleted diet without L-threonine supplementation. There was no significant difference for pancreas relative weight in birds fed course or fine diets. Gizzard relative weight increased in birds fed mash diets.

The results of blood biochemical parameters are shown in Table 4. Results showed that Glucose, cholesterol, HDL and LDL concentrations did not affect by dietary treatments (P >0.05). Triglyceride and VLDL concentration increased in birds fed pellet diets without L-threonine supplementation.

Discussion

Growth performance was not affected by particle size in present study. The effects of particle size on broiler performance have been studied in a number of researches [8]. The results of Behnke [8] indicated that particle size increased the surface area of the grain thus allowing for greater interaction with digestive enzymes [8]. Most researchers suggested that reducing mean particle size of cereal grains results in marked improvements in nutrient digestibility and efficiency of growth [9]. In contrast Kilburn and Edwards [10] suggested that large particle size of soybean meal in diets of broilers was more efficiently than fine particle size. Possibly these larger particles of soybean meal passed through the digestive tract at a slower rate and improved the nutrient utilization.

Interaction between particle size and feed physical form in broiler diets for weight gain and feed intake is well documented [4,11,12]. Svihus et al. [11] reported that feeding pelleted diets increased weight gain and feed intake and improved the feed conversion ratio compared with those birds maintained on mash diets. The results of present study support these findings. Amerah et al. [4] indicated

Table 2 Effects of dietary treatments on growth performance (gr/day) of broilers

Treatment	Weight gain			Feed intake			FCR[1]		
	1-21	21-42	1-42	1-21	21-42	1-42	1-21	21-42	1-42
Feed form									
Mash	56.7[b]	66.0	61.4[b]	91.7[b]	121.9[b]	106.8[b]	1.61	1.86	1.74
Pellet	61.5[a]	70.4	65.9[a]	101.8[a]	131.8[a]	116.7[a]	1.65	1.89	1.77
Feed size									
Course	59.4	66.7	63.1	97.5	127.2	112.0	1.64	1.91	1.78
Fine	58.8	69.6	64.2	95.9	126.5	111.6	1.63	1.83	1.73
L-threonine									
+	60.3[a]	66.9	63.6	97.3	124.6[b]	110.9	1.61	1.88	1.75
-	57.9[b]	69.5	63.7	96.2	129.2[a]	112.7	1.66	1.86	1.76
SEM	0.69	1.79	0.89	0.61	1.36	0.90	0.02	0.04	0.02
Interaction effects					Probability value				
A × B	0.67	0.93	0.94	0.20	0.71	0.86	0.78	0.87	0.93
A × C	0.56	0.10	0.15	0.01	0.57	0.16	0.34	0.03	0.01
B × C	0.24	0.17	0.35	0.17	0.71	0.83	0.60	0.20	0.25
A × B × C	0.25	0.10	0.04	0.69	0.66	0.63	0.34	0.09	0.03

[a-b]Means without a common superscripts in per column are significantly different (P < 0.05).
[1]FCR: Feed conversion ratio.
A: feed form; B: feed size; C: L-threonine.

Table 3 Effects of dietary treatments on carcass characteristics and internal organs of broilers

Treatment	Breast	Thigh	Liver	Pancreas	Gizzard	Heart	Intestine
			(% of live weight)				(cm)
Feed form							
Mash	25.55	18.76	2.22[b]	0.21	1.19[a]	0.42	211.62
Pellet	26.64	18.25	2.44[a]	0.20	0.99[b]	0.44	215.66
Feed size							
Course	25.92	18.49	2.26	0.19	1.10	0.44	217.12
Fine	26.27	18.52	2.39	0.21	1.09	0.42	210.16
L-threonine							
+	26.04	18.49	2.22[b]	0.20	1.06	0.43	215.62
-	26.15	18.52	2.44[a]	0.21	1.12	0.43	211.66
SEM	0.38	0.23	0.06	0.01	0.03	0.01	2.96
Interaction effects							
				Probability value			
A × B	0.44	0.18	0.51	0.61	0.79	0.48	0.30
A × C	0.03	0.14	0.08	0.15	0.65	0.24	0.001
B × C	0.33	0.59	0.23	0.38	0.77	0.08	0.21
A × B × C	0.14	0.43	0.16	0.49	0.30	0.32	0.06

[a-b]Means without a common superscripts in per column are significantly different (P < 0.05).
A: feed form; B: feed size; C: L-threonine.

Table 4 Effects of dietary treatments on blood biochemical characteristics of broilers

Treatment	Cholesterol	Triglyceride	Glucose	LDL	HDL	VLDL
			(mg/dl)			
Feed form						
Mash	128.8	94.7[b]	232.9	67.1	43.2	18.9[b]
Pellet	129.5	133.9[a]	258.8	68.8	41.5	26.6[a]
Feed size						
Course	137.7	119.2	259.6	68.7	43.6	23.7
Fine	120.6	109.4	232.5	67.2	41.0	21.8
L-threonine						
+	125.6	87.7[b]	237.3	68.2	42.7	17.5[b]
-	132.7	140.9[a]	254.5	67.6	42.0	28.1[a]
SEM	7.96	9.80	16.03	2.33	1.44	1.98
Interaction effects						
			Probability value			
A × B	0.53	0.38	0.96	0.44	0.77	0.37
A × C	0.10	1.00	0.83	0.06	0.15	0.97
B × C	0.45	0.68	0.60	0.86	0.60	0.66
A × B × C	0.41	0.13	0.80	0.67	0.84	0.12

[a-b]Means without a common superscripts in per column are significantly different (P < 0.05).
A: feed form; B: feed size; C: L-threonine.

that it may be due to increased nutritional density, improved starch digestibility resulting from chemical changes during pelleting, increased nutrient intake, changes in the physical form of the feed, reduced feed wastage, and decreased energy expenditure in eating. Nir et al. [3] reported that broilers fed wheat and sorghum mash diets with coarser particles had heavier body weights and better feed efficiency compared to those fed the finely ground diets.

The digestive tract development of poultry in known to be affected by feed particle size. Gabriel et al. [13] and Nir et al. [3] reported that relative intestinal weight was lower in birds fed coarse particle diets compared to those fed fine particle diets. The present data showed that the relative gizzard weight were higher in birds fed mash feeds than in those fed pelleted feeds. Similar findings have been reported by Nir et al. [3] and Amerah et al. [4]. These results may suggest that pelleting decreased the grinding requirement of the gizzard so that its function was reduced to that of transit.

Effect of L-threonine supplementation on broiler performance was not significant in present study. It is demonstrated that formulation of diets to contain up to 544 g of L-threonine/ton did not affect growth or carcass attributes of commercial broilers. It is well known that use of L-threonine decreased protein requirements of broilers and improved nitrogen efficiency utilization [14]. Rezaeipour et al. [5] indicated that dietary L-Threonine improved intestinal morphology and growth rate in broilers.

A dearth of information exists in terms of blood parameters in poultry; therefore, direct comparison cannot be made.

Conclusions

In conclusion, the data showed that feed form and dietary L-threonine had a greater influence on the different measured parameters than did the particle size. In summary, pellet form increased feed consumption and weight gain in whole of experimental period (1 to 42 days of age) while feed particle size and dietary L-threonine had no effect on broiler performance at the same time.

Competing interests
The authors declare that they have no competing interests.

Authors' contributions
VR and SG designed experiment, SG carried out animal trial and VR written manuscript. Both authors read and approved the final manuscript.

References

1. Goodband RD, Tokach MD, Nelssen JL: *MF-2050 Feed Manufacturing.* Manhattan: Department Grain Sci. Ind., Kansas State University; 2002.
2. Amerah AM, Ravindran V, Lentle RG, Thomas DG: **Influence of feed particle size and feed form on the performance, energy utilization, digestive tract development, and digesta parameters of broiler starters.** *Poult Sci* 2007, **86:**2615–2623.
3. Nir I, Hillel R, Ptichi I, Shefet G: **Effect of particle size on performance. 3. Grinding pelleting interaction.** *Poult Sci* 1995, **74:**771–783.
4. Amerah AM, Ravindran V, Lentle RG, Thomas DG: **Influence of feed particle size on the performance, energy utilization, digestive tract development, and digesta parameters of broiler starters fed wheat- and corn- based diets.** *Poult Sci* 2008, **87:**2320–2328.
5. Rezaeipour V, Fononi H, Irani M: **The effects of dietary L-threonine and** *Saccharomyces cerevisiae* **on performance, intestinal morphology and immune response of broiler chickens.** *S Afr J Anim Sci* 2012, **42**(3):266–273.
6. Dozier WA, Moran ET, Kidd MT: **Threonine requirement of broiler males from 42 to 56 days in a summer environment.** *J Appl Poult Res* 2000, **9:**460–467.
7. *SAS, Statistical Analysis Systems User's Guide, Version 8.02 Edition.* Cary, N.C. USA: SAS Institute, Inc; 2001.
8. Behnke KC: **Feed manufacturing technology: current issues and challenges.** *Anim Feed Sci Technol* 1996, **62:**49–57.
9. Ebrahimi R, Bojar pour M, Mokhtarzadeh S: **Effect of feed particle size on the performance and carcass characteristics of broilers.** *Int J Anim Vet Adv* 2010, **9**(10):1482–1484.
10. Kilburn J, Edwards HM: **The response of broiler to the feeding of mash or pelleted diets containing maize of varying particle sizes.** *Br Poult Sci* 2001, **42:**484–492.
11. Svihus B, Klovstad KH, Perez V, Zimonja O, Sahlstrom S, Schuller RB: **Physical and nutritional effects of pelleting of broiler chicken diets made from wheat ground to different coarsenesses by the use of roller mill and hammer mill.** *Anim Feed Sci Technol* 2004, **117:**281–293.
12. Peron A, Bastianelli D, Oury FX, Gomez J, Carre B: **Effects of food deprivation and particle size of ground wheat on digestibility of food components in broilers fed on a pelleted diet.** *Br Poult Sci* 2005, **46:**223–230.
13. Gabriel I, Mallet S, Leconte M: **Differences in the digestive tract characteristics of broiler chickens fed on complete diet or on whole wheat added to pelleted protein concentrate.** *Br Poult Sci* 2003, **44:**283–290.
14. Kidd MT, Kerr BJ: **Threonine responses in commercial broilers at 30 to 42 days.** *J Appl Poult Res* 1997, **6:**362–367.

28

Effects of a lipid-encapsulated zinc oxide supplement on growth performance and intestinal morphology and digestive enzyme activities in weanling pigs

Insurk Jang[1], Chang Hoon Kwon[2], Duck Min Ha[1], Dae Yun Jung[1], Sun Young Kang[1], Man Jong Park[1], Jeong Hee Han[2], Byung-Chul Park[3] and Chul Young Lee[1*]

Abstract

This study compared the effects of varying lipid content and dietary concentration of a lipid-encapsulated (LE) ZnO product to those of native ZnO and thereby to find insights into optimal lipid coating and dosage of the Zn supplement. A total of 192 21-d-old weanling pigs were allotted to 48 pens, after which each six pens received a ZnO-free basal diet supplemented with 125 ppm ZnO (100 ppm Zn; BASAL), 2,500 ppm Zn as native ZnO (HIGH), or 100 or 200 ppm Zn as LE ZnO (LE-100 or LE-250) containing 8%, 10%, or 12% lipid [LE-8%, LE-10%, or LE-12%, respectively; 2×3 factorial arrangement within the LE-ZnO diets (LE-ALL)] for 14 d. Forty pigs were killed at the end for histological and biochemical examinations. None of ADG, ADFI, gain:feed, and fecal consistency score differed between the LE-ALL and either of the BASAL and HIGH groups. Hepatic and serum Zn concentrations were greater (p <0.05) in the HIGH vs. LE-ALL group, but did not differ between LE-ALL and BASAL, between LE-100 and -250, or among LE-8%, -10%, and -12% groups. Villus height (VH), crypt depth (CD), and the VH:CD ratio in the duodenum, jejunum, and ileum did not differ between the LE-ALL and either of the BASAL and HIGH groups, except for a greater CD in the duodenum in the LE-ALL vs. HIGH group. Additionally, VH and CD in the duodenum and VH:CD in the jejunum were greater in the LE-250 vs. LE-100 group. Specific activities of sucrase, maltase, and leucine aminopeptidase in these intestinal regions and those of amylase and trypsin in the pancreas were not influenced by the lipid content or dietary concentration of LE ZnO and also did not differ between the LE-ALL and either of the BASAL and HIGH groups, except for a greater pancreatic amylase activity in the former vs. HIGH group. In conclusion, the present results indicate that the LE ZnO, regardless of its lipid percentage or supplementation level examined in this study, has no significant effect on growth performance, fecal consistency, or digestive enzyme activities of weanling pigs under the experimental conditions.

Keywords: Weanling pig, Zinc oxide, Dietary supplement, Growth, Diarrhea, Villus structure, Digestive enzyme

Background

Zinc oxide (ZnO) is commonly supplemented to the pig starter diet to 2,000 to 3,000 ppm to prevent the post-weaning diarrhea as well as to enhance transiently retarded growth of post-weaning pigs [1-3]. However, supplementation of ZnO at this pharmacological level poses substantial concerns about environmental pollution, because dietary ZnO is mostly unabsorbed and therefore excreted into the environment via feces [4,5]. This has led to limiting the in-feed ZnO concentration within 150 ppm by legislation in the European Union. Accordingly, some manufacturers have lately introduced new ZnO products which are more active or more efficiently delivered to the intestine than conventionally used ZnO, thereby suggesting the possibility of reducing the amount of ZnO added to the diet by substitution of the former for the latter.

A few types of non-covalent ZnO-carrier conjugates have been manufactured as a means of increasing the

* Correspondence: cylee@gntech.ac.kr
[1]The Regional Animal Industry Center, Gyeongnam National University of Science and Technology, Jinju 660-758, Republic of Korea
Full list of author information is available at the end of the article

efficiency of ZnO delivery. In this regard, Hu et al. [6-8] have reported that 600 to 900 ppm supplementation of Zn as ZnO supported on zeolite or 500 ppm of Zn as a ZnO-smectite conjugate was as effective as 2,000 to 2,250 ppm of Zn as native ZnO in enhancing growth performance and digestive function of weanling pigs. HiZox (Animine, France) is also a ZnO product whose surface area is maximized to increase the bioavailability of the compound. As for its relative potency, Morales et al. [9] have reported that post-weaning pigs fed a diet supplemented with 110 ppm of Zn as HiZox exhibited a greater ADG, a greater G:F ratio, and a better health status than those fed the same diet supplemented with 2,500 ppm of Zn as ZnO; however, this needs to be confirmed.

The ZnO particle has also been coated with an enteric substance to increase the delivery of the mineral to the intestine [10-12]. Shield Zn (CTCBIO, Inc., Seoul) is a lipid-encapsulated ZnO product, which has been manufactured based on a rationale that the encapsulated ZnO particle reaches the intestine efficiently and is subsequently released upon digestion of the lipid capsule by lipase because unlike the inorganic ZnO, the mineral component of the product is not released as Zn^{2+} under acidic pH in the stomach owing to the outer enteric coating [12]. In this connection, we have found that dietary supplementation of 72 ppm of Zn as the LE ZnO (100 ppm) was as effective as 2,000 ppm of Zn as native ZnO (2,500 ppm) in alleviating reduced growth, diarrhea, and impaired integrity of intestinal mucosal structure in weanling pigs artificially infected with enterotoxigenic *Escherichia* coli (ETEC) K88 [11]. The present study was undertaken to investigate the effects of the LE ZnO relative to those of native ZnO as well as the effects of the lipid content and dietary concentration of the LE ZnO on growth and the measures of digestive function as an initial step to finding the optimal lipid coating and dosage of the Zn supplement in naïve weanling pigs.

Methods
Animals
The protocol for the present experiment was approved by the Institutional Animal Care and Use Committee (IACUC) of Gyeongnam National University of Science and Technology. The (Yorkshire × Landrace) × Duroc piglets were divided into the high-, medium- and low-body weight categories at weaning at 21 days of age. Ninety-six high- and 96 medium-body weight weanling pigs were randomly allotted to 48 pens by body weight to forty-eight 1-m^2 nursery pens, with two females and two castrated males housed per pen equipped with a feeder and a nipple waterer. Each six pens received a ZnO-free basal nursery diet supplemented with 125 ppm of native ZnO (100 ppm ZnO; BASAL), 3,125 ppm of native ZnO (HIGH), 100 or

200 ppm Zn as LE ZnO (LE-100 or LE-250) containing 8%, 10%, or 12% lipid [LE-8%, LE-10%, or LE-12%, respectively; 2 × 3 factorial arrangement within the LE-ZnO diets (LE-ALL)] for 14 d (Table 1). The basal diet was formulated to contain a low percentage of crude protein to reduce the post-weaning diarrhea resulting from undigested proteins [10,13]. The ambient temperature was maintained at 30°C up to d 7 and then lowered to 29°C. Fecal consistency was scored on d 1, 4, 7, and 14 according to an arbitrary 3-point integer scale as described by Heo et al. [13] and Zhao et al. [14]: 1, well-formed feces; 2, sloppy feces; 3, diarrhea.

Collection of blood samples and intestinal tissues
A total of 40 pigs, which consisted of 16 pigs from the BASAL and HIGH groups (8 pigs each) and 24 pigs from the 6 LE groups (4 pigs each), were sacrificed as described previously [15]. Blood, pancreas, and intestinal tissues at the regions of the duodenum, jejunum, and ileum were collected also as described [11,15].

Determination of Zn
Five grams of hepatic tissue or 1 mL of serum was digested with 10 mL of 70% nitric acid at 150°C to complete solublization, filtered and diluted with 100 mL of distilled water. The Zn content in the digested and diluted solution was determined using an inductively coupled plasma atomic emission spectrophotometer (model 5300DV, Perkin-Elmer, Waltham, MA, USA).

Histological and biochemical determinations
The intestinal tissue was fixed, embedded in paraffin, mounted on the glass slide, stained, and subjected to microscopic determination of the villus height and crypt depth as described previously [11,15].

The intestinal mucosa and pancreatic tissue were homogenized and stored at −70°C until used. The protein content of the homogenate was determined using the bicinchronic acid protein assay kit (Pierce, Rockford, IL, USA); the specific activities of sucrase, maltase, leucine

Table 1 Calculated chemical composition of the basal diet[1] (as-fed basis)

Item	Content
DE, MCal/kg	3.34
Crude protein (%)	16.50
Ether extract (%)	3.91
Lysine (%)	1.21
Zn (ppm)	100, 250, or 2,500

[1]The composition of ingredients, which was grains-soy-whey-based, was reported previously [11]. Experimental diets were manufactured by supplementing the basal diet containing no Zn additive with 125 or 3,125 ppm of native ZnO or 139 to 355 ppm of Shield Zn® (CTCBIO, Seoul) encapsulated with 8%, 10%, or 12% lipid (w/w) to provide 100, 250, or 2,500 ppm of Zn.

aminopeptidase, and trypsin were determined as described previously [15-17].

Statistical analysis

All data were analyzed using the MIXED procedure of SAS (SAS Inst. Inc., Cary, NC, USA). The pen was the experimental unit in all variables, except for postmortem measurements in which the piglet was regarded as the experimental unit. The model included each dietary treatment as the main effect. In the analysis of repeated measurements, effects of the day and its interaction with the main effect were also included in the model. Accordingly, effects of the diet and the day including its interaction with the diet were tested using the experimental unit and experimental unit × day as error terms, respectively. In addition, effects of supplementation of LE ZnO vs. native ZnO as well as those of the lipid percentage and supplementation level of LE ZnO were analyzed by the contrast.

Results

Growth performance including the average daily gain (ADG), average daily feed intake (ADFI), and gain:feed (G:F) ratio during the 14-d experimental period did not differ between the LE-ALL group and either of the BASAL and HIGH groups (Table 2). Moreover, within the LE-ALL group animals, the performance parameters were not different between the LE-10% group and either

of the LE-8% and LE-12% groups or between the LE-100 and LE-250 groups.

The majority of the piglets exhibited well-formed feces throughout the experimental period, although the fecal consistency score (FCS) was greater ($P < 0.01$) on d 14 than at any other time point (Table 2). However, the FCS did not differ between the LE-ALL group and either of the LBASAL and HIGH groups, between the LE-10% group and either of the LE-8% and -12% groups, or between the LE-100 and -250 groups.

The Zn concentration in the liver did not differ between the LE-ALL and BASAL groups (Table 3). However, hepatic ZnO concentration was 11-fold greater in the HIGH vs. LE-ALL group whereas within the LE-ALL group, it was not influenced by the lipid percentage or supplementation level of the LE ZnO. Serum Zn concentration, which did not differ between the LE-ALL and BASAL groups, was marginally greater in the HIGH group vs. LE-ALL (2.45 vs. 1.49 µg/mL; $P < 0.01$). However, within the LE-ALL group, serum Zn concentration was not influenced by the lipid percentage or supplementation level of the ZnO product.

The villus height (VH) and crypt depth (CD) in the LE-ALL group were not different from those in the BASAL and HIGH groups in the duodenum (Table 3). Within the LE-ALL group, these morphology variables did not change due to a change in the lipid percentage of the LE ZnO, but increased when the supplementation

Table 2 Effects of supplementations of native ZnO and lipid-encapsulated ZnO (LE ZnO) on growth and fecal consistency of weanling pigs[1]

Variable	Dietary supplementation (ppm as Zn)										Contrast P-value				
	Native ZnO		LE ZnO								LE ZnO vs. Native ZnO		Within LE ZnO		
	100	2,500	8% lipid[2]		10% lipid[2]		12% lipid[2]		SEM	P-value	100	2,500	10% lipid vs.		100 vs. 250
			100	250	100	250	100	250					8%	12%	
Growth performance															
Initial BW (kg)	6.69	6.98	6.76	6.67	6.71	6.65	6.74	6.73	0.15	0.85	0.92	0.05	0.78	0.67	0.63
Final BW (kg)	9.63	9.89	9.65	9.35	9.52	9.18	9.65	9.73	0.20	0.32	0.65	0.14	0.52	0.15	0.34
ADG (g)	210	207	207	192	201	180	208	214	11	0.49	0.50	0.60	0.49	0.11	0.34
ADFI (g)	355	367	355	329	344	327	348	369	13	0.24	0.49	0.13	0.60	0.09	0.51
G:F	0.591	0.569	0.582	0.586	0.581	0.538	0.603	0.579	0.025	0.76	0.63	0.74	0.34	0.21	0.30
Fecal consistency score[3]															
Day 1	1.00	1.00	1.00	1.00	1.00	1.00	1.00	1.00							
Day 4	1.00	1.13	1.08	1.00	1.00	1.13	1.04	1.04	0.07[a]						
Day 7	1.04	1.00	1.04	1.00	1.04	1.13	1.13	1.17							
Day 14	1.17	1.00	1.17	1.17	1.25	1.35	1.17	1.08							
Overall[4]	1.05	1.03	1.07	1.04	1.07	1.15	1.08	1.07	0.04	0.59	0.35	0.11	0.07	0.27	0.63

[1] Data are means of 6 pens in each dietary group.
[2] Denotes the percentage of lipid (w/w) encapsulating the ZnO particle.
[3] Scored according to a 3-notch integer scale: 1, well formed; 2, sloppy; 3, diarrhea.
[4] P-values for the day and day × treatment were <0.01 and 0.71, respectively.
[a] Applies to all day × treatment combinations.
BW, body weight; ADG, average daily gain; ADFI, average daily feed intake; G:F, gain:feed.

Table 3 Effects of dietary supplementations of native ZnO and lipid-encapsulated ZnO (LE ZnO) on hepatic and circulating Zn concentrations, intestinal villus structure, and specific activities of digestive enzymes in weanling pigs

Enzyme	Native ZnO[1]			Dietary supplementation (ppm as Zn) — LE ZnO								Contrast P-value				
				8% lipid[2]		10% lipid[2]		12% lipid[2]		SEM	P-value	LE ZnO vs. Native ZnO		Within LE ZnO		
														10% lipid vs.		100 vs.
	100	2,500	SEM	100	250	100	250	100	250			100	2,500	8%	12%	250
Zn concentration																
Liver (µg/g)	34.9	412.1	13.6	33.3	55.4	27.8	33.5	36.3	39.0	19.2	<0.01	0.87	<0.01	0.48	0.72	0.52
Serum (µg/ml)	1.78	2.45	0.14	1.53	1.63	1.52	1.50	1.14	1.62	0.22	<0.01	0.10	<0.01	0.75	0.57	0.31
Villus structure																
Duodenum																
VH (µm)	292	284	14	299	365	299	315	277	323	20	0.07	0.21	0.10	0.22	0.71	<0.01
CD (µm)	230	227	11	240	282	238	276	231	256	16	0.07	0.07	0.05	0.78	0.41	<0.01
VH:CD	1.27	1.26	0.04	1.25	1.29	1.26	1.15	1.20	1.26	0.05	0.56	0.37	0.66	0.21	0.60	0.98
Jejunum																
VH (µm)	280	252	16	254	315	240	266	275	263	21	0.31	0.54	0.38	0.15	0.46	0.17
CD (µm)	230	206	15	234	262	225	238	251	218	20	0.49	0.64	0.08	0.44	0.89	0.87
VH:CD	1.21	1.16	0.03	1.09	1.20	1.07	1.12	1.11	1.20	0.04	0.03	0.02	0.30	0.21	0.13	<0.01
Ileum																
VH (µm)	245	209	13	215	247	245	234	230	232	19	0.56	0.48	0.11	0.67	0.67	0.62
CD (µm)	228	186	13	205	221	238	224	238	180	19	0.15	0.73	0.05	0.33	0.24	0.24
VH:CD	1.10	1.21	0.07	1.30	1.16	1.03	1.05	1.09	1.31	0.11	0.36	0.54	0.50	0.08	0.15	0.71
Enzyme activity (µmol end product · mg protein⁻¹ · min⁻¹)																
Sucrase																
Duodenum	0.043	0.034	0.008	0.076	0.039	0.040	0.049	0.041	0.044	0.012	0.25	0.61	0.14	0.26	0.89	0.39
Jejunum	0.874	0.537	0.225	0.319	0.401	0.160	0.628	0.509	0.170	0.318	0.59	0.06	0.51	0.91	0.86	0.79
Ileum	0.098	0.117	0.032	0.105	0.099	0.088	0.107	0.094	0.119	0.046	1.00	0.91	0.68	0.93	0.86	0.74
Maltase																
Duodenum	5.18	4.41	1.10	8.15	4.56	5.98	7.71	5.34	7.80	1.56	0.36	0.28	0.10	0.76	0.86	0.88
Jejunum	15.87	12.70	3.48	8.35	12.95	10.28	14.10	13.88	5.38	4.92	0.76	0.22	0.64	0.76	0.61	0.99
Ileum	4.20	5.39	1.49	5.11	4.02	2.91	6.27	4.13	3.95	2.11	0.96	0.91	0.57	0.99	0.80	0.69
LAP																
Duodenum	2.00	2.30	0.36	2.19	1.91	1.99	2.72	1.90	3.20	0.50	0.53	0.45	0.96	0.55	0.70	0.17
Jejunum	3.15	2.73	0.27	2.37	2.53	3.17	2.51	3.25	2.20	0.38	0.30	0.14	0.86	0.31	0.77	0.11
Ileum	1.76	1.93	0.29	2.04	1.71	1.68	1.94	1.61	1.81	0.41	0.99	0.90	0.69	0.88	0.81	0.89
Amylase (PAN)	241.2	142.1	23.0	262.1	185.1	224.3	240.7	237.6	231.4	31.4	0.06	0.67	<0.01	0.78	0.95	0.39
Trypsin (PAN)	0.156	0.107	0.022	0.149	0.127	0.150	0.184	0.160	0.197	0.032	0.43	0.86	0.06	0.37	0.71	0.53

[1] Data are means of 8 animals in each group.
[2] Denotes the percentage of lipid (w/w) encapsulating the ZnO particle. Data are means of 4 animals in each group.
VH, villus height; CD, crypt depth; LAP, leucine aminopeptidase; PAN, pancreas.

level was increased from 100 ppm Zn to 250 ppm (P <0.01). However, the VH:CD ratio in the duodenum did not differ between the LE-ALL group and either of the BASAL and HIGH groups, between the LE-10% group and either of the LE-8% and -12% groups, or between the LE-100 and -250 groups. In the jejunum, the VH and CD did not differ between the LE-ALL group and either of the BASAL and HIGH groups and

also were not influenced by the lipid percentage or supplementation level of the LE ZnO. However, the VH:CD ratio was greater (P <0.05) in the BASAL group (1.21) than in the LE-ALL group (1.13), within which it was greater (P <0.01) in the LE-250 vs. LE-100 group (1.17 vs. 1.09). In the ileum, the CD was greater in the LE-ALL vs. HIGH group (218 vs. 186 µm); otherwise, the villus morphology variables did not differ between the

LE-ALL group and either of the BASAL and HIGH groups, between the LE-100 and -250 groups, or among the LE-8%, -10%, and -12% groups.

Specific activities of sucrase, maltase, and leucine aminopeptidase in the mucosa were not affected by any treatment factor examined in the present study in any intestinal segment (Table 3). Specific activities of amylase and trypsin in the pancreas also did not change due to any treatment factor, except for a greater amylase activity in the LE-ALL vs. HIGH group.

Discussion

The present results indicated that neither supplementation level of native ZnO or LE ZnO nor lipid percentage of the latter has any significant effect on ADG, ADFI, G: F, or FCS. However, this does not necessarily mean that neither pharmacological supplementation of native ZnO nor the basal-level supplementation of LE ZnO has any significant effect on these performance parameters, because even the known growth-enhancing effect of the former [2,3,18,19] was not apparent under the present experimental conditions. As a matter of fact, we have observed a growth-enhancing effect of the LE ZnO supplemented at a basal level (100 ppm as Zn concentration; unpublished results). The lack of growth-enhancing effects of the LE ZnO and HIGH treatments in the present study is therefore reflective of the known fact that growth enhancers including ZnO are less effective under experimental settings than under production conditions [20,21]. In this context, the present results were also reminiscent of the lack of effects of the in-feed antibiotics, the known growth enhancers in swine [20,21], on growth performance of post-weaning pigs placed in small experimental pens in our previous study [15]. As such, more studies are warranted to determine the optimal usage of the LE ZnO as a growth-enhancing dietary supplement for weanling pigs under production conditions.

The several-fold and marginally greater Zn concentrations in the HIGH vs. BASAL group in the liver and serum, respectively, were consistent with published results [22-24]. Moreover, neither circulating Zn concentration was different between the LE ZnO and BASAL groups as in a report of Kim et al. [10], nor hepatic Zn concentration was influenced by the lipid coating. These results imply that the supplementation level of ZnO is well reflected into the Zn concentration in the liver where many heavy metals are stored and that the absorption rate of Zn at the intestine is not influenced by lipid coating of the ZnO particle.

The lack of effect of the LE ZnO or HIGH treatment on the intestinal villus structure was quite different from the increase in the VH and/or VH:CD ratio as well as the decrease in CD in response to either treatment in weanling pigs artificially infected with ETEC K88 in our previous study [11]. Furthermore, the positive effect of pharmacological ZnO on the integrity of the villus structure often observed in weanling pigs [25-27] also was not apparent in the present study. These results are thus interpreted to suggest that the beneficial effect of either Zn supplement on the integrity of the villus may be apparent only when the villus structure of the piglet is substantially damaged by any causative like the microbial infection. In this context, the increased VH in the duodenum and increased VH:CD in the jejunum in response to the LE-250 treatment vs. LE-100 observed in the present study, albeit intriguing, needs to be rigorously confirmed to make any firm conclusion as to the dose effect of the LE ZnO on the villus structure in naïve weanling pigs.

Results on the digestive enzyme activities in the intestinal mucosa and pancreatic tissue also indicated that these are not affected by the supplementation level of the LE ZnO or native ZnO or by the lipid percentage of the former, although the pancreatic amylase activity was greater in the LE-ALL vs. HIGH group. Similarly, in the study of Hedemann et al. [28], effects of the high-ZnO (2,500 ppm as Zn) diet on pancreatic and intestinal enzyme activities in weanling pigs were inconclusive or equivocal.

Conclusions

The present results indicated that neither the high-ZnO supplementation nor the physiolgical supplementation of the LE ZnO containing 8 to 12% lipid to 100 or 250 ppm Zn has any significant effect on growth performance, fecal consistency, villus morphology, or digestive enzyme activities of the piglets under the experimental conditions. More studies are necessary, however, to determine the effects of the LE ZnO relative to those of native ZnO on these measures in weanling pigs under production conditions.

Competing interests
The authors declare that they have no competing interests.

Authors' contributions
BCP and CYL designed the experiment and analyzed the data. DMH, DYJ, MJP, and CYL managed the experimental animals. CHK and JHH analyzed the intestinal morphology. SYK and IJ determined the enzyme activities. All authors read and approved the final manuscript.

Acknowledgements
This work was supported by CTCBIO, Inc. and the Regional Animal Industry Center at Gyeongnam National University of Science and Technology.

Author details
[1]The Regional Animal Industry Center, Gyeongnam National University of Science and Technology, Jinju 660-758, Republic of Korea. [2]College of Veterinary Medicine and Institute of Veterinary Science, Kangwon National University, Chuncheon 200-701, Republic of Korea. [3]R & D Institute, Sunjin Co., Ltd, 517-3 Doonchon-dong, Kangdong-gu, Seoul 134-060, Republic of Korea.

References

1. Heo JM, Opapeju FO, Pluske JR, Kim JC, Hampson DJ, Nyachoti CM: Gastrointestinal health and function in weaned pigs: a review of feeding strategies to control post-weaning diarrhoea without using in-feed antimicrobial compounds. *J Anim Physiol Anim Nutr* 2013, **97**:207–237.
2. Hill GM: Minerals and Mineral Utilization in Swine. In *Sustainable Swine Nutrition.* Edited by Chiba LI. Oxford: John Wiley & Sons, Inc; 2013:173–195.
3. Sales J: Effect of pharmacological concentrations of dietary zinc oxide on growth of post-weaning pigs: a meta-analysis. *Biol Trace Elem Res* 2013, **152**:343–349.
4. Poulsen HD, Larsen T: Zinc excretion and retention in growing pigs fed increasing levels of zinc oxide. *Livest Prod Sci* 1995, **43**:235–242.
5. Jondreville C, Revy PS, Dourmad JY: Dietary means to better control the environmental impact of copper and zinc by pigs from weaning to slaughter. *Livest Prod Sci* 2003, **84**:147–156.
6. Hu CH, Gu LY, Luan ZS, Song J, Zhu K: Effects of montmorillonite-zinc oxide hybrid on performance, diarrhea, intestinal permeability and morphology of weanling pigs. *Anim Feed Sci Technol* 2012, **177**:108–115.
7. Hu C, Song J, Li Y, Luan Z, Zhu K: Diosmectite-zinc oxide composite improves intestinal barrier function, modulates expression of pro-inflammatory cytokines and tight junction protein in early weaned pigs. *Br J Nutr* 2013, **110**:681–688.
8. Hu CH, Xiao K, Song J, Luan ZS: Effects of zinc oxide supported on zeolite on growth performance, intestinal microflora and permeability, and cytokines expression of weaned pigs. *Anim Feed Sci Technol* 2013, **181**:65–71.
9. Morales J, Cordero G, Pineiro C, Durosoy S: Zinc oxide at low supplementation improves productive performance and health status of piglets. *J Anim Sci* 2012, **90**(Suppl 4):436–438.
10. Kim JC, Hansen CF, Mullan BP, Pluske JR: Nutrition and pathology of weaner pigs: nutritional strategies to support barrier function in the gastrointestinal tract. *Anim Feed Sci Technol* 2012, **173**:3–16.
11. Kwon CH, Lee CY, Han SY, Kim SJ, Park BC, Jang I, Han JH: Effects of dietary supplementation of lipid-encapsulated zinc oxide on colibacillosis, growth and intestinal morphology in weaned piglets challenged with enterotoxigenic Escherichia coli. *Anim Sci J* 2014, **85**:805–813.
12. Shen J, Chen Y, Wang Z, Zhou A, He M, Mao L, Zou H, Peng Q, Xue B, Zhang X, Wu S, Lv Y: Coated zinc oxide improves intestinal immunity function and regulates microflora composition in weaned piglets. *Br J Nutr* 2014, **111**:2123–2134.
13. Heo JM, Kim JC, Hansen CF, Mullan BP, Hampson DJ, Pluske JR: Feeding a diet with decreased potein content reduces indices of protein fermentation and the incidence of postweaning diarrhea in weaned pigs challenged with an enterotoxigenic strain of *Escherichia coli*. *J Anim Sci* 2009, **87**:2833–2843.
14. Zhao PY, Jung JH, Kim IH: Effect of mannan oligosaccharides and fructan on growth performance, nutrient digestibility, blood profile, and diarrhea score in weanling pigs. *J Anim Sci* 2012, **90**:833–839.
15. Lee CY, Lim JW, Ko YH, Kang SY, Park MJ, Ko T, Lee JH, Hyun Y, Jeong KS, Jang IS: Intestinal growth and deveopment of weanling pigs in response to dietary supplementation of antibiotics, phytogenic products and brewer's yeast plus *Bacillus* spores. *J Anim Sci Technol* 2011, **53**:227–235.
16. Jang IS, Ko YH, Yang HY, Ha JS, Kim JY, Kim JY, Kang SY, Yoo DH, Nam DS, Kim DH, Lee CY: Influence of essential oil components on growth performance and the functional activity of the pancreas and small intestine in broiler chickens. *Asian-Aust J Anim Sci* 2004, **17**:394–400.
17. Jang IS, Ko YH, Kang SY, Lee CY: Effect of a commercial essential oil on growth performance, digestive enzyme activity and intestinal microflora population in broiler chickens. *Anim Feed Sci Technol* 2007, **134**:304–315.
18. Hahn JD, Baker DH: Growth and plasma zinc responses of young pigs fed pharmacologic levels of zinc. *J Anim Sci* 1993, **71**:3020–3024.
19. Hill GM, Mahan DC, Carter SD, Cromwell GL, Ewan RC, Harrold RL, Lewis AJ, Miller PS, Shurson GC, Veum TL: Effect of pharmacological concentrations of zinc oxide with or without the inclusion of an antibacterial agent on nursery pig performance. *J Anim Sci* 2001, **79**:934–941.
20. Cromwell GL: Antimicrobial and promicrobial agents. In *Swine Nutrition.* 2nd edition. Edited by Chiba LI. Oxford: John Wiley & Sons, Inc; 2001:401–426.
21. Wenk C: Recent advances in animal feed additives such as metabolic modifiers, antimicrobial agents, probiotics, enzymes and highly available minerals - review. *Asian-Aust J Anim Sci* 2000, **13**:86–95.
22. Jensen-Waern M, Melin L, Lindberg R, Johannisson A, Petersson L, Wallgren PP: Dietary zinc oxide in weaned pigs – effects on performance, tisue concentrations, morphology, neutrophil functions and faecal microflora. *Res Vet Sci* 1998, **64**:225–231.
23. Carlson D, Beattie JH, Poulsen HD: Assessment of zinc and copper status in weaned piglets in relation to dietary zinc and copper supply. *J Anim Physiol Anim Nutr* 2007, **91**:19–28.
24. Davin R, Manzanilla EG, Klasing KC, Perez JF: Effect of weaning and in-feed high doses of zinc oxide on zinc levels in different body compartments of piglets. *J Anim Physiol Anim Nutr* 2013, **97**:6–12.
25. Li BT, Van Kessel AG, Caine WR, Huang SX, Kirkwood RN: Small intestinal morphology and bacterial populations in ileal digesta and feces of newly weaned pigs receiving a high dietary level of zinc oxide. *Can J Anim Sci* 2001, **81**:511–516.
26. Li X, Yin J, Li D, Chen X, Zang J, Zhou X: Dietary supplementation with zinc oxide increases IGF-I and IGF-I receptor gene expression in the small intestine of weanling piglets. *J Nutr* 2006, **136**:1786–1791.
27. Owusu-Asiedu A, Nyachoti CM, Marquardt RR: Response of early-weaned pigs to an enterotoxigenic Escherichia coli (K88) challenge when fed diets containing spray-dried porcine plasma or pea protein isolate plus egg yolk antibody, zinc oxide, fumaric acid, or antibiotic. *J Anim Sci* 2003, **81**:1790–1798.
28. Hedemann MS, Jensen BB, Poulsen HD: Influence of dietary zinc and copper on digestive enzyme activity and intestinal morphology in weaned pigs. *J Anim Sci* 2006, **84**:3310–3320.

29

A whole genomic scan to detect selection signatures between Berkshire and Korean native pig breeds

Zewdu Edea and Kwan-Suk Kim[*]

Abstract

Background: Scanning of the genome for selection signatures between breeds may play important role in understanding the underlie causes for observable phenotypic variations. The discovery of high density single nucleotide polymorphisms (SNPs) provide a useful starting point to perform genome–wide scan in pig populations in order to identify loci/candidate genes underlie phenotypic variation in pig breeds and facilitate genetic improvement programs. However, prior to this study genomic region under selection in commercially selected Berkshire and Korean native pig breeds has never been detected using high density SNP markers. To this end, we have genotyped 45 animals using Porcine SNP60 chip to detect selection signatures in the genome of the two breeds by using the F_{ST} approach.

Results: In the comparison of Berkshire and KNP breeds using the FDIST approach, a total of 1108 outlier loci (3.48%) were significantly different from zero at 99% confidence level with 870 of the outlier SNPs displaying high level of genetic differentiation ($F_{ST} \geq 0.490$). The identified candidate genes were involved in a wide array of biological processes and molecular functions. Results revealed that 19 candidate genes were enriched in phosphate metabolism (GO: 0006796; *ADCK1, ACYP1, CAMK2D, CDK13, CDK13, ERN1, GALK2, INPP1; MAK, MAP2K5, MAP3K1, MAPK14, P14KB, PIK3C3, PRKC1, PTPRK, RNASEL, THBS1, BRAF, VRK1*). We have identified a set of candidate genes under selection and have known to be involved in growth, size and pork quality (*CART, AGL, CF7L2, MAP2K5, DLK1, GLI3, CA3* and *MC3R*), ear morphology and size (*HMGA2* and *SOX5*) stress response (*ATF2*, MSRB3, *TMTC3* and *SCAF8*) and immune response (*HCST* and *RYR1*).

Conclusions: Some of the genes may be used to facilitate genetic improvement programs. Our results also provide insights for better understanding of the process and influence of breed development on the pattern of genetic variations.

Keywords: Korean native pig, Genome-wide, SNP, Selection signature

Background

Pigs have long been of great economic importance to many farmers in the world. Molecular evidence supports independent domestication of pig in Asia and Europe from wild boar sub-species [1,2]. As compared to their wild ancestor, domestic pig breeds display a wide range of phenotypic variations that have been manipulated and shaped during the course of domestication and breed development for a wide range of traits. Some pig breeds, particularly commercial breeds have been intensively selection for better growth, meat quality and fertility traits may have resulted in loss of genetic diversity. To the contrast, most traditional breeds are reared by small-holder farmers and less subjected to selection pressure and harbor higher genetic diversity for adaptation under marginal environments.

The superiority of some commercial pig breeds for growth and carcass traits over traditional breeds, have led them to be the breed of choice and their continuous utilization in improvement of native populations through crossbreeding. In Asia, it has been known that commercial breeds have contributed to the genetic pool of most

* Correspondence: kwanskim@chungbuk.ac.kr
Department of Animal Science, Chungbuk National University, Cheongju 361-763, Korea

indigenous breeds [3]. Likewise, for the last two decades, western pig breeds have been imported into the Korea peninsula and crossed with the Korean native pigs (KNP) in order to improve growth and carcass related traits [4]. As a result the number of Korean native pigs decreased noticeably following the introduction of improved breeds. Although commercial breeds are superior in terms of growth and feed efficiency traits, the Korean native pig harbors unique genetic material for product quality and better adaptation to low management levels [5]. Despite the indiscriminate crossbreeding, little is known regarding the genome composition difference between the Korean native and European (Berkshire) pig breeds.

There is a growing interest in spotting genomic regions or genes that have been under selection. F_{ST} statistic is among the most widely used measures to identify genomic regions or loci that display high differentiation between populations [6]. Genomic regions or loci that show significantly high F_{ST} values compared with neutral loci offer evidence for positive selection. Until recently, there has been little success of detecting genomic regions under selection in livestock species attributed to lack of high density molecular markers. However, through the advancement of high-throughput sequencing technology, thousands of single nucleotide polymorphisms have been discovered and open opportunities to facilitate and transform livestock genetic improvement programs. In pig, several thousands of SNPs spinning the whole genome has been discovered using next generation technologies [7]. The availability and discovery of such large number of SNPs provide a useful starting point to perform genome–wide scan in pig populations in order to identify candidate genes underlining phenotypic variations between breeds. However, prior to this study genomic region under selection in commercially selected pig breed like Berkshire and Korean native pig has never been detected using high density SNP markers. Scanning of the genome for selection signature between highly selected and traditional breeds may play important role in identifying genes underlying for phenotypic variation. In addition, it can be used to facilitate genetic improvement and conservation programs. To this end, we have genotyped 45 animals are using Porcine SNP 60 BeadChip to identify loci variants showing directional selection in comparing European (Berkshire) and Korean native pig breeds using the F_{ST} approach.

Methods
Pig breeds, sampling and genotyping
Samples were collected from unrelated Berkshire (n = 29) and from Korea native pig (KNP, n = 16) breed. Briefly, Korean native pig was phenotypically discriminated as long black coarse hairs, long straight noses, greatly

protruded mouth and straightly upright ears. The breed is known for its high prolificacy, better meat quality (high redness and intramuscular fat [5] and strong adaptability under low management conditions, but showed a slower growth rate, small adult body weight, smaller litter sizes, and lower carcass yield [8]. On the other hand, Berkshire pig breed is characterized by medium to large body size, fast growth rate, early maturing, and large litter size, medium and erect ears.

DNA samples of Korean native pig were obtained from National Institute of Animal Science (NIAS) and that of Berkshire were obtained from Dasan Breeding Farm in Korea. Sample collection procedures were approved by the National Institute of Animal Science (NIAS). During sample collection animals were treated humanely. All animals were genotyped performed using the Illumina Pocrine SNP60 BeadChip [9]. Common monomorphic SNPs for all of the breeds were discarded from further analyses. SNPs were filtered with criteria of call rate (≥90%), minor allele frequency (MAF ≥5%) and Hardy-Weinberg equilibrium (HWE ≥0.001). Thus finally about 31,755 SNPs were considered for the study.

Statistical analysis
Genetic variations
Genetic diversity was assessed for each breed by calculating observed and expected heterozygosities using Arlequin software [10]. Principal component analysis was performed to illustrate the pattern of individual clustering using SNP and Variation Suite version 7 [11]. PCA assigns individuals to their population of origin using a common clustering algorithm Patterson et al. [12]. In the principal component analysis, the first principal component (PC1) accounts for the greater variation followed by principal component (PC2).

Detection of outlier loci or signature of selection
Detection of outlier loci was based on calculation of fixation index (F_{ST}) at different significance levels as a measure of genetic differentiation for each locus between Berkshire and KNP following the FDIST approach proposed by [13] as implemented in Arlequin software [10]. Briefly, the FDIST program calculate genetic differentiation index (F_{ST}) for each loci and then uses coalescent simulation to generate the null distribution of F_{ST} values based on the infinite island model [13]. Within this framework, we ran 20,000 coalescent simulations to obtain the P – values of locus-specific F_{ST} conditioned on observed levels of heterozygosity with default settings. This method provides evidence for divergent selection by looking for outlier loci with FST values higher than expected, controlling for heterozygosity. The corresponding candidate genes for outlier SNPs (P <0.01)

were annotated with the pig genome analysis data repository [14].

Biological process and molecular functional analyses of the candidate genes under selection

To known the biological process and molecular functional of each candidate genes, we assessed their Gene Ontology (GO) and classification using a web-based Database for Annotation, Visualization, and Integrated Discovery (DAVID) tools [15]. Furthermore, enrichment analysis was performed to identify biological processes and molecular functions over-represented by Fisher Exact test (EASE score). Any GO terms that have a larger than expected subset of selected genes were considered over-represented and gave insight into the functional characteristics of the annotated genes.

Haplotype blocks detection

To investigate whether any of the significantly differentiated loci or genes (P <0.01) are in strong linkage disequilibrium, we further analyzed LD and haplotype blocks for the two breeds following the [16] method using the SNP and Variation Suite version 7 [11]. According to this method, SNP pairs to be in strong linkage disequilibrium (LD) if the one-sided upper 95% confidence bound on D' is.0.98 (that is, consistent with no historical recombination) and the lower bound is above 0.7.

Results

Genetic diversity and population structure

The average observed heterozygosity was 0.321 ± 0.171 in Berkshire and 0.326 ± 0.173 in Korean native pig, whereas the expected heterozygosity was found to be 0.319 ± 0.156 and 0.336 ± 0.153 for Berkshire and Korean native pig, respectively. The average within – breed fixation index (F_{IS}) was shown deficiency of heterozygosity (0.029) in Korean native pig whereas it was negative (−0.008) in Berkshire. To illustrate the pattern of individual animals clustering, we performed principal component analyses (PCA). Principal component one (PC1) and principal component 2 (PC2) accounted for 82.33% and 17.67% of the total variance, respectively (Figure 1) and clearly separated individuals according to their breed group.

Genetic differentiation, outlier loci and candidate genes under selection

In this study, we are primarily intended to identify outlier loci in the comparison of two pig breeds (Berkshire and Korean native pigs). Level of differentiations between the breeds was measured by fixation indices. The overall F_{ST} was 0.157 with about 29% (9127) of the loci having an F_{ST} value below zero or equal to zero. The highest genetic differentiation between the two breeds

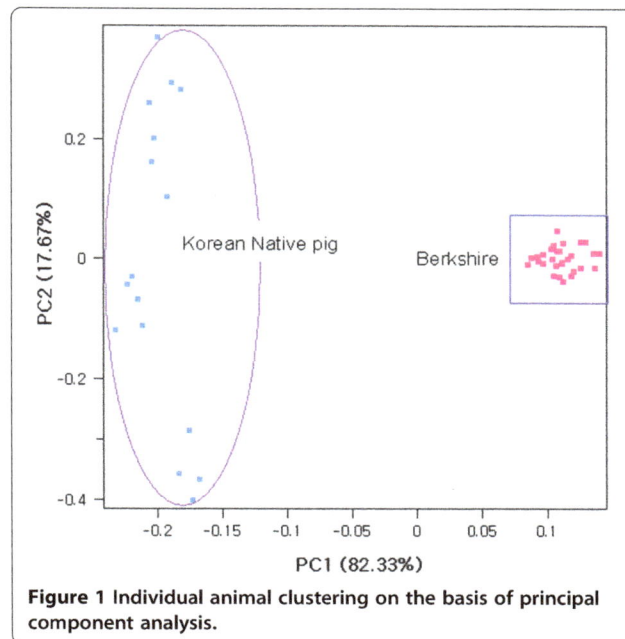

Figure 1 Individual animal clustering on the basis of principal component analysis.

was observed on chromosome 16 where 7 SNPs or loci (rs81228734, rs81458940, rs81459172, rs81459185, rs 81459183, rs81297918 and rs81459195) displayed an F_{ST} value of 1. In the comparison of Berkshire and KNP breeds, using the FDIST approach indicated that a total of 1108 loci (3.48%) were significantly different from zero at 99% confidence level with 870 of the outlier SNPs displaying a high level of genetic differentiation ($F_{ST} \geq 0.48$) (see Additional file 1) and revealing that the loci are potentially under directional selection. The distribution of F_{ST} as a function of expected heterozygosity based on the 31755 loci is presented in Figure 2. The F_{ST} value plot by chromosome is given in Additional file 2.

Haplotype blocks

The distribution of haplotype blocks in Korean native pig and Berkshire are shown in Additional file 1: Table S1. The overall the distribution of haplotype blocks was higher in Korean native pig where a total of 76 variable size blocks was identified. In the contrast, in the Berkshire only 32 haplotype blocks were detected. The number of haplotype blocks detected in the Korean native pig population ranged from 10 in chromosome 5 to none in chromosome 6 and 9. In contrast, the Berkshire population had the highest number of haplotype blocks (7) in chromosome 7 and no haplotype block was identified for pairs of SNPs in chromosome 10, 11, 14, 17 and 18. The size of each block varied from 5.716 kb to 158.824 kb in KNP and ranged from 11. 916 kb to 158.270 kb in Berkshire samples.

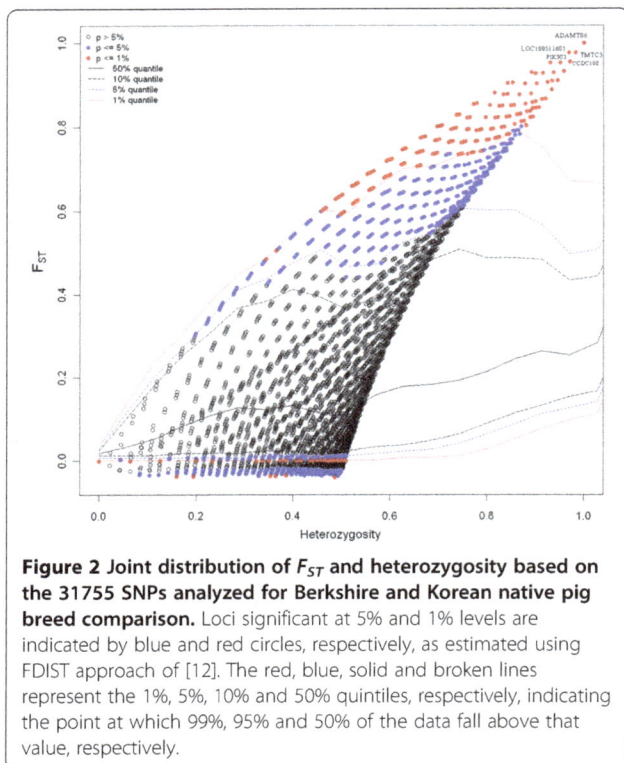

Figure 2 Joint distribution of F_{ST} and heterozygosity based on the 31755 SNPs analyzed for Berkshire and Korean native pig breed comparison. Loci significant at 5% and 1% levels are indicated by blue and red circles, respectively, as estimated using FDIST approach of [12]. The red, blue, solid and broken lines represent the 1%, 5%, 10% and 50% quintiles, respectively, indicating the point at which 99%, 95% and 50% of the data fall above that value, respectively.

The first block, in chromosome 1 in Korean native pig covered 116.539 kb and contained IGF2R gene whereas the second block of the same chromosome spanned about 55.54 kb and covers SCAF8 gene. Two blocks in chromosome 5 (first and second blocks) which spanned 51.785 kb and 137.242 kb, respectively encompassed WIFI candidate gene which is known to be associated with bone development. Similarly, the first block in chromosome 5 spanned about 51.785 kb and encompassed important gene (MSRB3) which is known to be related to cold and heat stresses. In chromosome 17, the third haplotype block covers MC3R gene previously known to be associated with body weight, adipose mass and feed conversion efficiency. The two breeds shared 12 common haplotye blocks. Of these blocks, block 69 in Korean native pig and block 28 in Berkshire spanned 158.27 kb and includes HOXA10 gene that play an important role in morphogenesis (Additional file 3). We compared the pattern of F_{ST} and haplotype blocks distribution. The highest genetic differentiation (F_{ST} =1.00) values were observed for SNPs positioned within chromosome 16, but there was no any halotype block detected for these particular SNPs. Common for both breeds, the largest haplotype blocks were detected on chromosome 15 (87,700,078 - 87,163,500) with average F_{ST} value of 0.82, while the smallest haplotype (12.848 kb) was detected for chromosome 13 with these SNPs displaying an average genetic differentiation of 0.00.

Gene ontology (GO) term analyses of the candidate genes under selection

The candidate genes were analyzed for their gene ontology (GO) and tested for enrichment based on a Fisher Exact test to identify biological process and molecular functions most pertinent by our genes list. The biological process and molecular functions of the candidate gene are present in Additional files 3 and 4, respectively. The identified candidate genes involved in a wide array of biological processes and molecular functions. Results revealed that 19 candidate genes were enriched in phosphate (GO: 0006796) and phosphorus metabolisms (GO: 0006793). GO term enriched in ion transport contained 16 candidate genes, while 5 (GLI3, CHD7, FBN2, HOXA10 and NR2F2) and 4 (GLI3, CHD7, FBN2 and HOXA10) candidate genes were involved in limb development (GO: 0060173) and morphogenesis (GO: 0035108), respectively. In addition, our functional analysis demonstrated that higher numbers of the candidate genes with significant enrichment had molecular functions related to ion binding (48). We also detected a significant encirclement for candidate genes involved in nucleotide and nucleoside bindings. Molecular function enriched with ATP bindings contained 23 candidate genes (ATP2A3, ADCK1, CAMK2D, CHD7, CDKL3, DGKB, ERN1, GALK2, KIF6, MAK, MAGI3, MAP2K5, MAP3K1, MAP3K14, MYO1B, PI4KB, PIK3C3, PRKCI, RNASEL, LOC392335, LOC441420, MYO5B, UBE2W, BRAF and VRK1) under selection. About 15 and 12 potential candidate genes were involved in protein kinase and protein serine/threonine kinase activity, respectively.

Discussion

The post-genomics era have opened opportunity in the scanning of whole for election signature in most commercial livestock species. Selection signatures may be used to identify genes or chromosomal regions that are possible targets of positive selection. The detection of selection signatures for local adaptation and phenotypic variations in Korean and western pig breeds is yet lacking. In this study, we compared two phenotypically distinct pig breeds in order to identify a sub-set loci significantly differentiated by employing an F_{ST} test in porcine 60 K SNP chip. We detected 1108 outlier loci (p <0.01) showing signature of selection and some of associated with genes known to be associated with production traits, ear size and morphology and diseases.

Growth and carcass qualities are important traits influencing the pig industry and these traits have been received considerable attention in breed improvement programs. Modern pigs have selected for lower levels of fat and fast lean growth [17]. In support with these facts, we identified potential candidate genes associated with growth, fat composition and feed conversion efficiency. Some of the genes include: *CART, AGL, CF7L2, MAP2K5, DLK1* and *MC3R*. For example study by [18] revealed that chromosomal region harboring the *CART* gene is a promising QTL in pig production traits (abdominal fat, weight and back fat thickness). In addition, this gene plays a crucial role in a variety of physiological processes, including food intake and body weight regulation [19]. Recent genome-wide association studies (GWAS) found that genetic polymorphisms in the *AGL* gene has shown to be association with growth and carcass traits in the crossbred population of Landrace and Jeju (Korea) Black pig [20]. We also detected signature of selection at *TCF7L2* loci which found to be associated with fat deposition traits in pigs [17]. This gene is known to locate on chromosome 14 where chromosome – wide significant trait loci for last ribs back fat and carcass weight were detected in Berkshire and Yorkshire crosses population [21].

Another important candidate gene under selection is *MAP2K5*. *MAP2K5* associated with body mass index and obesity in human [22]. Furthermore this gene is a component of the MAPK-family intracellular signaling pathways, responding to extracellular growth factors 2 (*IGF2*) [23]. Interestingly, we detected selection signatures at *DLK1* which is one of among imprinted genes in the callipyge locus (*CLPG*) region and associated with fat deposition, lean muscle mass and prenatal and postnatal growth rates in pigs [24]. In the comparison of the small sized KNP against the medium sized Berkshire, we identified selection footprints in *HMGA2* and *CA3* genes which were previously known to be associated with meat quality traits [25].

We have additionally identified two candidate genes, *GLI3* and *MC3R*, that known to influence body weight and growth traits. *GLI3* is associated with growth traits [26]. Melanocortin-3 receptor (*MC3R*) was previously reported to affect adipose mass in mice [27]. This gene is also associated with feed conversion and body weight in broiler [28] and with body weight in cattle [29]. KNP grow slower compared to the faster growing ability of European commercial pig breed (Berkshire). These genes may be involved for observable phenotypic variations in terms of growth traits in the two breeds and could serve as a molecular marker in the breeding programs. Therefore, our study provides evidence that these candidate genes detected here are likely under selection for better carcass quality traits and may be used for marker-assisted selection in beef cattle breeding program.

Among the potential candidate genes displaying signature of selection signature are *ATF2, MSRB3, TMTC3* and *SCAF8* genes which were involved in stress responses [30,31]. Particularly the *MSRB3* gene is known to play a key role in protection mechanisms against cold and heat stresses [32]. Considering the extreme environmental temperature where KNP is originated and

developed, this gene is likely under selection for cold resistance.

Breed difference for disease resistance quite obvious between improved and native breeds. KNP is known to have adaptability under low management systems. We identified genes known to playing physiological functions of either inhabiting or activating immune response. These include pig immune receptor (*HCST* or *DAP10*) which was detected predominantly in lymphohematopoietic tissues [33]. Studies in humans and mice demonstrated that *DAP10* and *DAP12* can either activate or inhibit immune responses [34] implying that they play an important role in innate immune responses. Malignant hyperthermia (MH) causes major economic losses in the swine industry. Interestingly, here we detected signature of selection in *RYR1* gene which is an essential gene in swine. A single point mutation in the *RYR1* gene was found to be correlated with MH in breeds of swine [35].

Pigs have undergone morphological evolution through the course of domestication and breed development. For instance, strong selection signatures have been detected in loci harboring quantitative trait loci that explain morphological changes in the domestic pig [36]. In line with this evidence, our GO classification analysis indicated that some of the candidate genes under selection were known to be associated with limb development and morphogenesis. Among others, ear size and morphology are important conformation characteristics for breed discrimination. In this study we detected to two potential candidate genes (*HMGA2* and *SOX5*) known to have a key role in affecting ear size and morphology. *HMGA2-LPP* fusion protein promotes chondrogenesis [37]. More interestingly, studies revealed that *HMGA2*-deficient mice develop smaller ears [38] and in dogs, it may be involved in differences in the size and type of ears [39]. Furthermore, *SOX5* plays a role in chondrogenesis [40]. Considering the distinct variations in ear morphology and size displayed by the study populations, the detected genes are potentially under selection for the observable differences. Considerable variation was observed regarding to the number and distribution of haplotype blocks in the two breeds. The relatively higher number of haplotype blocks detected in the KNP population is consistence with the demographic history of the breed [4]. The bottle neck associated with reduction in KNP population may lead to greater LD.

Modern pig breeds have been selected for reproduction traits. KNP is known for its high prolificacy and here we identified some potential genes having role in reproduction or fertility. Possible role of phospholipase B in sperm maturation and activation was investigated in guinea pig [41]. *KIF6* or *KRP3* gene has involved in spermatid maturation mediated by possible interaction with the Ran GTPase [42]. Previously *PDE3A* is identified as the major cAMP-degrading PDE in the oocyte and regulates the resumption of meiosis [43].

Conclusions

In this study we identified several candidate genes which have known associated with pork production (growth, size, and pork quality), morphology, stress and immune response. Some of the genes may be used to facilitate genetic improvement programs. Our results also provide insights for better understanding of the process and influence of breed development on the pattern of genetic variations. As the current annotation of pig genome is not conclusive, it is worth noting that many of the outlier loci or genes without GO terms may have relevant biological meanings and functions.

Additional files

Additional file 1: Loci displaying selection signature at 99% confidence level and candidate genes in the comparison of Berkshire and Korean native pig breeds.

Additional file 2: F_{ST} **plot by chromosome in the comparison of Koran native and Berkshire pig breeds.**

Additional file 3: Biological process of the candidate genes under selection based on gene ontology (GO) analysis.

Additional file 4: Molecular functions of the candidate genes under selection based on gene ontology (GO) analysis.

Abbreviations
DAVID: Database for Annotation, Visualization, and Integrated Discovery; DNA: Deoxyribonucleic acid; EASE: Expression analysis systematic explorer; F_{IS}: Within − breed fixation index; F_{ST}: Fixation index; GO: Gene ontology; ILRI: International Livestock Research Institute; HWE: Hardy-Weinberg equilibrium; KNP: Korean native pig; MAF: Minor allele frequency; NIAS: National Institute of Animal Science; PC: Principal component; PCA: Principal component analysis; SNP: Single nucleotide polymorphism.

Competing interests
The authors declare that they have no competing interests.

Authors' contributions
ZE analyzed data and wrote this paper; KSK conceived this research and edited the paper. Both authors read and approved the final manuscript.

Authors' information
E.Z PhD student at Chungbuk National University and graduate fellow at ILRI; K.S.K professor of Animal Science at Chungbuk National University.

Acknowledgements
This work was supported by a grant entitled "Development of Genetic Improvement Systems in Pigs Using Genomic and Reproductive Technologies" from the Korea Institute of Planning and Evaluation for Technology of Food, Agriculture, Forestry and Fisheries and 2-7-10 Agenda Research (PJ00670701) from the National Institute of Animal Science.

References
1. Giuffra E, Kijas JM, Amarger V, Carlborg O, Jeon JT, Andersson L: **The origin of the domestic pig: Independent domestication and subsequent introgression.** *Genetics* 2000, 154:1785–1791.
2. Larson G, Dobney K, Albarella U, Fang M, Matisoo-Smith E, Robins J, Lowden S, Finlayson H, Brand T, Willerslew E, Rowley-Conwy P, Andersson L, Cooper A: **Worldwide phylogeography of wild boar reveals multiple centers of pig domestication.** *Science* 2005, 307:1618–1621.
3. Yang SL, Wang ZG, Liu B, Zhang GX, Zhao SH, Yu M, Fan B, Li MH, Xiong TA, Li K: **Genetic variation and relationships of eighteen Chinese indigenous pig breeds.** *Genet Sel Evol* 2003, 35:657–671.
4. Kim TH, Kim KS, Choi BH, Yoon DH, Jang GW, Lee KT, Chung HY, Lee HY, Park HS, Lee JW: **Genetic structure of pig breeds from Korea and China using microsatellite loci analysis.** *J Anim Sci* 2005, 83:2255–2263.
5. Park BY, Kim NK, Lee CS, Hwang IH: **Effect of fiber type on postmortem proteolysis in longissimus muscle of Landrace and Korean native black pigs.** *Meat Sci* 2007, 77:482–491.
6. Porto-Neto LR, Lee SH, Lee HK, Gondro C: **Detection of signatures of selection using Fst.** *Methods Mol Biol* 2013, 1019:423–436.
7. Ramos AM, Crooijmans RP, Affara NA, Amaral AJ, Archibald AL, Beever JE, Bendixen C, Churcher C, Clark R, Dehais P, Hansen MS, Hedegaard J, Hu ZL, Kerstens HH, Law AS, Megens H-J, Milan D, Nonneman DJ, Rohrer GA, Rothschild MF, Smith TPL, Schnabel RD, Van Tassell CP, Taylor JF, Wiedmann RT, Schook LB, Groene MAM: **Design of a high density SNP genotyping assay in the pig using SNPs identified and characterized by next generation sequencing technology.** *PLoS One* 2009, 4:e6524.
8. Hwang IH, Park BY, Cho SH, Kim JH, Lee JM: **Identification of muscle proteins related to objective meat quality in Korean native black pig.** *Asian-Australas J Anim Sci* 2004, 17:1599–1607.
9. Illumina. [http://www.illumina.com]
10. Excoffier L, Lischer H: **Arlequin: An integrated software package for population genetics data analysis.** *Evol Bioinform* 2011, 1:47–50.
11. SNP and Variation Suite Version 7. [http://www.goldenhelix.com]
12. Patterson N, Price A, Reich D: **Population Structure and Eigen analysis.** *PLoS Genet* 2006, 2006(2):e190.
13. Beaumont MA, Nichols RA: **Evaluating loci for use in the genetic analysis of population structure.** *Proc R Soc B* 1996, 263:1619–1626.
14. ORG Data Repository. [http://www.animalgenome.org/repository/pig/]
15. Database for Annotation, Visualization, and Integrated Discovery (DAVID) tools. [http://david.abcc.ncifcrf.gov/]
16. Gabriel SB, Schaffner SF, Nguyen H, Moore JM, Roy J, Blumenstiel B, Higgins J, DeFelice M, Lochner A, Faggart M, Liu-Cordero SN, Rotimi C, Adeyemo A, Cooper R, Ward R, Lander ES, Daly MJ, Altshuler D: **The Structure of Haplotype Blocks in the Human Genome.** *Science* 2002, 296:2225. doi:10.1126/science.1069424.
17. Du ZQ, Fan B, Zhao X, Amoako R, Rothschild MF: **Association analyses between type 2 diabetes genes and obesity traits in pigs.** *Obesity* 2009, 17(17):323–329.
18. Stachowiak M, Cieslak J, Skorczyk A, Nowakowska J, Szczerbal I, Szydlowski M, Switonski M: **The pig CART (cocaine- and amphetamine-regulated transcript) gene and association of its microsatellite polymorphism with production traits.** *J Anim Breed Genet* 2009, 126:37–42.
19. Vicentic A, Jones DC: **The CART system in appetite and drug addiction.** *J Pharmacol Exp Ther* 2007, 320:499–506.
20. Han SH, Shin KY, Lee SS, Ko MS, Jeong DK, Oh HS, Yang BC, Cho IC: **SINE indel polymorphism of AGL gene and association with growth and carcass traits in Landrace x Jeju Black pig F(2) population.** *Mol Biol Rep* 2010, 37:467–471.
21. Malek M, Dekkers JC, Lee HK, Baas TJ, Prusa K, Huff-Lonergan E, Rothschild MF: **A molecular genome scan analysis to identify chromosomal regions influencing economic traits in the pig. II. Meat and muscle composition.** *Mamm Genome* 2001, 12:637–645.
22. Rask-Andersen M, Jacobsson JA, Moschonis G, Ek AE, Chrousos GP, Marcus C, Manios Y, Fredriksson R, Schiöth HB: **The MAP2K5-linked SNP rs2241423 is associated with BMI and obesity in two cohorts of Swedish and Greek children.** *BMC Med Genet* 2012, 13:36. doi:10.1186/1471-2350-13-36.
23. Carter EJ, Cosgrove RA, Gonzalez I, Eisemann JH, Lovett FA, Cobb LJ, Pell JM: **MEK5 and ERK5 are mediators of the pro-myogenic actions of IGF-2.** *J Cell Sci* 2009, 122:3104–3112.
24. Kim KS, Kim JJ, Dekkers JCM, Rothschild MF: **Polar overdominant inheritance of a DLK1 polymorphism is associated with growth and fatness in pigs.** *Mamm Genome* 2003, 15:555–559.
25. Wimmers K, Murani E, Te Pas MF, Chang KC, Davoli R, Merks JW, Henne H, Muraniova M, da Costa N, Harlizius B, Schellander K, Böll I, Braglia S, de Wit AA, Cagnazzo M, Fontanesi L, Prins D, Ponsuksili S: **Associations of functional candidate genes derived from gene-expression profiles of**

prenatal porcine muscle tissue with meat quality and muscle deposition. *Anim Genet* 2007, **38**:474–484.

26. Huang YZ, Wang KY, He H, Shen QW, Lei CZ, Lan XY, Zhang CL, Chen H: **Haplotype distribution in the GLI3 gene and their associations with growth traits in cattle.** *Gene* 2013, **513**:141–146.

27. Butler AA, Kesterson RA, Khong K, Cullen MJ, Pelleymounter MA, Dekoning J, Baetscher M, Cone RD: **A unique metabolic syndrome causes obesity in the melanocortin-3 receptor-deficient mouse.** *Endocrinology* 2000, **141**:3518–3521.

28. Sharma P, Bottje W, Okimoto R: **Polymorphisms in uncoupling protein, melanocortin 3 receptor, melanocortin 4 receptor, and pro-opiomelanocortin genes and association with production traits in a commercial broiler line.** *Poult Sci* 2008, **87**:2073–2086.

29. Luoreng ZM, Wang XP, Ma Y, Li F, Guo DS, Li N, Wang JR: **Three novel SNPs in the coding region of the Bovine MC3R gene and their aassociations with ggrowth traits.** *Biochem Genet* 2014, **52**:116–124.

30. Liu S, Wang F, Yan L, Zhang L, Song Y, Xi S, Jia J, Sun G: **Oxidative stress and MAPK involved into ATF2 expression in immortalized human urothelial cells treated by arsenic.** *Arch Toxicol* 2013, **87**:981–989.

31. Racapé M, Duong Van Huyen JP, Danger R, Giral M, Bleicher F, Foucher Y, Pallier A, Pilet P, Tafelmeyer P, Ashton-Chess J, Dugast E, Pettré S, Charreau B, Soulillou JP, Brouard S: **The involvement of SMILE/TMTC3 in endoplasmic reticulum stress response.** *PLoS One* 2011, **6**:e19321.

32. Lim DH, Han JY, Kim JR, Lee YS, Kim HY: **Methionine sulfoxide reductase B in the endoplasmic reticulum is critical for stress resistance and aging in Drosophila.** *Biochem Biophys Res Commun* 2012, **419**:20–269.

33. Yim D, Jie HB, Sotiriadis J, Kim YS, Kim KS, Rothschild MF, Lanier LL, Kim YB: **Molecular cloning and characterization of pig immunoreceptor DAP10 and NKG2D.** *Immunogenetics* 2001, **53**:243–249.

34. Lanier LL: **DAP10- and DAP12-associated receptors in innate immunity.** *Immunol Rev* 2009, **227**:150–160.

35. Fujii J, Otsu K, Zorzato F, Leon SD, Khanna VK, Weiler JE, O'Brien PJ, MacLannan DH: **Identification of a mutation in porcine ryanodine receptor associated with malignant hyperthermia.** *Science* 1991, **253**:448–451.

36. Rubin C, Megens HJ, Barrio AM, Maqbool K, Sayyab S, Schwochow D, Wang C, Carlborg Ö, Jern P, Jørgensen CB, Archibald AL, Fredholm M, Groenen MA, Andersson L: **Strong signatures of selection in the domestic pig genome.** *PNAS* 2012, **2012**(109):19529-1936.

37. Kubo T, Matsui Y, Goto T, Yukata K, Yasui N: **Over expression of HMGA2- LPP fusion transcripts promotes expression of the alpha 2 type XI collagen gene.** *Biochem Biophys Res Commun* 2006, **340**:476–481.

38. Xiang X, Benson KF, Chada K: **Mini-mouse: disruption of the pygmy locus in a transgenic insertional mutant.** *Science* 1990, **247**:967–969.

39. Boyko AR, Quignon P, Li L, Schoenebeck JJ, Degenhardt JD, Lohmueller KE, Zhao K, Brisbin A, Parker HG, von Holdt BM, Cargill M, Auton A, Reynolds A, Elkahloun AG, Castelhano M, Mosher DS, Sutter NB, Johnson GS, Novembre J, Hubisz MJ, Siepel A, Wayne RK, Bustamante CD, Ostrander EA: **A simple genetic architecture underlies morphological variation in dogs.** *PLoS Biol* 2010, **8**:e1000451.

40. Smits P, Li P, Mandel J, Zhang Z, Deng JM, Behringer RR, de Crombrugghe B, Lefebvre V: **The transcription factors L-Sox5 and Sox6 are essential for cartilage formation.** *Dev Cell* 2001, **1**:277–290.

41. Delagebeaudeuf C, Gassama-Diagne A, Nauze M, Ragab A, Li RY, Capdevielle J, Ferrara P, Fauvel J, Chap H: **Ectopic epididymal expression of guinea pig intestinal phospholipase B. Possible role in sperm maturation and activation by limited proteolytic digestion.** *J Biol Chem* 1998, **273**:13407–13414.

42. Zou Y, Millette CF, Sperry AO, KRP3A and KRP3B: **Candidate motors in spermatid maturation in the seminiferous epithelium.** *Biol Reprod* 2002, **66**:843–855.

43. Sasseville M, Côté N, Guillemette C, Richard FJ: **New insight into the role of phosphodiesterase 3A in porcine oocyte maturation.** *BMC Dev Biol* 2006, **6**:47.

30

Effects of dietary *omega*-3 polyunsaturated fatty acids on growth and immune response of weanling pigs

Qizhang Li, Joel H Brendemuhl, Kwang C Jeong and Lokenga Badinga[*]

Abstract

The recognition that *omega*-3 polyunsaturated fatty acids (*n*-3 PUFA) possess potent anti-inflammatory properties in human models has prompted studies investigating their efficacy for animal growth and immunity. This study examined the effect of feeding an *n*-3 PUFA-enriched diet on growth and immune response of weanling piglets. Newly weaned pigs (averaging 27 ± 2 days of age and 8.1 ± 0.7 kg of body weight) were assigned randomly to receive a control (3% vegetable oil, n = 20) or *n*-3 PUFA-supplemented (3% marine *n*-3 PUFA, n = 20) diet for 28 day after weaning. Female pigs consuming the *n*-3 PUFA-enriched diet were lighter at week 4 post-weaning than those fed the vegetable oil supplement. Weanling pigs gained more weight, consumed more feed and had better growth to feed ratios between days 14 and 28 than between days 0 and 14 post-weaning. Plasma insulin-like growth factor I (IGF-I) decreased between days 0 (87.2 ± 17.0 ng/mL) and 14 (68.3 ± 21.1 ng/mL) after weaning and then increased again by day 28 (155.2 ± 20.9 ng/mL). In piglets consuming the vegetable oil-enriched diet, plasma tumor necrosis factor alpha (TNF-α) increased from 37.6 ± 14.5 to 102.9 ± 16.6 pg/mL between days 0 and 14 post-weaning and remained high through day 28 (99.0 ± 17.2 pg/mL). The TNF-α increase detected in the piglets fed vegetable oil was not observed in the piglets fed *n*-3 PUFA. Results indicate that weaning induces considerable immune stress in piglets and that this stress can be mitigated by dietary supplementation of *n*-3 PUFA.

Keywords: n-3 PUFA, Growth, Immunity, Pig

Background

Nutritional, environmental and immune challenges associated with weaning may lead to considerable economic losses to pork producers. This period is generally characterized by decreased voluntary feed intake, altered gut integrity and increased concentrations of inflammatory cytokines in blood [1-3]. These nutritional and physiological abnormalities often result in diarrhea and depression of growth in newly weaned piglets. Restrictions of antibiotic usage in swine have compelled the industry to find alternatives that offer both performance enhancement and protection from disease [4,5]. In this regard, Liu et al. [6] reported that dietary fish oil reduced the release of pro-inflammatory cytokines in weaned pigs challenged with *Escherichia coli* lipopolysaccharide. A more recent study indicated that prenatal exposure to long-chain *n*-3 PUFA increased postnatal glucose absorption in piglets [7]. Although exact mechanisms by which dietary *n*-3 PUFA modulate immune and metabolic functions in pigs are yet to be fully elucidated, the above study would indicate that dietary *n*-3 PUFA may help the piglets adapt quickly to the rapidly changing diet at weaning [7].

Currently, there is very little information regarding the use of *n*-3 PUFA in the diets of pigs raised under minimal disease and stress conditions. To test the hypothesis that nutritional management strategies that attenuate intestinal inflammation may partition nutrients to skeletal muscle for optimal growth, this study was designed to examine the effects of dietary *n*-3 PUFA on growth and immune response of weanling pigs raised without an added bacterial or environmental challenge.

* Correspondence: lbadinga@ufl.edu
Department of Animal Sciences, Institute of Food and Agricultural Sciences, University of Florida, Gainesville 32611, USA

Results and discussion

Weaning imposes tremendous stress on piglets and is accompanied by marked changes in gastrointestinal physiology, microbiology and immunology [8]. The biochemical and histological changes that occur in the small intestine cause excessive secretion of pro-inflammatory cytokines and induce severe intestinal inflammation. *Omega*-3 PUFA are known to possess anti-inflammatory properties in humans [9,10], swine [6,11] and chickens [12]. To test the hypothesis that nutritional management strategies that attenuate intestinal inflammation may repartition nutrients to tissue accretion, we examined the effects of dietary *n*-3 PUFA on growth and immune response of weanling pigs (Figure 1) raised without an added bacterial or environmental challenge.

Inclusion of 3% *n*-3 PUFA in the weanling piglet's diet did not result in significantly improvement of average daily gain (ADG), average daily feed intake (ADFI) or growth to feed ratio (G: F) in weanling pigs. These findings are consistent with an earlier study [13] which detected no effects of dietary flax seed meal (rich in alpha linolenic acid) on basal body weight gain, feed intake or feed efficiency in weanling pigs. Additional studies using control diets with less 18: 3*n*-3 are needed to examine true effects of long-chain *n*-3 PUFA on growth and feed intake responses. Female piglets consuming the *n*-3 PUFA-supplemented diet were lighter at week 4 post-weaning than those consuming the vegetable oil-enriched diet. Whether or not this phenomenon was due to alteration in body composition as a result of feeding *n*-3 PUFA to nursery pigs was not documented. In rodents [14-16] and humans [17,18], diets rich in *n*-3 PUFA lower fat stores and increase lean tissue mass. It is, therefore, possible that the smaller body weight of female piglets

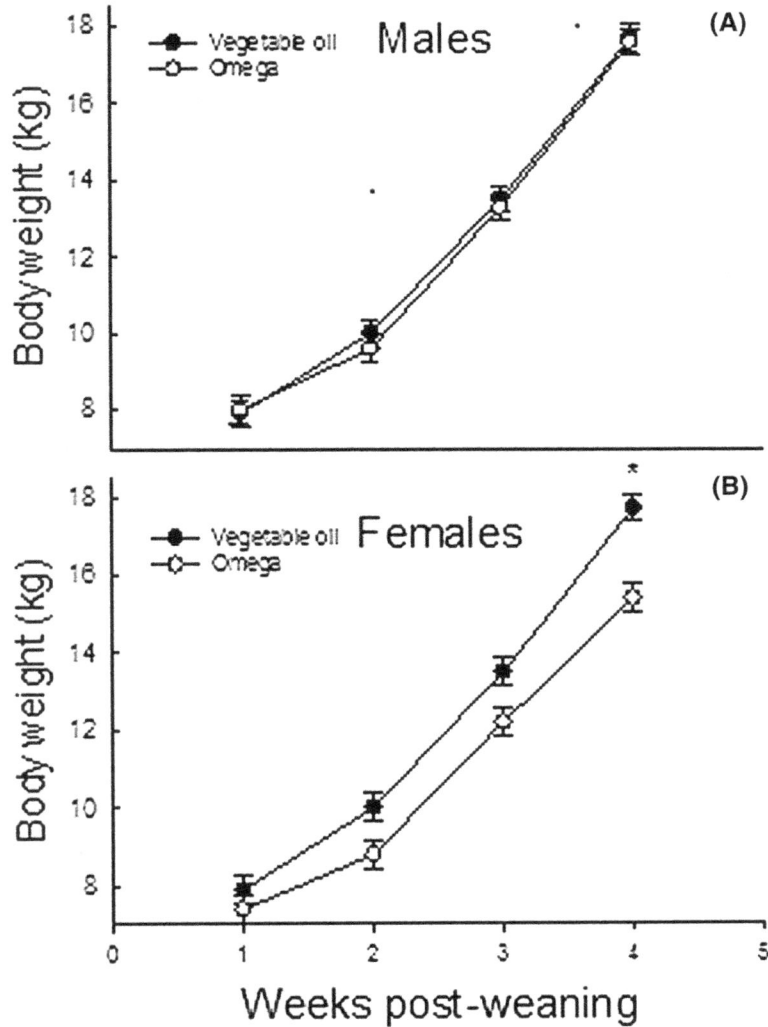

Figure 1 Body weights of male (A) and female (B) pigs during four weeks after weaning. A diet x gender x week interaction was detected ($P < 0.04$) for body weight. Asterisk indicates significant difference ($P < 0.01$) at the specified week.

consuming *n*-3 PUFA detected at week 4 post-weaning was due to a decrease in fat accretion at the expense of lean tissue. The improvement of body weight gain detected in experimental animals between days 14 and 28 post-weaning likely resulted from an increase in feed intake and a decrease in basal inflammatory challenges during the second phase of growth (Figure 2).

Concentrations of IGF-I in plasma decreased immediately following weaning and increased again by day 28 post-weaning (Figure 3). These findings indicate that weaning may cause a significant metabolic stress in weanling pigs and that this stress decreases with increasing weeks after weaning. There is little information on the effect of dietary *n*-3 PUFA on peripheral concentrations of growth factors in the pig. In the present study, inclusion of 3% *n*-3 PUFA into the piglet's diet had no detectable effects on plasma IGF-I concentration during the first four weeks after weaning. These observations are consistent with previous studies [6,13] which showed no beneficial effects of dietary fish oil on basal IGF-I concentration in weaned pigs. Thissen and Verniers [19] reported that IL-6 and TNF-α decreased both growth hormone (GH) and IGF-I mRNA in rat hepatocyte primary cultures. We did not examine GH or IGF-I transcript modulation by inflammatory cytokines, and therefore, whether or not the lack of *n*-3 PUFA effects on plasma IGF-I concentration detected in the present study was indicative of cytokine-mediated uncoupling of GH and IGF-I gene expression in weanling pigs warrants further investigation.

Tumor necrosis factor-α, a cytokine produced primarily by monocytes and macrophages, is thought to be one of the principal mediators of inflammation [20]. In the present study, plasma TNF-α concentrations were lower in weanling piglets supplemented with *n*-3 PUFA than those fed the vegetable oil supplement (Figure 3). These findings are consistent with previous in vitro [21-23] and in vivo [11,24,25] studies and suggest that *n*-3 PUFA inclusion in the diet could mitigate the immune stress in weanling pigs. Whereas exact mechanisms of *n*-3 PUFA suppression of TNF-α are yet to be fully elucidated, we speculate that suppression of TNF-α production by *n*-3 PUFA may be attributed, in part, to their inhibitory effects on NF-κB activation and or translocation to the nucleus [9,22,23]. Nuclear factor-κB are normally confined in the cytoplasm through their association with IκB. When cells are activated by inflammatory stimuli, the IκB are rapidly phosphorylated and degraded to free the NF-κB. The free NF-κB then migrate to the nucleus where they bind to cognate DNA binding sites and activate inflammatory gene transcription [9]. Any factor that prevents IκB phosphorylation and, thus, NF-κB activation, will decrease pro-inflammatory gene expression in the nucleus. Additionally, long-chain PUFA serve

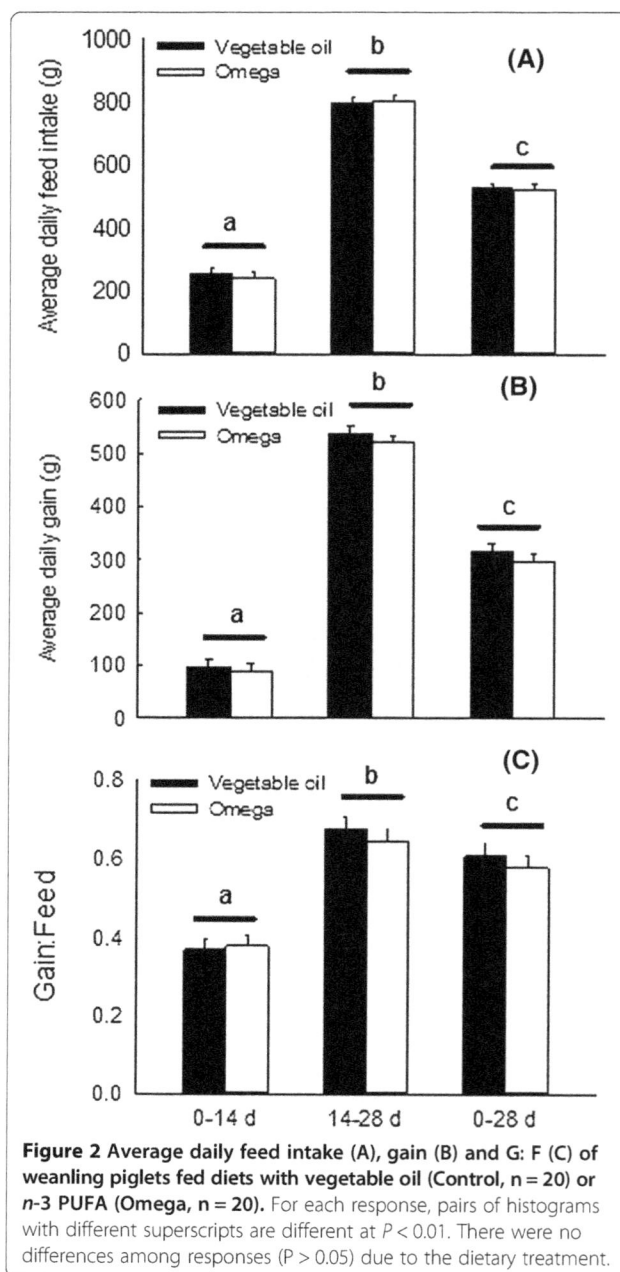

Figure 2 Average daily feed intake (A), gain (B) and G: F (C) of weanling piglets fed diets with vegetable oil (Control, n = 20) or *n*-3 PUFA (Omega, n = 20). For each response, pairs of histograms with different superscripts are different at *P* < 0.01. There were no differences among responses (P > 0.05) due to the dietary treatment.

as ligands for peroxisome proliferator-activated receptors (PPAR), which are known to inhibit nuclear translocation of NF-κB [9]. Thus, activation of PPAR may be another intracellular mechanism by which marine *n*-3 PUFA regulate NF-κB activation and TNF-α production in animal models [9].

Hematological traits of swine are influenced by a variety of environmental and physiological factors including diet, age, gender and housing [26,27]. In the present study, most of the blood characteristics examined did not differ among pigs fed the two diets (Table 1). Blood samples for complete blood cell counts were collected at

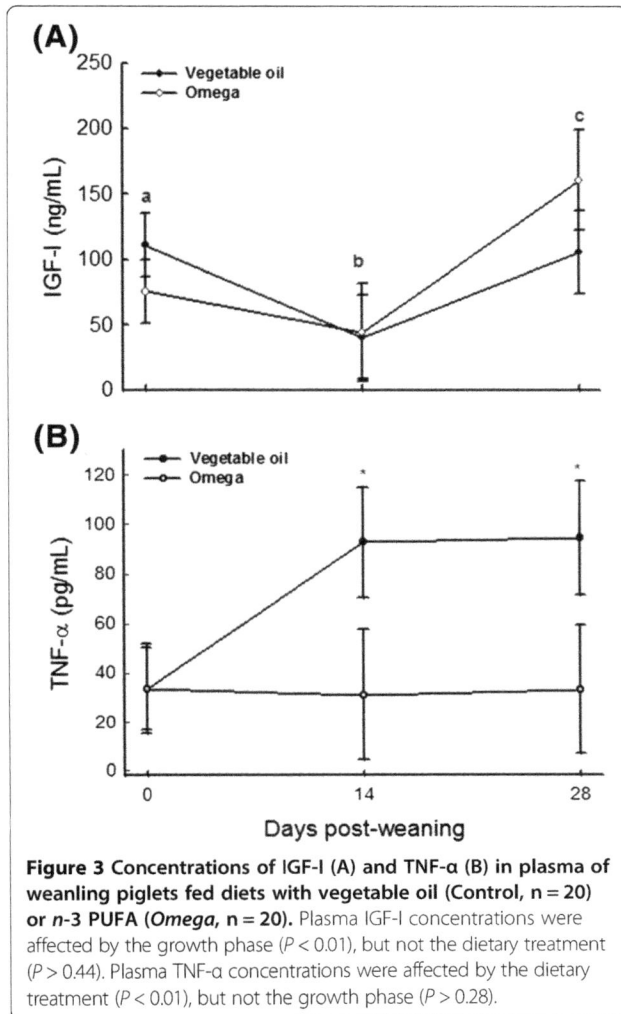

Figure 3 Concentrations of IGF-I (A) and TNF-α (B) in plasma of weanling piglets fed diets with vegetable oil (Control, n = 20) or n-3 PUFA (Omega, n = 20). Plasma IGF-I concentrations were affected by the growth phase (P < 0.01), but not the dietary treatment (P > 0.44). Plasma TNF-α concentrations were affected by the dietary treatment (P < 0.01), but not the growth phase (P > 0.28).

Table 1 Hematological traits of weanling pigs fed diets with vegetable oil or long-chain *omega*-3 fatty acids[a]

Trait	Experimental diets[b]		SEM	P[c]
	Control	Omega		
WBC[d] $\times 10^3/mm^3$	14.2	15.7	1.2	0.44
Lymphocytes, %	42.3	41.5	4.0	0.90
Neutrophils, %	52.8	51.0	3.8	0.75
Eosinophils, %	1.5	2.8	0.4	0.07
Monocytes, %	2.5	4.3	0.7	0.13
RBC[e] $\times 10^3/mm^3$	7.0	6.6	0.3	0.45
Hemoglobin	10.4	9.2	0.7	0.28
Hematocrit, %	34.1	31.4	2.1	0.41
Platelets $\times 10^3/mm^3$	378.5	674.0	80.4	0.04

[a]Means represent 4 pigs per dietary treatment.
[b]Diets were: Control (3% vegetable oil) and *omega* (3% Gromega Ultra 345, provided by JBS United, Inc., Sheridan, IN).
[c]P-values for control compared to Omega diet.
[d]White blood cells.
[e]Red blood cells.

challenges. The observation that female piglets consuming the n-3 PUFA-supplemented diet were lighter at week 4 post-weaning than those consuming the vegetable oil-enriched diet (Figure 2) may be indicative of a decrease in fat accretion at the expense of lean tissue. Additionally, dietary n-3 PUFA may improve the immune status of weanling pigs, as reflected by considerably lower plasma TNF-α in pigs consuming n-3 PUFA than those fed vegetable oil. The gradual increase in body weight, feed intake and feed efficiency following weaning likely reflects a progressive adaptation to post-weaning diets and a gradual improvement of the gastrointestinal microbiota.

Methods

Animals, diets and experimental design

The animal protocol for this research was approved by the institutional Animal Research Committee of the University of Florida. To avoid potential differences due to farrowing season, the study was conducted using 40 piglets born within one week at the Swine Research Unit of the University of Florida (Gainesville, FL) during the month of March 2013. Forty crossbred pigs (averaging 27 ± 2 days of age and 8.1 ± 0.7 kg of body weight) were balanced for initial body weight and gender across two treatment groups in a complete randomized block design. Experimental animals were fed either a control (3% vegetable oil, n = 20) or n-3 PUFA (3% marine n-3 PUFA; *Gromega 345*, JBS United, Inc, Sheridan IN, n = 20)-supplemented diet for four weeks after weaning. The vegetable oil was purchased from Sysco Corporations (Houston, TX) and contained approximately 22% total fat. Omega-3 fatty polyunsaturated fatty acids used in this study were provided by JBS United (Sheridan, Indiana)

4 weeks after weaning, and it is possible that by this sampling time, the weanling piglets had already recovered from most physiological and dietary challenges normally associated with weaning in pigs. Alternatively, the piglets used in this study were raised in a clean environment and, thus, may not have acquired the "normal" gastrointestinal microflora that would cause clinical diseases. This hypothesis was further supported by our inability to detect salmonella and enterotoxigenic *E. coli* in fecal samples collected at week 4 post-weaning (data not shown).

Conclusions

In the pig, the period following weaning is generally characterized by sub-optimal growth, deteriorated feed efficiency, and a high incidence of diarrhea. Results of this study provided no evidence for n-3 PUFA modulation of growth of male weanling pigs raised in the absence of significant immunological and environmental

and contained a minimum of 39% crude fat. Complete ingredient compositions and FA profiles of experimental diets are summarized in Tables 2 and 3, respectively. Pigs were housed in pens (groups of 5 animals per pen; pen size = 2.4 m × 1.8 m) and kept on the same diet for the entire experimental period. Body weight and feed consumption were recorded weekly throughout the 4-week experiment. These observations were used to calculate ADG, ADFI, and G: F.

Blood collection and analysis

On days 0, 14 and 28 of the experiment, jugular venous blood samples (8 ml from each experimental pig) were collected into evacuated heparinized tubes (BD Franklin Lakes, NJ) and centrifuged (3,000 × g for 15 min) to separate plasma. The plasma samples were stored at −80°C until analysis. Concentrations of IGF-I and TNF-α in plasma were analyzed using commercially available ELISA kits (R&D Systems, Inc., Minneapolis, MN). Hormone and cytokine analyses were performed in single assays and intra-assay CV were 4.0 and 4.7% for IGF-I and TNF-α, respectively. The least detectable concentrations were 0.06 ng/mL and 5.50 pg/ml. On day 27 of the experiment, additional blood samples were collected for complete blood cell counts, and hematological traits were determined as described by Quiroz-Rocha et al. [28].

Fecal evaluation

Two fecal consistency scores were assigned to each pen on weeks 1, 2, 3, and 4 post-weaning. The scale used to assess fecal consistency was based on a numerical scale

Table 2 Ingredient and calculated compositions of experimental diets

Composition	Experimental diets[a]	Omega
	Control	
Ingredient:		
Corn, %	61.90	61.90
Soybean meal, %	25.00	25.00
Vegetable oil, %	3.00	-
Gromega Ultra 345, %	-	3.00
Min-Vit Premix, %	10.00	10.00
L-Lysine.HCL, %	0.10	0.10
Calculated composition:		
ME, kcal/kg	3282.38	3282.38
CP, %	19.53	19.53
CF, %	3.39	3.39
Lysine, %	1.40	1.40
Calcium, %	0.78	0.97
Phosphorus, %	0.63	0.63

[a]Diets were: Control (3% vegetable oil) and omega (3% Gromega Ultra 345, provided by JBS United, Inc., Sheridan, IN).

Table 3 Fatty acid profile (g/100 g of total fat) of experimental diets[a]

Fatty acid	Experimental diets[b]	*Omega*
	Control	
C14:0	0.21	2.61
C15:0	0.00	2.61
C16:0	14.68	19.76
C16:1, 9c	0.32	2.98
C17:0	0.13	0.42
C17:1	0.00	0.39
C18:0	4.26	4.71
C18:1, 9c	24.72	23.89
C18:2n-6	49.79	37.04
C18:3n-3	4.53	2.15
C18:4n-3	0.00	0.47
C20:0	0.38	0.40
C20:1n-9	0.00	0.71
C20:5n3	0.00	1.30
C22:0	0.41	0.26
C22:5n-3	0.00	0.26
C22:6n-3	0.00	0.96
C24:0	0.26	0.34
Σ n-6	49.79	37.04
Σ n-3	4.53	5.54
Σn-6 /Σn-3	10.99	6.69
Σ SFA	20.33	28.79
Σ UFA	79.36	70.15

[a]Fatty acid analysis was performed by the University of Missouri Analytical Laboratory.
[b]Diets were: Control (3% vegetable oil) and *omega* (3% Gromega Ultra 345, provided by JBS United, Inc., Sheridan, IN).

Table 4 Fecal consistency scores[a] of weanling pigs fed diets with vegetable oil or long-chain *omega*-3 fatty acids[b]

Week post-weaning	Diets[c]	Omega	SEM	P[d]
	Control			
1	2.6	2.5	0.2	0.67
2	1.9	1.9	0.2	1.00
3	2.0	1.9	0.2	0.68
4	1.3	1.1	0.2	0.68

[a]The scale used for assessing fecal consistency was based on a numerical scale of 1 to 3, where 1 represented a normal (hard) feces, 2 represented a soft moist feces, and 3 represented diarrhea (watery liquid).
[b]Means represent average fecal scores for 4 pens per dietary treatment.
[c]Diets were: Control (3% vegetable oil) and *omega* (3% Gromega Ultra 345, provided by JBS United, Inc., Sheridan, IN).
[d]P-values for control compared to *Omega* diet.

of 1 to 3, where 1 represented a normal (hard) feces, 2 represented a soft (moist) feces, and 3 represented diarrhea (watery liquid). The weekly score for each pen was calculated by averaging the two fecal consistency scores (Table 4).

Statistical analysis

Effects of diets on growth, IGF-I, TNF-α and fecal characteristics were analyzed using the MIXED procedure of Statistical Analysis System (version 9.3) with repeated measures [29]. For individual measurements (body weights), fixed effects included diet, gender, diet × gender interaction, week after weaning, diet × week interaction, gender × week interaction and diet × gender × week interaction. The pig, nested within gender and diet, was considered a random variable, and therefore the pig variance was used to test the effects of diet, gender, and diet x gender interaction. Initial weights were used as covariates in these analyses. A similar model was used to test the effect of diet on plasma IGF-I and TNF-α concentrations, except that week after weaning was replaced by day of blood sample collection. For collective measurements (feed intake, average daily gain, feed efficiency, and fecal consistency score), the statistical model included the effect of diets, pen (diet), week relative to weaning, diet × week interaction. In these models, pen was used as experimental unit to test the main effect of diet. Single blood samples were collected for complete blood cell counts, and, therefore, the statistical models for hematological traits contained only the main effect of diet. For all responses, significant differences between means were declared at $P < 0.05$.

Abbreviations

ADFI: Average daily feed intake; ADG: Average daily gain; G: F: Gain to feed ratio; GH: Growth hormone; IGF-I: Insulin-like growth factor I; PPAR: Peroxisome proliferator-activated receptors; RBC: Red blood cells; TNF-α: Tumor necrosis factor alpha; WBC: White blood cells.

Competing interests

The authors declare that they have no competing interests.

Authors' contributions

QL carried out the experiment and was responsible for field data processing and laboratory analysis. JHB was involved in the design and organization of the experiment at the University of Florida Swine Unit. KCJ performed bacterial analysis of feces and contributed substantially to the writing of the manuscript. LB was involved in the design, analysis and execution of the experiment and had primary responsibility for data processing and writing of the manuscript. All authors read and approved the final manuscript.

Acknowledgements

The authors thank JBS United, Inc (Sheridan, IN, USA) for kindly providing *Gromega Ultra 345* for this research and the University of Florida Swine Unit crew for their help with the field work. The present study was supported partially by the Department of Animal Sciences of the University of Florida.

References

1. Le Dividich J, Sève B: Effects of underfeeding during the weaning period on growth, metabolism, and hormonal adjustments in the piglet. *Domest Anim Endocrinol* 2000, 19:63–74.
2. Montagne L, Boudry G, Favier C, Le Huërou-Luron I, Lallès JP, Sève B: Main intestinal markers associated with the changes in gut architecture and function in piglets after weaning. *Br J Nutr* 2007, 97:45–57.
3. Pié S, Lallès JP, Blazy F, Laffitte J, Sève B, Oswald IP: Weaning is associated with an up-regulation of expression of inflammatory cytokines in the intestine of piglets. *J Nutr* 2004, 134:641–647.
4. Cromwell GL: Why and how antibiotics are used in swine production. *Anim Biotechnol* 2002, 13:7–27.
5. Vondruskova H, Slamova R, Trckova M, Zraly Z, Pavlik I: Alternatives to antibiotic growth promoters in prevention of diarrhoea in weaned piglets: a review. *Vet Medic* 2010, 5:199–224.
6. Liu YL, Li DF, Gong LM, Yi GF, Gaines AM, Carroll JA: Effects of fish oil supplementation on the performance and immunological, adrenal, and somatotropic responses of weaned pigs after an Escherichia coli lipopolysaccharide challenge. *J Anim Sci* 2003, 81:2758–2765.
7. Gabler NK, Radcliffe JS, Spencer JD, Webel DM, Spurlock ME: Feeding long-chain *n*-3 polyunsaturated fatty acids during gestation increases intestinal glucose absorption potentially via the acute activation of AMPK. *J Nutr Biochem* 2009, 20:17–25.
8. Heo JM, Opapeju FO, Pluske JR, Kim JC, Hampson DJ, Nyachoti CM: Gastrointestinal health and function in weaned pigs: a review of feeding strategies to control post-weaning diarrhoea without using in-feed antimicrobial compounds. *J Anim Physiol Anim Nutr* 2012, 97:207–237.
9. Calder PC: Omega-3 fatty acids and inflammatory processes. *Nutrients* 2010, 2:355–374.
10. Calder PC: Omega-3 polyunsaturated fatty acids and inflammatory processes: nutrition or pharmacology? *Br J Clin Pharmacol* 2012, 75:645–662.
11. Carroll JA, Gaines AM, Spencer JD, Allee GL, Kattesh HG, Roberts MP, Zannelli ME: Effect of menhaden fish oil supplementation and lipopolysaccharide exposure on nursery pigs. I. Effects on the immune axis when fed diets containing spray-dried plasma. *Domest Anim Endocrinol* 2003, 24:341–351.
12. Korver DR, Klasing KC: Dietary fish oil alters specific inflammatory immune responses in chicks. *J Nutr* 1997, 127:2039–2046.
13. Eastwood L, Kish PR, Beaulieu AD, Leterme P: Nutritional value of flaxseed meal for swine and its effects on the fatty acid profile of the carcass. *J Anim Sci* 2009, 87:3607–3619.
14. Baillie RA, Takada R, Nakamura M, Clarke SD: Coordinate induction of peroxisomal acyl-CoA oxidase and UCP-3 by dietary fish oil: a mechanism for decreased body fat deposition. *Prostaglandins Leukot Essent Fatty Acids* 1999, 60:351–356.
15. Belzung F, Raclot T, Groscolas R: Fish oil n-3 fatty acids selectively limit the hypertrophy of abdominal fat depots in growing rats fed high-fat diets. *Am J Physiol* 1993, 264:R1111–R1118.
16. Hill JO, Peters JC, Lin D, Yakubu F, Greene H, Swift L: Lipid accumulation and body fat distribution is influenced by type of dietary fat fed to rats. *Int J Obes Relat Metab Disord* 1993, 17:223–236.
17. Jump DB, Clark SD, Thelen A, Liimatta M: Coordinate regulation of glycolytic and lipogenic gene expression by polyunsaturated fatty acids. *J Lipid Res* 1994, 35:1076–1084.
18. Noreen EE, Sass MJ, Crowe ML, Pabon VA, Brandauer J, Averill LK: Effects of supplemental fish oil on resting metabolic rate, body composition, and salivary cortisol in healthy adults. *J Int Soc Sports Nutr* 2010, 7:31–37.
19. Thissen JP, Verniers J: Inhibition by interleukin-1β and tumor necrosis factor-α of the insulin-like growth factor I messenger ribonucleic acid response to growth hormone in rat hepatocyte primary culture. *Endocrinology* 1997, 138:1078–1084.
20. Bemelmans MH, van Tits LJ, Buurman WA: Tumor necrosis factor: Function, release and clearance. *Crit Rev Immunol* 1996, 16:1–11.
21. Lo CJ, Chiu KC, Fu M, Lo R, Helton S: Fish oil decreases tumor necrosis factor gene transcription by altering the NF-κB activity. *J Surg Res* 1999, 82:216–221.
22. Novak TE, Babcock TA, Jho DH, Helton WS, Espat NJ: NF-κB inhibition by ω-3 fatty acids modulates LPS-stimulated macrophage TNF-α transcription. *Am J Physiol Lung Cell Mol Physiol* 2003, 284:L84–L89.
23. Zhao Y, Joshi-Barve S, Barve S, Chen LH: Eicosapentaenoic acid prevents LPS-induced TNF-α expression by preventing NF-kB activation. *J Am Coll Nutr* 2004, 23:71–78.

24. Gaines AM, Carroll JA, Yi GF, Allee GL, Zannelli ME: **Effect of menhaden fish oil supplementation and lipopolysaccharide exposure on nursery pigs. II. Effects on the immune axis when fed simple or complex diets containing no spray-dried plasma.** *Domest Anim Endocrinol* 2003, **24:**353–365.

25. Malekshahi Moghadam A, Saedisomeolia A, Djalali M, Djazayery A, Pooya S, Sojoudi F: **Efficacy of omega-3 fatty acid supplementation on serum levels of tumour necrosis factor-alpha, C-reactive protein and interleukin-2 in type 2 diabetes mellitus patients.** *Singapore Med J* 2012, **53:**615–619.

26. Friendship RM, Lumsden JH, McMillan I, Wilson MR: **Hematology and biochemistry reference values for Ontario swine.** *Can J Comp Med* 1984, **48:**390–393.

27. Wilson GD, Harvey DG, Snook CR: **A review of factors affecting blood biochemistry in the pig.** *Br Vet J* 1972, **128:**596–610.

28. Quiroz-Rocha GF, LeBlanc SJ, Duffield TF, Wood D, Leslie KE, Jacobs RM: **Reference limits for biological and hematological analytes of dairy cows one week before and one week after parturition.** *Can Vet J* 2009, **50:**383–388.

29. Littell RC, Henry PR, Ammerman CB: **Statistical analysis of repeated measures data using SAS procedures.** *J Anim Sci* 1998, **76:**1216–1231.

31

Capacitation and acrosome reaction differences of bovine, mouse and porcine spermatozoa in responsiveness to estrogenic compounds

Do-Yeal Ryu[†], Ye-Ji Kim[†], June-Sub Lee, Md Saidur Rahman, Woo-Sung Kwon, Sung-Jae Yoon and Myung-Geol Pang[*]

Abstract

Background: Endocrine disruptors are exogenous substance, interfere with the endocrine system, and disrupt hormonal functions. However, the effect of endocrine disruptors in different species has not yet been elucidated. Therefore, we investigated the possible effects of 17ß-estradiol (E2), progesterone (P4), genistein (GEN) and 4-tert-octylphenol (OP), on capacitation and the acrosome reaction in bovine, mouse, and porcine spermatozoa. In this *in vitro* trial, spermatozoa were incubated with 0.001-100 µM of each chemical either 15 or 30 min and then assessed capacitation status using chlortetracycline staining.

Results: E2 significantly increased capacitation and the acrosome reaction after 30 min, while the acrosome reaction after 15 min incubation in mouse spermatozoa. Simultaneously, capacitation and the acrosome reaction were induced after 15 and 30 min incubation in porcine spermatozoa, respectively. Capacitation was increased in porcine spermatozoa after 15 min incubation at the lowest concentration, while the acrosome reaction was increased in mouse spermatozoa after 30 min ($P < 0.05$). E2 significantly increased the acrosome reaction in porcine spermatozoa, but only at the highest concentration examined ($P < 0.05$). P4 significantly increased the acrosome reaction in bovine and mouse spermatozoa treated for 15 min ($P < 0.05$). The same treatment significantly increased capacitation in porcine spermatozoa ($P < 0.05$). P4 significantly increased capacitation in mouse spermatozoa treated for 30 min ($P < 0.05$). GEN significantly increased the acrosome reaction in porcine spermatozoa treated for 15 and 30 min and in mouse spermatozoa treated for 30 min ($P < 0.05$). OP significantly increased the acrosome reaction in mouse spermatozoa after 15 min ($P < 0.05$). Besides, when spermatozoa were incubated for 30 min, capacitation and the acrosome reaction were higher than 15 min incubation in E2 or GEN. Furthermore, the responsiveness of bovine, mouse and porcine spermatozoa to each chemical differed.

Conclusions: In conclusion, all chemicals studied effectively increased capacitation and the acrosome reaction in bovine, mouse, and porcine spermatozoa. Also we found that both E2 and P4 were more potent than environmental estrogens in altering sperm function. Porcine and mouse spermatozoa were more responsive than bovine spermatozoa.

Keywords: Capacitation, Acrosome reaction, Estrogen, Endocrine disruptor, Spermatozoa

* Correspondence: mgpang@cau.ac.kr
[†]Equal contributors
Department of Animal Science and Technology, Chung-Ang University, 4726
Seodong-daero, Anseong 456-756, Gyeonggi-Do, Republic of Korea

Background

Estrogens play key roles in regulating various physiological phenomena related to normal growth, development, and reproduction in mammals [1]. Although estrogens have been considered to be female reproductive hormones, recent evidence has indicated that they play an important role in the development and regulation of the male reproductive system [2]. Estrogens were detected in male serum as well as in the male gonad [3].

Recent reports have shown that endocrine disruptors are exogenous substances that interfere with the endocrine system and might disrupt hormone function in various wildlife species. Moreover, it has also been suggested that these compounds may be responsible for a variety of reproductive disturbances in men, including possible declines in sperm concentration [4,5]. Many non-steroid compounds in the environment have been found to exhibit estrogenic activity [6,7], where these include natural phytoestrogens, pesticides, and industrial products with various homologies to estradiol that can bind with estrogen receptors, acting either as agonists or antagonists [8]. Most studies of xenobiotics have focused on long-term developmental effects on the testis, the male reproductive tract and semen quality. When xenobiotics are used at relatively high dosages, they disrupt spermatogenesis and hence decrease overall male fertility [9].

Luconi et al. [2] have identified and characterized novel nongenomic estrogen receptors on the cell membrane of human spermatozoa that interfere with the effects of progesterone (P4). Thus, spermatozoa may represent a suitable model to study the possible effects of estrogenic xenobiotics on the function of spermatozoa. Adeoya-Osiguwa et al. [10] recently reported evidence that very low dosages of several xenobiotics have direct effects on the function of spermatozoa, significantly accelerating capacitation and the acrosome reaction in mice. However, there are no data on these effects in domestic animals. Examination of the physiological and nonphysiological effects of estrogenic xenobiotics on sperm function requires a high-quality *in vitro* test system [11].

To look for any significant effects of treatment length and concentration on the rate of capacitation and acrosome reaction, we examined the effects of 17β-estradiol (E2), P4, and two estrogenic compounds, namely genistein (GEN) and 4-tert-octylphenol (OP) on bovine, mouse and porcine spermatozoa.

Methods

All procedures were performed according to guidelines for the ethical treatment of animals and approved by Institutional Animal Care and Use Committee of Chung-Ang University (Approval no. 12-0021).

Medium and reagents

Throughout this study, bovine, mouse and porcine spermatozoa were treated in modified TCM 199 that comprised TCM 199 with Earle's salts containing 10% heat-inactivated fetal calf serum (v/v), 0.91 mM sodium pyruvate, 3.05 mM D-glucose, 2.92 mM calcium lactate, 50 IU/l penicillin G, and 30 μg/ml streptomycin sulfate. A stock solution of 1000 μM E2 and P4 (Sigma-Aldrich, St Louis, MO, USA) was prepared in dimethyl sulfoxide (DMSO) and stored at −20°C. Working stock solutions were prepared daily by first diluting the initial stock solution in 10% DMSO: 0.9% NaCl (1:1). This solution was used for subsequent dilutions of the standard medium. The other stock solutions (100 μM) of GEN (Sigma-Aldrich, St Louis, MO, USA), P4 and OP (Sigma-Aldrich, St Louis, MO, USA) were prepared in absolute ethanol and stored at −20°C. The working stock solutions were prepared daily using standard medium as diluent.

Preparation of spermatozoa

Frozen bovine semen and liquid porcine semen were obtained from the National Agriculture Cooperation Federation (Goyang, Gyeonggi, Korea) and Yonam Genetics, Inc. (Chonan, Chungnam, Korea), respectively. Epididymal mouse sperm cells were collected from 9-week-old male ICR mice (Central Lab. Animal Inc, Seoul, Korea). For frozen-thawed bovine and liquid porcine semen, sperm cells were centrifuged at $500 \times g$ for 3 min and the sperm pellets were diluted with modified TCM 199 with or without chemicals. Subsequently, suspensions were centrifuged at $500 \times g$ for 3 min and sperm pellets were diluted with modified TCM 199 solution containing one of the chemicals examined. These were incubated for 15 min or 30 min in an atmosphere of 5% CO_2 at 39°C. For mouse spermatozoa, caudal epididymal spermatozoa from three mature ICR males were released into sterile plastic dishes containing modified TCM 199 (0.1% BSA). Suspensions were then allowed to disperse for 5 min on a warming tray and motile sperm cells were collected. Sperm cells were diluted with modified TCM 199 containing one of the chemicals examined. These samples were incubated for 15 min or 30 min in an atmosphere of 5% CO_2 at 37°C. The analysis of each condition for each animal was replicated 3 times.

Combined Hoechst 33258/chlortetracycline fluorescence assessment of spermatozoa (H33258/CTC)

The dual staining method performed was based on that described by Perez et al. [12], with some modifications. Briefly, 135 μl of semen (2×10^8 cell/ml) was added to 15 μl of H33258 solution (10 μg H33258/ml D-PBS) and incubated at 37°C for 10 min in a light-shielded water bath. Excess dye was removed by layering the mixture over 250 μl of 2% (w/v) polyvinylpyrrolidone (PVP) in PBS that had been centrifuged at $400 \times g$ for 10 min. The supernatant was discarded and the pellet was resuspended in 700 µl of PBS and 500 µl of this solution was added to 500 µl of a freshly prepared CTC solution (1.3 mg CTC in 5 ml buffer: 20 µM Tris, 130 µM NaCl, 5 µM cystein). After 20 sec, the reaction was stopped by the addition of 10 µl of 12.5% (v/v) glutaraldehyde solution in 1 M Tris buffer and maintained at 4°C in the dark until evaluation (within 24 h of preparation). Samples were observed with a Nikon microphot-FXA under epifluorescence illumination using UV BP 340-380/LP 425 and BP 450-490/LP 515 excitation/emission filters for H33258 and CTC, respectively. Spermatozoa were classified as shown in Table 1: dead (D type, when nuclei showed bright blue fluorescence over the sperm head), live noncapacitated (F type, bright green fluorescence distributed uniformly over the entire sperm head, with or without stronger fluorescent line at the equatorial segment), live capacitated (B type, green fluorescence over acrosomal region and dark postacrosome), or live acrosome-reacted (AR type, spermatozoa showing a mottled green fluorescence over head, green fluorescence only in post acrosomal region or no fluorescence on the head) [13]. All spermatozoa had bright green fluorescent midpieces. Two slides per sample were evaluated, with at least 100 spermatozoa per slide.

Statistics

The data were analyzed using ANOVA performed with SPSS software (v. 12.0; Chicago, Illinois, USA). This test compares responses within replicates for a significant difference to be obtained, a consistent and reasonable magnitude is required between control and treated samples. A value of $P < 0.05$ was considered statistically significant.

Results

This study investigated the possible effects of E2, P4, and two estrogenic environmental estrogens, GEN, and OP, on capacitation and acrosome reaction in bovine, mouse, and porcine spermatozoa *in vitro*. Bovine and porcine spermatozoa suspensions were incubated with 0.001-100 µM E2, P4, GEN, and OP for either 15 or 30 min at 39°C and then assessed using CTC fluorescence. Mouse spermatozoa were incubated at 37°C.

Table 1 Determination of capacitation status patterns

Patterns (%)	Description
F	Full fluorescence characteristic of ejaculated spermatozoa: characteristic of uncapacitated spermatozoa
B	Banded, indicative of capacitated spermatozoa and fluorescence only in the post acrosomal region: characteristic of capacitated spermatozoa
AR	Spermatozoa showing a mottled green fluorescence over head, green fluorescence only in post acrosomal region or no fluorescence on the head: typical acrosome-reacted spermatozoa

Effects of E2 on bovine, mouse, and porcine spermatozoa

A concentration-dependent pattern of capacitation was observed in mouse spermatozoa. E2 (0.001-100 µM) significantly increased the acrosome reaction in mouse spermatozoa and capacitation in porcine spermatozoa after 15 min incubation ($P < 0.05$, Figure 1B and C). Following 30 min exposure, E2 significantly increased both capacitation and the acrosome reaction in mouse spermatozoa, and significantly increased the acrosome reaction in porcine spermatozoa ($P < 0.05$) (Figure 2B and C). No detectable effects were observed in bovine spermatozoa treated for 15 min or 30 min (Figures 1A and 2A). The acrosome reaction in mouse spermatozoa was responded to E2 at higher concentrations (10–100 µM) after 15 min incubation and responded to 0.001 µM after 30 min incubation. Capacitation in mouse spermatozoa was increased significantly at higher concentrations (10–100 µM) after 30 min incubation ($P < 0.05$) (Figures 1B and 2B). Capacitation was increased in porcine spermatozoa after 15 min incubation in E2 at a concentration of 0.001 µM. This effect decreased gradually with increasing doses (Figure 1C). E2 significantly increased the acrosome reaction in porcine spermatozoa, but only after 30 min incubation at the highest concentration ($P < 0.05$) (Figures 1C and 2C).

Effects of P4 on bovine, mouse, and porcine spermatozoa

After a 15-minute exposure to P4, a concentration-dependent effect on acrosome reaction was observed. P4 (0.001-100 µM) significantly increased the acrosome reaction in both bovine and mouse spermatozoa treated for 15 min ($P < 0.05$). This condition also significantly increased capacitation ($P < 0.05$) (Figure 3A and B). No detectable acrosome reaction was observed in porcine spermatozoa (Figure 3C). P4 significantly increased capacitation in a dose-dependent manner in mouse spermatozoa treated for 30 min ($P < 0.05$) (Figure 4B), while no detectable effects were observed in bovine or porcine spermatozoa (Figure 4A and C).

In bovine spermatozoa, the acrosome reaction was significantly increased after 15 min incubation at higher concentrations ($P < 0.05$) (Figure 3A) (10–100 µM). The mouse spermatozoa first exhibited a response at 0.001 µM (Figure 3B). However, no significant effects on acrosome reaction were observed after 30 min incubation in any treated spermatozoa (Figure 4).

Effects of GEN on bovine, mouse, and porcine spermatozoa

GEN (0.001-100 µM) significantly increased the acrosome reaction in porcine spermatozoa treated for 15 min ($P < 0.05$) (Figure 5C), while no detectable effects were observed in mouse spermatozoa (Figure 5B). GEN also significantly increased the acrosome reaction in both mouse and porcine

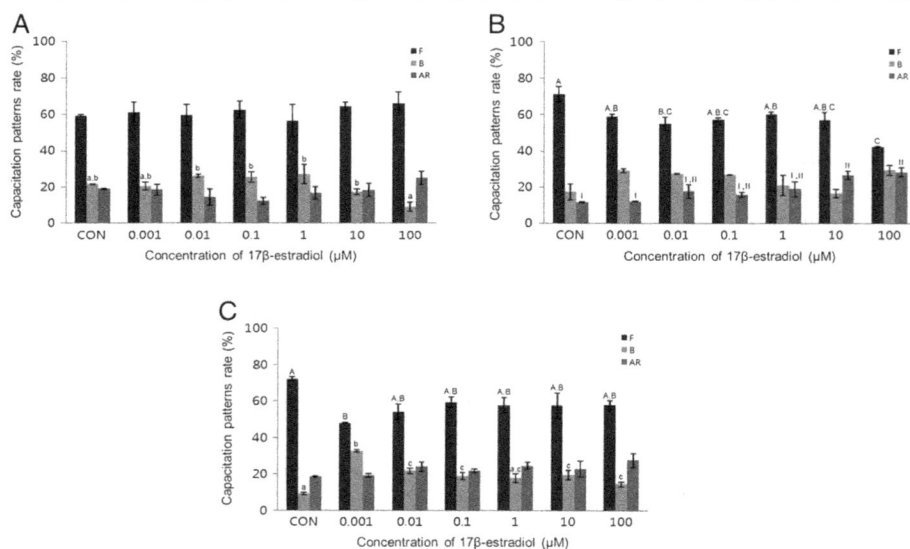

Figure 1 Effects of 15 min of incubation with 17β-estradiol (E2) on capacitation status. (A) Change of sperm capacitation status of bovine spermatozoa in the absence or presence of E2 (0.001 to 100 μM). **(B)** Change of sperm capacitation status of mouse spermatozoa in the absence or presence of E2 (0.001 to 100 μM). **(C)** Change of sperm capacitation status of porcine spermatozoa under in the absence or presence of E2 (0.001 to 100 μM). Capacitation status was distinguished F, B and AR pattern (Black Bar: F pattern, Grey Bar: B pattern, Dark-grey Bar: AR pattern). Data represent mean ± SEM, n =3. [A, B, C] Values with different superscripts were significantly different compared to control and the F pattern group, by ANOVA (P <0.05). [a, b, c] Values with different superscripts were significantly different compared to control and the B pattern group, by ANOVA (P <0.05). [I, II] Values with different superscripts were significantly different compared to control and the AR pattern group, by ANOVA (P <0.05).

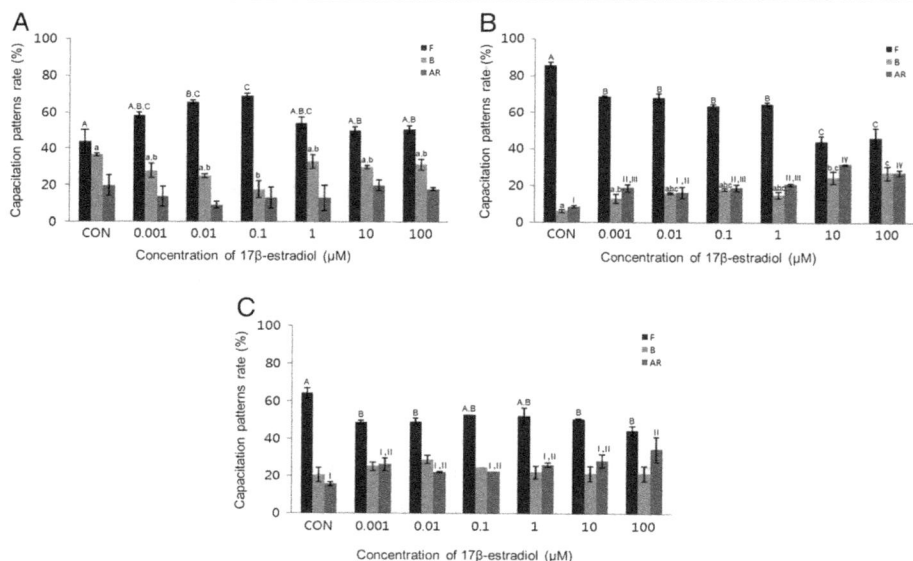

Figure 2 Effects of 30 min of incubation with 17β-estradiol (E2) on capacitation status. (A) Change of sperm capacitation status of bovine spermatozoa in the absence or presence of E2 (0.001 to 100 μM). **(B)** Change of sperm capacitation status of mouse spermatozoa in the absence or presence of E2 (0.001 to 100 μM). **(C)** Change of sperm capacitation status of porcine spermatozoa in the absence or presence of E2 (0.001 to 100 μM). Capacitation status was distinguished F, B and AR pattern (Black Bar: F pattern, Grey Bar: B pattern, Dark-grey Bar: AR pattern). Data represent mean ± SEM, n =3. [A, B, C] Values with different superscripts were significantly different compared to control and the F pattern group, by ANOVA (P <0.05). [a, b, c] Values with different superscripts were significantly different compared to control and the B pattern group, by ANOVA (P <0.05). [I, II, III, IV] Values with different superscripts were significantly different compared to control and the AR pattern group, by ANOVA (P <0.05).

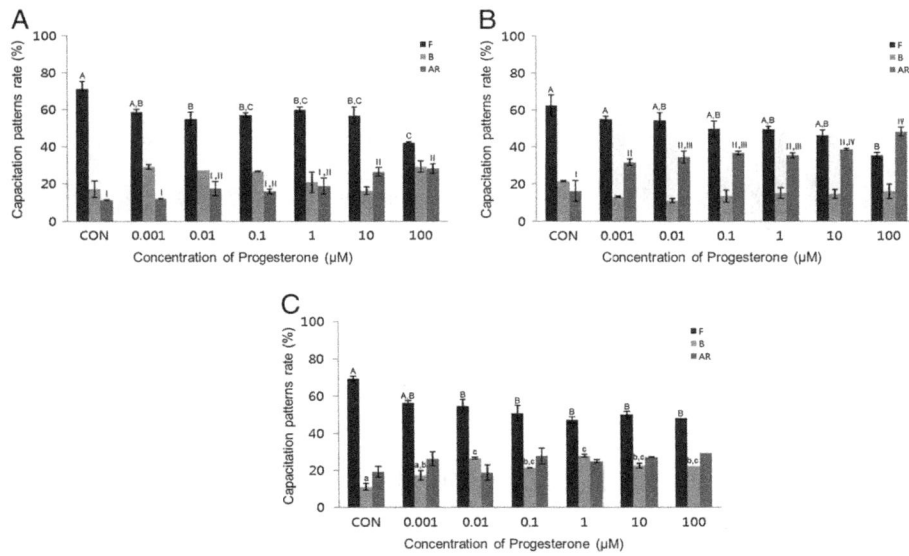

Figure 3 Effects of 15 min of incubation with progesterone (P4) on capacitation status. (A) Change of sperm capacitation status of bovine spermatozoa in the absence or presence of P4 (0.001 to 100 μM). **(B)** Change of sperm capacitation status of mouse spermatozoa in the absence or presence of P4 (0.001 to 100 μM). **(C)** Change of sperm capacitation status of porcine spermatozoa in the absence or presence of P4 (0.001 to 100 μM). Capacitation status was distinguished F, B and AR pattern (Black Bar: F pattern, Grey Bar: B pattern, Dark-grey Bar: AR pattern). Data represent mean ± SEM, n =3. [A, B, C] Values with different superscripts were significantly different compared to control and the F pattern group, by ANOVA (P <0.05). [a, b, c] Values with different superscripts were significantly different compared to control and the B pattern group, by ANOVA (P <0.05). [I, II, III, IV] Values with different superscripts were significantly different compared to control and the AR pattern group, by ANOVA (P <0.05).

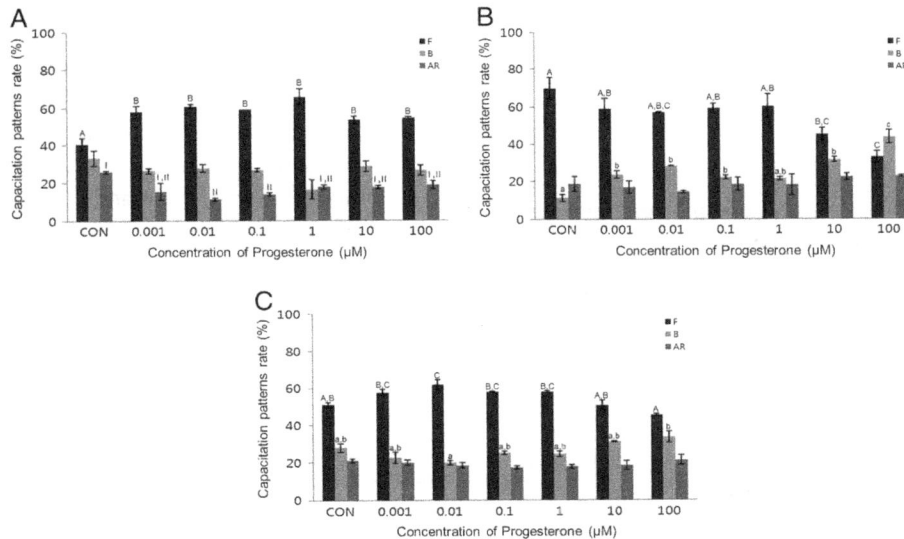

Figure 4 Effects of 30 min of incubation with progesterone (P4) on capacitation status. (A) Change of sperm capacitation status of bovine spermatozoa in the absence or presence of P4 (0.001 to 100 μM). **(B)** Change of sperm capacitation status of mouse spermatozoa in the absence or presence of P4 (0.001 to 100 μM). **(C)** Change of sperm capacitation status of porcine spermatozoa in the absence or presence of P4 (0.001 to 100 μM). Capacitation status was distinguished F, B and AR pattern (Black Bar: F pattern, Grey Bar: B pattern, Dark-grey Bar: AR pattern). Data represent mean ± SEM, n =3. [A, B, C] Values with different superscripts were significantly different compared to control and the F pattern group, by ANOVA (P <0.05). [a, b, c] Values with different superscripts were significantly different compared to control and the B pattern group, by ANOVA (P <0.05). [I, II] Values with different superscripts were significantly different compared to control and the AR pattern group, by ANOVA (P <0.05).

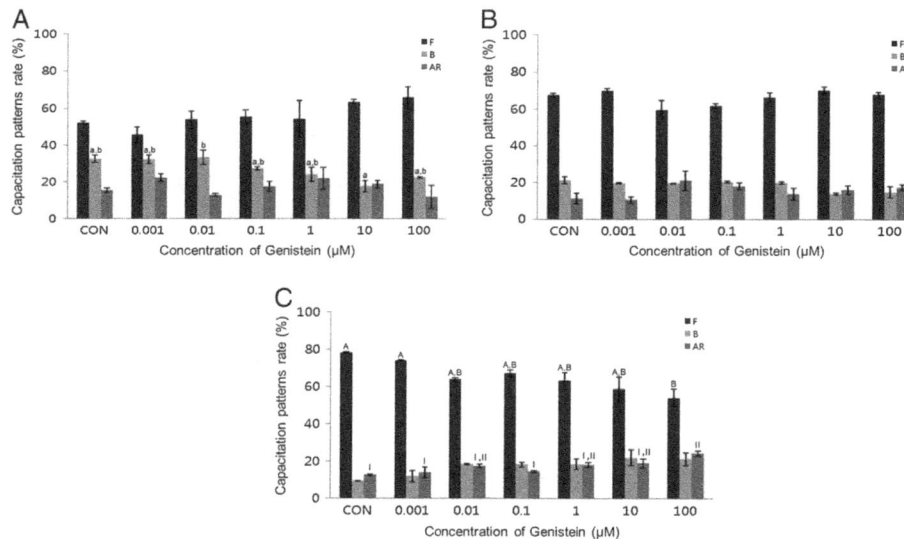

Figure 5 Effects of 15 min of incubation with genistein (GEN) on capacitation status. (A) Change of sperm capacitation status of bovine spermatozoa in the absence or presence of GEN (0.001 to 100 μM). **(B)** Change of sperm capacitation status of mouse spermatozoa in the absence or presence of GEN (0.001 to 100 μM). **(C)** Change of sperm capacitation status of porcine spermatozoa in the absence or presence of GEN (0.001 to 100 μM). Capacitation status was distinguished F, B and AR pattern (Black Bar: F pattern, Grey Bar: B pattern, Dark-grey Bar: AR pattern). Data represent mean ± SEM, n =3. [A, B] Values with different superscripts were significantly different compared to control and the F pattern group, by ANOVA (P <0.05). [a, b] Values with different superscripts were significantly different compared to control and the B pattern group, by ANOVA (P <0.05). [I, II] Values with different superscripts were significantly different compared to control and the AR pattern group, by ANOVA (P <0.05).

spermatozoa treated for 30 min (P <0.05) (Figure 6B and C). A concentration-dependent effect on the acrosome reaction was observed in both mouse and porcine spermatozoa. Capacitation was increased in bovine spermatozoa after 15 min incubation at a concentration of 0.001 μM. Upon treatment with 0.1 μM, this effect gradually decreased with increasing doses of GEN (Figure 5A). No detectable effect was observed after 30 min incubation.

Effects of OP on bovine, mouse, and porcine spermatozoa

OP (0.001-100 μM) significantly increased the acrosome reaction in mouse spermatozoa after 15 min (P <0.05) (Figure 7B). This effect was dose-dependent manner. OP treatment also increased capacitation in porcine spermatozoa incubated for 15 min (Figure 7C). OP increased the acrosome reaction in bovine spermatozoa treated for 30 min, however these differences were not significant except at 100 μM. No detectable effects were observed in mouse, porcine, or bovine spermatozoa (Figure 8).

Discussion

The present study addressed the question whether estrogens and endocrine disruptors interfere with bovine, mouse, and porcine spermatozoa function. Four chemicals, namely, E2, P4, GEN, and OP were evaluated at concentrations from 0.001 to 100 μM. Uncapacitated spermatozoa were treated for either 15 or 30 min. We

then assessed capacitation and the acrosome reaction using CTC analysis.

Interestingly, spermatozoa from mice, porcine, and bovine were responded to E2 in a different way. Mouse and porcine spermatozoa responded at different concentrations and times. Our results from mouse spermatozoa are in accordance with [10]. Unfortunately, we could not compare our findings from bovine and porcine with data from others because, to date, no such data has been collected by others. In various somatic cell systems, E2 has been reported to modulate Ca^{2+} fluxes, generate cyclic nucleotides, activate various kinases and modulate ion channels [14]. cAMP plays a role in spermatozoa physiology, many treatments that accelerate capacitation cause an increase in cAMP. cAMP plays a role in spermatozoa physiology many treatments that acceleration of capacitation cause by increasing in cAMP. Moreover, continuous stimulation of cAMP production appears to be associated with acrosome loss [10], these observations demonstrate the role of E2 in some of the physiological changes that take place in spermatozoa. Estrogens are classically thought to act by binding to estrogen receptors, ESR1 and ESR2 [15]. Notably, several studies have reported the presence of ERs on human [16,17] and rat [18] cell membranes. GEN and other estrogenic components are able to bind to both estrogen receptors, ESR1 and ESR2 [19-21]. These chemicals may exert their effects by binding to the same receptors. P4 significantly increased

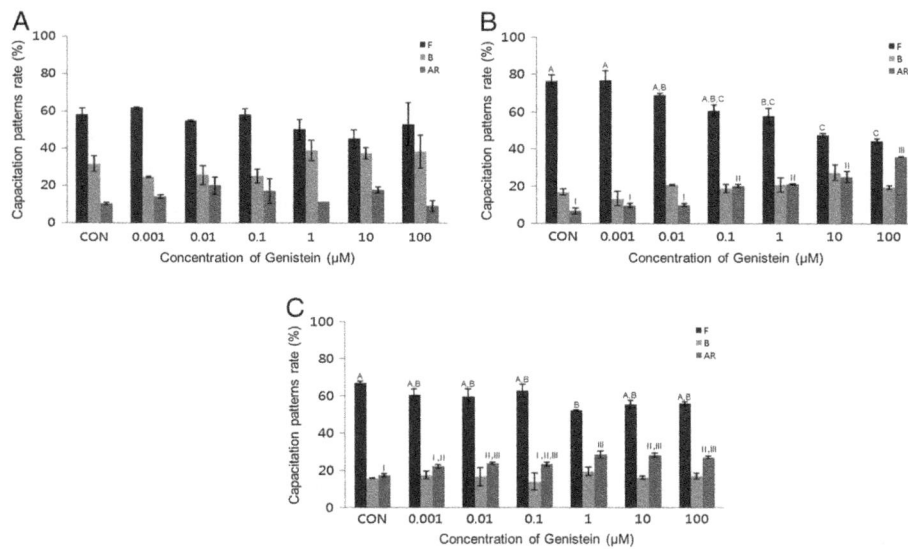

Figure 6 Effects of 30 min of incubation with genistein (GEN) on capacitation status. (A) Change of sperm capacitation status of bovine spermatozoa in the absence or presence of GEN (0.001 to 100 μM). **(B)** Change of sperm capacitation status of mouse spermatozoa in the absence or presence of GEN (0.001 to 100 μM). **(C)** Change of sperm capacitation status of porcine spermatozoa in the absence or presence of GEN (0.001 to 100 μM). Capacitation status was distinguished F, B and AR pattern (Black Bar: F pattern, Grey Bar: B pattern, Dark-grey Bar: AR pattern). Data represent mean ± SEM, n =3. [A, B, C] Values with different superscripts were significantly different compared to control and the F pattern group, by ANOVA (P <0.05). [I, II, III] Values with different superscripts were significantly different compared to control and the AR pattern group, by ANOVA (P <0.05).

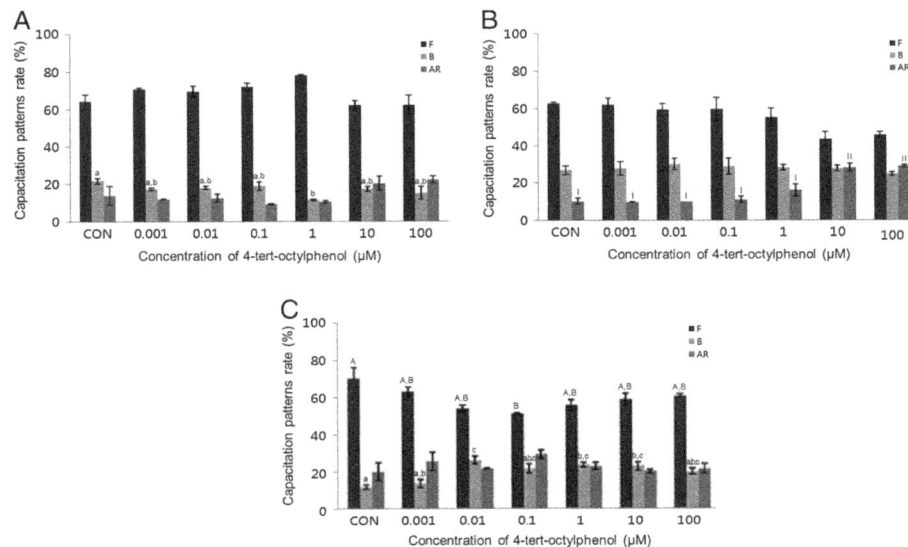

Figure 7 Effects of 15 min of incubation with 4-tert-octylphenol (OP) on capacitation status. (A) Change of sperm capacitation status of bovine spermatozoa in the absence or presence of OP (0.001 to 100 μM). **(B)** Change of sperm capacitation status of mouse spermatozoa in the absence or presence of OP (0.001 to 100 μM). **(C)** Change of sperm capacitation status of porcine spermatozoa in the absence or presence of OP (0.001 to 100 μM). Capacitation status was distinguished F, B and AR pattern (Black Bar: F pattern, Grey Bar: B pattern, Dark-grey Bar: AR pattern). Data represent mean ± SEM, n =3. [A, B] Values with different superscripts were significantly different compared to control and the F pattern group, by ANOVA (P <0.05). [a, b, c] Values with different superscripts were significantly different compared to control and the B pattern group, by ANOVA (P <0.05). [I, II] Values with different superscripts were significantly different compared to control and the AR pattern group, by ANOVA (P <0.05).

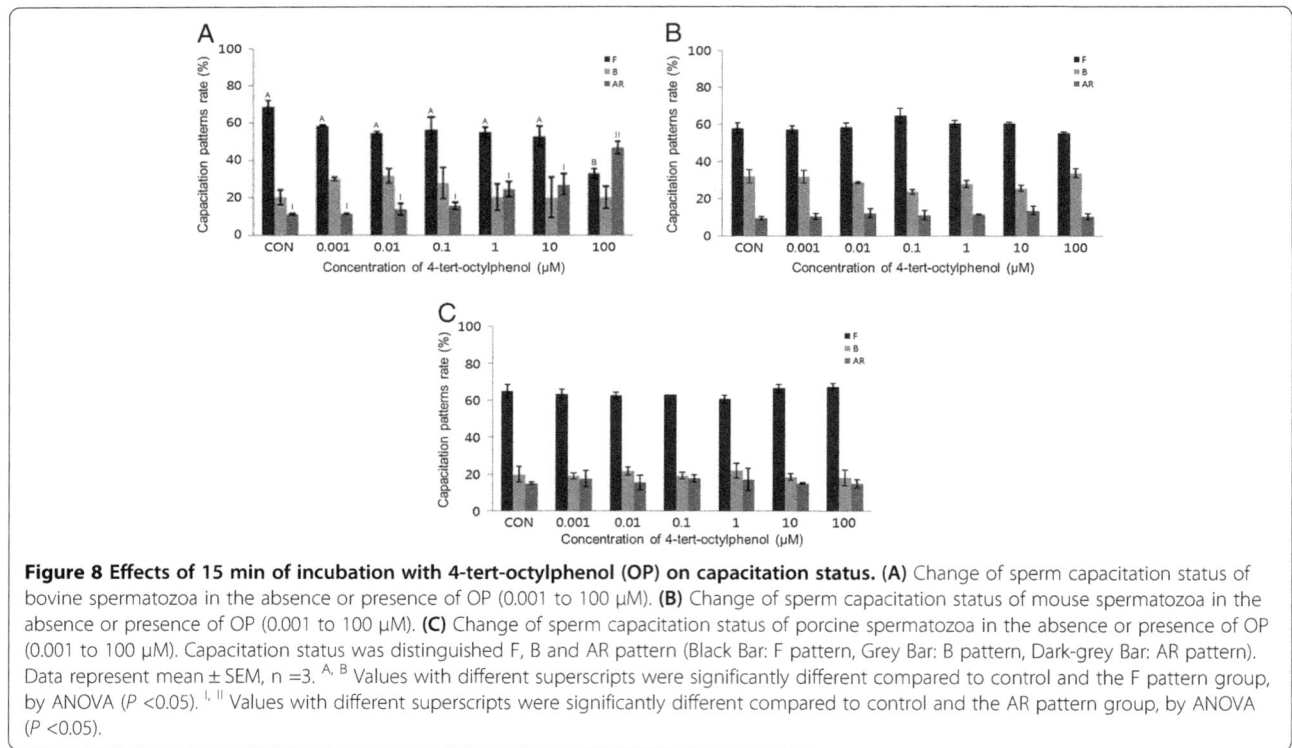

Figure 8 Effects of 15 min of incubation with 4-tert-octylphenol (OP) on capacitation status. (A) Change of sperm capacitation status of bovine spermatozoa in the absence or presence of OP (0.001 to 100 μM). **(B)** Change of sperm capacitation status of mouse spermatozoa in the absence or presence of OP (0.001 to 100 μM). **(C)** Change of sperm capacitation status of porcine spermatozoa in the absence or presence of OP (0.001 to 100 μM). Capacitation status was distinguished F, B and AR pattern (Black Bar: F pattern, Grey Bar: B pattern, Dark-grey Bar: AR pattern). Data represent mean ± SEM, n =3. [A, B] Values with different superscripts were significantly different compared to control and the F pattern group, by ANOVA (P <0.05). [I, II] Values with different superscripts were significantly different compared to control and the AR pattern group, by ANOVA (P <0.05).

the acrosome reaction in both bovine and mouse spermatozoa treated for 15 min (P <0.05) (Figure 3A and B), while significantly increased capacitation was observed in porcine spermatozoa (P <0.05) (Figure 3C). P4 significantly increased capacitation with dose-dependent manner in mouse spermatozoa treated for 30 min (P <0.05) (Figure 4B), while no detectable effects were observed in either bovine or porcine spermatozoa. Spermatozoa from mice, porcine, and bovine were responded to P4 in the same way as they responded to E2. P4 seems to be involved in the physiological induction of the AR [22]. The changes induced by P4 require intercellular mechanisms dependent on protein kinase C and the Ca^{2+} channel. Unlike Ca^{2+} ionophores, P4 is a physiological inducer of the AR, because it does not bypass normal regulatory mechanisms [23]. In most species, P4 concentration within the reproductive tract fluids is still unknown [24,25]. In canine spermatozoa, the exposure of P4 binds with its receptors finally correlated with the maturation state of spermatozoa [26]. In stallions, the proportion of spermatozoa with exposed P4 receptors seems to correlate with the fertility of a given animal [27]. However, in bulls, P4 is not involved in the capacitation process, but rather initiates the AR, when spermatozoa are under capacitation conditions [28]. Others have reported that P4 significantly enhances sperm capacitation but does not significantly increase the AR of heparin-capacitated spermatozoa [29]. Therefore, our results indicate that P4 can induce the acrosome reaction in bovine and mouse spermatozoa, as

well as stimulating capacitation in mouse and porcine spermatozoa. These findings are in contrast to other reports on bull spermatozoa that said P4 has a potential role in capacitation but not in AR. However, a significant time dependent increases of AR was reported in boar spermatozoa induced by P4 [30]. P4 also failed to accelerate capacitation or acrosome loss in human sperm suspensions pre-incubated for 1 h prior to a 30-min P4 treatment [10]. GEN significantly increased the acrosome reaction in both mouse and porcine spermatozoa treated for 30 min (P <0.05) (Figure 6B and C). A concentration-dependent effect on the acrosome reaction was observed in both mouse and porcine spermatozoa. GEN accelerated capacitation but this effect was not significant. Porcine spermatozoa responded to GEN in the same way as mouse spermatozoa did. In porcine spermatozoa, the acrosome reaction was increased at a concentration of 0.01 μM, but mouse spermatozoa required at least 0.1 μM of GEN to elicit a significantly difference. Our results from mouse and porcine spermatozoa are in accordance with those of Fraser et al. [9], who reported that GEN accelerated capacitation and increased the acrosome reaction in human and mouse spermatozoa. While other investigators [19-21] have reported that GEN is able to bind to both estrogen receptors (ERs)-alpha and ER-beta, it has lower affinity than E2. Therefore, these observations were not in accordance with our data, which revealed similar effects for GEN and E2. However, other studies have reported that E2 and GEN both stimulate uncapacitated spermatozoa and that capacitation is associated with an increase in protein phosphorylation in humans [31,32]. Daidzein, like GEN, is an isoflavone associated with soy isoflavones and has been proven to have the same effect as GEN on uncapacitated mouse spermatozoa. Thus, soy isoflavones-containing products have at least two components with the potential to alter sperm function [33].

Finally, OP significantly increased the acrosome reaction in mouse spermatozoa, in a dose-dependent manner ($P < 0.05$) (Figures 7B and 8B), and significantly increased capacitation in porcine spermatozoa treated for 15 min ($P < 0.05$) (Figure 7C). Bovine spermatozoa showed significant increases in the AR following 30-minute incubation at the highest concentration of OP studied ($P < 0.05$) (Figure 8A). However, no detectable effects were observed in mouse or porcine spermatozoa following incubation for 30 min. While OP has been reported to be estrogenic in fish and mammalian cells and mimics the effect of E2 by binding to the estrogen receptor [34], our data revealed that OP had low estrogenic activity in all spermatozoa tested. This was in accordance with Aydogan et al. [34], who reported that the estrogenic activity of OP was approximately 10^{-3} to 10^{-7} M relative to E2. The low efficiency of OP may result from the lower affinity with which it binds to ERs [19-21]. Luconi et al. [2] also reported that octyphenol polyethoxylate had no detectable effect in humans.

Conclusions

All chemicals studied effectively altered capacitation and the acrosome reaction in bovine, mouse, and porcine spermatozoa. However, when spermatozoa were incubated for 15 or 30 min in all chemicals studied, capacitation status and acrosome reaction were significantly different in the responsiveness of bovine, mouse and porcine spermatozoa to E2, P4, GEN, and OP were significantly different. Porcine spermatozoa were more responsive than the other spermatozoa. Therefore, we suggest that porcine spermatozoa can be used as a suitable tool for *in vitro* screening of potential endocrine disruptors.

Competing interests
The authors declare that they have no competing interests.

Authors' contributions
Conceived and designed the experiments: DYR, YJK, MGP. Performed the experiments: DYR, YJK, JSL, MSR. Analyzed the data: DYR, YJK, MSR, WSK, SJY, MGP. Contributed reagents/materials/analysis tools: DYR, YJK, MGP. Wrote the paper: DYR, YJK, MGP. All authors read and approved the final manuscript.

Acknowledgement
This study was financially supported by the Cooperative Research Program for Agriculture Science and Technology Development (Project no. PJ008415), Rural Development Administration, Republic of Korea.

References

1. Stormshak F, Bishop CV: **Estrogen and progesterone signaling: genomic and nongenomic actions in domestic ruminants.** *J Anim Sci* 2008, **86**:299–315.
2. Luconi M, Muratori M, Forti G, Baldi E: **Identification and characterization of a novel functional estrogen receptor on human sperm membrane that interferes with progesterone effects.** *J Clin Endocrinol Metab* 1999, **84**:1670–1678.
3. Couse JF, Korach KS: **Estrogen receptor null mice: what have we learned and where will they lead us?** *Endocr Rev* 1999, **20**:358–417.
4. Sharpe RM: **Hormones and testis development and the possible adverse effects of environmental chemicals.** *Toxicol Lett* 2001, **120**:221–232.
5. Skakkebaek NE, Rajpert-De Meyts E, Main KM: **Testicular dysgenesis syndrome: an increasingly common developmental disorder with environmental aspects.** *Hum Reprod* 2001, **16**:972–978.
6. Sohoni P, Sumpter JP: **Several environmental oestrogens are also anti-androgens.** *J Endocrinol* 1998, **158**:327–339.
7. Akingbemi BT, Hardy MP: **Oestrogenic and antiandrogenic chemicals in the environment: effects on male reproductive health.** *Ann Med* 2001, **33**:391–403.
8. Luconi M, Bonaccorsi L, Forti G, Baldi E: **Effects of estrogenic compounds on human spermatozoa: evidence for interaction with a nongenomic receptor for estrogen on human sperm membrane.** *Mol Cell Endocrinol* 2001, **178**:39–45.
9. Fraser LR, Beyret E, Milligan SR, Adeoya-Osiguwa SA: **Effects of estrogenic xenobiotics on human and mouse spermatozoa.** *Hum Reprod* 2006, **21**:1184–1193.
10. Adeoya-Osiguwa SA, Markoulaki S, Pocock V, Milligan SR, Fraser LR: **17beta-Estradiol and environmental estrogens significantly affect mammalian sperm function.** *Hum Reprod* 2003, **18**:100–107.
11. Park YJ, Mohamed e-SA, Kwon WS, You YA, Ryu BY, Pang MG: **Xenoestrogenic chemicals effectively alter sperm functional behavior in mice.** *Reprod Toxicol* 2011, **32(4)**:418–424.
12. Perez LJ, Valcarcel A, de Las Heras MA, Mouse DF, Baldassarre H: **In vitro capacitation and induction of acrosomal exocytosis in ram spermatozoa as assessed by the chlortracycline assay.** *Theriogenology* 1996, **45**:1037–1046.
13. Maxwell WM, Johnson LA: **Physiology of spermatozoa at high dilution rates: the influence of seminal plasma.** *Theriogenology* 1999, **52**:1353–1362.
14. Kelly MJ, Levin ER: **Rapid action of plasma membrane estrogen receptor.** *Trends Endocrinol Metab* 2001, **12**:152–156.
15. Sirianni R, Chimento A, Ruggiero C, De Luca A, Lappano R, Andò S, Maggiolini M, Pezzi V: **The novel estrogen receptor, G protein-coupled receptor 30, mediates the proliferative effects induced by 17 beta-estradiol on mouse spermatogonial GC-1 cell line.** *Endocrinology* 2008, **149**:5043–5051.
16. Misao R, Niwa K, Morishita S, Fujimoto J, Nakanishi Y, Tamaya T: **Immunohistochemical detection of estrogen and progesterone receptors in spermatozoa of infertile men.** *Int J Fertil* 1997, **42**:421–425.
17. Durkee TJ, Mueller M, Zinaman M: **Identification of estrogen receptor protein and messenger ribonucleic acid in human spermatozoa.** *Am J Obstet Gynecol* 1998, **178**:1288–1297.
18. Saberwal GS, Sharma MK, Balasinor N, Choudhary J, Juneja HS: **Estrogen receptor, calcium mobilization and rat sperm motility.** *Mol Cell Biochem* 2002, **237**:11–20.
19. Kuiper GG, Carlsson B, Grandien K, Enmark E, Haggblad J, Nilsson S, Gustafsson JA: **Comparison of the ligand binding specificity and transcript tissue distribution of estrogen receptors alpha and beta.** *Endocrinology* 1998, **138**:863–870.
20. Milligan SR, Kalita JC, Heyerick A, Rong H, De Cooman L, De Keukeleire D: **Identification of a potent phytoestrogen in hops (Humulus lupulus L.) and beer.** *J Clin Endocrinol Metab* 1999, **83**:2249–2252.
21. Milligan SR, Kalita JC, Pocock V, Van de Kauter V, Stevens JF, Deinzer ML, Rong H, De Keukeleire D: **The endocrine activities of 8-prenylnaringenin and related hop (Humulus lupulus l.) flavonoids.** *J Clin Endocrinol Metab* 2000, **85**:4912–4915.
22. Neild DN, Gadella BM, Aguero A, Stout TAE, Colenbrander B: **Capacitation, acrosome finction and chromatin structure in stallion sperm.** *Anim Reprod Sci* 2005, **89**:47–56.

23. Witte TS, Schäfer-Somi S: Involvement of cholesterol, calcium and progesterone in the induction of capacitation and acrosome reaction of mammalian spermatozoa. *Anim Reprod Sci* 2007, **102**:181–193.
24. Saaranen MJ, Calvo L, Dennison L, Banks S, Bustillon AD, Dorfmann M, Goldstein M, Thorsell L, Schulmann JD, Sherins RJ: Acrosome reaction inducing activity in follicular fluids correlates with progesterone concentration but not with oocyte maturity or fertilizability. *Hum Reprod* 1993, **8**:1448–1454.
25. Libersky EA, Boatman DE: Progesterone concentrations in serum, follicular fluid, and oviductal fluid of the golden hamster during periovulatory period. *Biol Reprod* 1995, **53**:477–482.
26. Sirivaidyapong S, Bevers MM, Gadella BM, Colenbrander B: Induction of the acrosome reaction in dog sperm cells is dependent on epididymal maturation: the generation of a functional progesterone receptor is involved. *Mol Reprod Dev* 2001, **58**:451–459.
27. Rathi R, Colenbrander B, Stout TAE, Bevers MM, Gadella BM: Progesterone induces acrosome reaction in stallion spermatozoa via a protein tyrosine kinase dependent pathway. *Mol Reprod Dev* 2003, **64**:120–128.
28. Therien I, Manjunath P: Effect of progesterone on bovine sperm capacitation and acrosome reaction. *Biol Reprod* 2003, **69**:1408–1415.
29. Lucoseviciute K, Zilinskas H, Januskauskas A: Effect of exogenous progesterone on post-thaw capacitation and acrosome reaction of bovine spermatozoa. *Reprod Dom Anim* 2004, **39**:154–161.
30. Wu JT, Chiang KC, Cheng FP: Expression of progesterone receptor(s) during capacitation and incidence of acrosome reaction induced by progesterone and zona proteins in boar spermatozoa. *Anim Reprod Sci* 2006, **93**:24–45.
31. Visconti PE, Moore GD, Bailey JL, Leclerc P, Connors SA, Pan D, Olds-Clarke P, Kopf GS: Capacitation of mouse spermatozoa. II: protein tyrosine phosphorylation and capacitation are regulated by a cAMP-dependent pathway. *Development* 1995, **121**:1139–1150.
32. Adeoya-Osiguwa SA, Fraser LR: Fertilization promoting peptide and adenosine, acting as first messengers, regulate cAMP production and consequent protein tyrosine phosphorylation in a capacitation-dependent manner. *Mol Reprod Dev* 2000, **57**:384–392.
33. Pukazhenthi BS, Wildt DE, Ottinger MA, Howard J: Inhibition of domestic cat spermatozoa acrosome reaction and zona pellucida penetration by tyrosine kinase inhibitors. *Mol Reprod Dev* 1998, **49**:48–57.
34. Aydogan M, Barlas N: Effect of maternal 4-tert-octylphenol exposure on the reproductive tract of male rats and adulthood. *Reprod Toxicol* 2006, **22**:455–460.

Permissions

The contributors of this book come from diverse backgrounds, making this book a truly international effort. This book will bring forth new frontiers with its revolutionizing research information and detailed analysis of the nascent developments around the world.

We would like to thank all the contributing authors for lending their expertise to make the book truly unique. They have played a crucial role in the development of this book. Without their invaluable contributions this book wouldn't have been possible. They have made vital efforts to compile up to date information on the varied aspects of this subject to make this book a valuable addition to the collection of many professionals and students.

This book was conceptualized with the vision of imparting up-to-date information and advanced data in this field. To ensure the same, a matchless editorial board was set up. Every individual on the board went through rigorous rounds of assessment to prove their worth. After which they invested a large part of their time researching and compiling the most relevant data for our readers.

The editorial board has been involved in producing this book since its inception. They have spent rigorous hours researching and exploring the diverse topics which have resulted in the successful publishing of this book. They have passed on their knowledge of decades through this book. To expedite this challenging task, the publisher supported the team at every step. A small team of assistant editors was also appointed to further simplify the editing procedure and attain best results for the readers.

Apart from the editorial board, the designing team has also invested a significant amount of their time in understanding the subject and creating the most relevant covers. They scrutinized every image to scout for the most suitable representation of the subject and create an appropriate cover for the book.

The publishing team has been an ardent support to the editorial, designing and production team. Their endless efforts to recruit the best for this project, has resulted in the accomplishment of this book. They are a veteran in the field of academics and their pool of knowledge is as vast as their experience in printing. Their expertise and guidance has proved useful at every step. Their uncompromising quality standards have made this book an exceptional effort. Their encouragement from time to time has been an inspiration for everyone.

The publisher and the editorial board hope that this book will prove to be a valuable piece of knowledge for researchers, students, practitioners and scholars across the globe.

List of Contributors

Davendra Kumar
Division of Animal Physiology and Biochemistry, Central Sheep and Wool Research Institute, Avikanagar, 304501 via Jaipur, Rajasthan, India

Syed Mohammed Khursheed Naqvi
Division of Animal Physiology and Biochemistry, Central Sheep and Wool Research Institute, Avikanagar, 304501 via Jaipur, Rajasthan, India

Sung-Hoon Lee
Livestock Experiment Station, Gyeongsangnamdo Livestock Promotion Research Institute, 251 Cheonghyun-ro, Sinan-myeon, Sancheong 666-962, Republic of Korea

Young-Kuk Joo
Livestock Experiment Station, Gyeongsangnamdo Livestock Promotion Research Institute, 251 Cheonghyun-ro, Sinan-myeon, Sancheong 666-962, Republic of Korea

Jin-Woo Lee
Livestock Experiment Station, Gyeongsangnamdo Livestock Promotion Research Institute, 251 Cheonghyun-ro, Sinan-myeon, Sancheong 666-962, Republic of Korea

Young-Joo Ha
Livestock Experiment Station, Gyeongsangnamdo Livestock Promotion Research Institute, 251 Cheonghyun-ro, Sinan-myeon, Sancheong 666-962, Republic of Korea

Joon-Mo Yeo
Department of Beef & Dairy Science, Korea National College of Agriculture and Fisheries, 212 Hyohaeng-ro, Bongdam-eup, Hwaseong 445-760, Republic of Korea

Wan-Young Kim
Department of Beef & Dairy Science, Korea National College of Agriculture and Fisheries, 212 Hyohaeng-ro, Bongdam-eup, Hwaseong 445-760, Republic of Korea

Kyung-Hoon Lee
Department of Animal Science and Technology, College of Animal Bioscience & Technology, Konkuk University, 120 Neungdong-ro, Seoul, Gwangjin-gu 143-701, South Korea

Tao Wang
Department of Animal Science and Technology, College of Animal Bioscience & Technology, Konkuk University, 120 Neungdong-ro, Seoul, Gwangjin-gu 143-701, South Korea

Yong-Cheng Jin
Department of Animal Science, College of Animal Science and Veterinary Medicine, Jilin University, Changchun 130062, P. R. China

Sang-Bum Lee
Department of Animal Science and Technology, College of Animal Bioscience & Technology, Konkuk University, 120 Neungdong-ro, Seoul, Gwangjin-gu 143-701, South Korea

Jin-Ju Oh
Department of Animal Science and Technology, College of Animal Bioscience & Technology, Konkuk University, 120 Neungdong-ro, Seoul, Gwangjin-gu 143-701, South Korea

Jin-Hee Hwang
Department of Animal Science and Technology, College of Animal Bioscience & Technology, Konkuk University, 120 Neungdong-ro, Seoul, Gwangjin-gu 143-701, South Korea

Ji-Na Lim
Department of Animal Science and Technology, College of Animal Bioscience & Technology, Konkuk University, 120 Neungdong-ro, Seoul, Gwangjin-gu 143-701, South Korea

Jae-Sung Lee
Department of Animal Science and Technology, College of Animal Bioscience & Technology, Konkuk University, 120 Neungdong-ro, Seoul, Gwangjin-gu 143-701, South Korea

Hong-Gu Lee
Department of Animal Science and Technology, College of Animal Bioscience & Technology, Konkuk University, 120 Neungdong-ro, Seoul, Gwangjin-gu 143-701, South Korea

Duck-Min Ha
Regional Animal Industry Center, Gyeongnam National University of Science and Technology, Jinju 660-758, Korea

Dae-Yun Jung
Regional Animal Industry Center, Gyeongnam National University of Science and Technology, Jinju 660-758, Korea

Man Jong Park
Regional Animal Industry Center, Gyeongnam National University of Science and Technology, Jinju 660-758, Korea

Byung-Chul Park
Sunjin Co., Ltd, 517-3 Doonchon-dong, Kangdong-gu, Seoul 134-060, Korea

C Young Lee
Regional Animal Industry Center, Gyeongnam National University of Science and Technology, Jinju 660-758, Korea

María Elena Fernández
Instituto de Genética Veterinaria (IGEVET), CCT La Plata – CONICET – Facultad de Ciencias Veterinarias, Universidad Nacional de La Plata, Calle 60 y 118 s/n, La Plata B1900AVW, CC 296, Argentina
Fellow of the Consejo Nacional de Investigaciones Científicas y Técnicas (CONICET), La Plata, Argentina

Daniel Estanislao Goszczynski
Instituto de Genética Veterinaria (IGEVET), CCT La Plata – CONICET – Facultad de Ciencias Veterinarias, Universidad Nacional de La Plata, Calle 60 y 118 s/n, La Plata B1900AVW, CC 296, Argentina
Fellow of the Consejo Nacional de Investigaciones Científicas y Técnicas (CONICET), La Plata, Argentina

Alberto José Prando
Departamento de Producción Animal, Facultad de Ciencias Veterinarias, Universidad Nacional de La Plata, La Plata, Argentina

Pilar Peral-García
Instituto de Genética Veterinaria (IGEVET), CCT La Plata – CONICET – Facultad de Ciencias Veterinarias, Universidad Nacional de La Plata, Calle 60 y 118 s/n, La Plata B1900AVW, CC 296, Argentina

Andrés Baldo
Departamento de Producción Animal, Facultad de Ciencias Veterinarias, Universidad Nacional de La Plata, La Plata, Argentina

Guillermo Giovambattista
Instituto de Genética Veterinaria (IGEVET), CCT La Plata – CONICET – Facultad de Ciencias Veterinarias, Universidad Nacional de La Plata, Calle 60 y 118 s/n, La Plata B1900AVW, CC 296, Argentina

Juan Pedro Liron
Instituto de Genética Veterinaria (IGEVET), CCT La Plata – CONICET – Facultad de Ciencias Veterinarias, Universidad Nacional de La Plata, Calle 60 y 118 s/n, La Plata B1900AVW, CC 296, Argentina

Damian Satterthwaite-Phillips
Illinois Natural History Survey, University of Illinois Urbana-Champaign, 1816 S. Oak Street, Champaign, IL 61820, USA

Jan Novakofski
Department of Animal Sciences, University of Illinois at Urbana-Champaign, 1503 S. Maryland Drive, Urbana, IL 61801, USA

Nohra Mateus-Pinilla
Illinois Natural History Survey, University of Illinois Urbana-Champaign, 1816 S. Oak Street, Champaign, IL 61820, USA

Dixie May
Instituto de Investigaciones en Ciencias Veterinarias, UABC, Mexicali, Baja California 21100, México

Jose F Calderon
Instituto de Investigaciones en Ciencias Veterinarias, UABC, Mexicali, Baja California 21100, México

Victor M Gonzalez
Instituto de Investigaciones en Ciencias Veterinarias, UABC, Mexicali, Baja California 21100, México

Martin Montano
Instituto de Investigaciones en Ciencias Veterinarias, UABC, Mexicali, Baja California 21100, México

Alejandro Plascencia
Instituto de Investigaciones en Ciencias Veterinarias, UABC, Mexicali, Baja California 21100, México

Jaime Salinas-Chavira
Facultad de Medicina Veterinaria y Zootecnia, UAT, Cd. Victoria, Tamaulipas 87000, México

Noemi Torrentera
Instituto de Investigaciones en Ciencias Veterinarias, UABC, Mexicali, Baja California 21100, México

Richard A Zinn
Department of Animal Science, University of California, Davis E. Holton Rd, El Centro, CA 92242, USA

Chai Hyun Lee
Department of Animal Resources Technology, Gyeongnam National University of Science and Technology, Jinju 660-758, South Korea

Dae-Yun Jung
Department of Animal Resources Technology, Gyeongnam National University of Science and Technology, Jinju 660-758, South Korea

Man Jong Park
The Regional Animal Industry Center, Gyeongnam National University of Science and Technology, Jinju 660-758, South Korea

C Young Lee
Department of Animal Resources Technology, Gyeongnam National University of Science and Technology, Jinju 660-758, South Korea
The Regional Animal Industry Center, Gyeongnam National University of Science and Technology, Jinju 660-758, South Korea

Jung-Hyun Kim
Department of Public Health, Graduate School of Public Health, Seoul National University, Seoul 151-742, Korea

Ki-Chang Nam
Department of Animal Science and Technology, Sunchon National University, Suncheon 540-742, Korea

Cheorun Jo
Department of Agricultural Biotechnology and Research Institute of Agriculture and Life Science, Seoul National University, Seoul 151-921, Korea

Dong-Gyun Lim
Department of Health Administration and Food Hygiene, Jinju Health College, Jinju 660-757, Korea

Sung Il Kim
Department of Animal Science, Gyeonbuk Provincial College, Yecheong-eup 757-807, Korea

Bo cheon Seo
Gyeongnam National University of Science and Technolony, Jinju 757-803, Korea

In Surk Jang
Gyeongnam National University of Science and Technolony, Jinju 757-803, Korea

Ouk Kim
Department of Animal Science, Dong-A University, Busan 602-714, Korea

Chang Bon Choi
Department of Biotechnolony, Yeungnam University, Gyeongsan 712-749, Korea

Keun Ki Jung
Moksan Hanwoo Reserch Institute, 108-7, Bujeok-ri, Apryang-myun, Gyeongsan, Gyeongsangbuk-do 712-821, Korea

Ismail M Hdud
School of Veterinary Medicine and Science, Faculty of Medicine and Health Sciences, University of Nottingham, Sutton Bonington Campus, Leicestershire LE12 5RD, UK
School of Veterinary Medicine and Science, Tripoli University, Tripoli, Libya

Paul T Loughna
School of Veterinary Medicine and Science, Faculty of Medicine and Health Sciences, University of Nottingham, Sutton Bonington Campus, Leicestershire LE12 5RD, UK
Medical Research Council-Arthritis Research UK Centre for Musculoskeletal Ageing Research, University of Nottingham, Leicestershire, UK

Eui-Ryong Chung
Division of Animal Science and Resources, College of Life Science and Natural Resources, Sangji University, 660 Usandong, Wonju, Gangwondo 220-702, South Korea

Gye-Woong Kim
Department of Animal Resources, Kong-Ju National University, # 54 Daehakro, Yesan, Chungnam 340-702, Korea

Jae-Young Yoo
Department of Obstetrics and Gynecology, Ewha Woman's University, Seoul 158-710, Korea

Hack-Youn Kim
Department of Animal Resources, Kong-Ju National University, # 54 Daehakro, Yesan, Chungnam 340-702, Korea

Roya Yavarifard
Department of Animal Science, Faculty of Agricultural Sciences, University of Guilan, P. O. Box: 41635? 1314, Rasht, Iran

Navid Ghavi Hossein-Zadeh
Department of Animal Science, Faculty of Agricultural Sciences, University of Guilan, P. O. Box: 41635? 1314, Rasht, Iran

Abdol Ahad Shadparvar
Department of Animal Science, Faculty of Agricultural Sciences, University of Guilan, P. O. Box: 41635? 1314, Rasht, Iran

Kwan-Ha Park
Department of Aquatic Life Medicine, Kunsan National University, Gunsan, Geonbuk 573-400, South Korea

Sanghoon Choi
Department of Aquatic Life Medicine, Kunsan National University, Gunsan, Geonbuk 573-400, South Korea

Seung-Hwan Lee
Hanwoo Experiment Station, National Institute of Animal Science, RDA, Pyeong-Chang 232-950, Korea

Byoung-Ho Park
Animal Genetic and Breeding Division, National Institute of Animal Science, Cheon-An, Korea

Aditi Sharma
Hanwoo Experiment Station, National Institute of Animal Science, RDA, Pyeong-Chang 232-950, Korea

Chang-Gwon Dang
Hanwoo Experiment Station, National Institute of Animal Science, RDA, Pyeong-Chang 232-950, Korea

Seung-Soo Lee
Animal Genetic and Breeding Division, National Institute of Animal Science, Cheon-An, Korea

Tae-Jeong Choi
Animal Genetic and Breeding Division, National Institute of Animal Science, Cheon-An, Korea

Yeon-Ho Choy
Animal Genetic and Breeding Division, National Institute of Animal Science, Cheon-An, Korea

Hyeong-Cheol Kim
Hanwoo Experiment Station, National Institute of Animal Science, RDA, Pyeong-Chang 232-950, Korea

Ki-Jun Jeon
Hanwoo Experiment Station, National Institute of Animal Science, RDA, Pyeong-Chang 232-950, Korea

Si-Dong Kim
Animal Genetic and Breeding Division, National Institute of Animal Science, Cheon-An, Korea

Seong-Heum Yeon
Hanwoo Experiment Station, National Institute of Animal Science, RDA, Pyeong-Chang 232-950, Korea

Soo-Bong Park
Animal Genetic and Breeding Division, National Institute of Animal Science, Cheon-An, Korea

Hee-Seol Kang
Hanwoo Experiment Station, National Institute of Animal Science, RDA, Pyeong-Chang 232-950, Korea

Seung-Chang Kim
Animal Genomics & Bioinformatics Division, National Institute of Animal Science, Rural Development Administration, Chuksan-gil 77, Kwonsun-gu, Suwon, Korea

Hong-Chul Jang
Animal Genomics & Bioinformatics Division, National Institute of Animal Science, Rural Development Administration, Chuksan-gil 77, Kwonsun-gu, Suwon, Korea

Sung-Dae Lee
Swine Science Division, National Institute of Animal Science, Rural Development Administration, Cheon-an, Chungnam 330-801, Korea

Hyun-Jung Jung
Swine Science Division, National Institute of Animal Science, Rural Development Administration, Cheon-an, Chungnam 330-801, Korea

Jun-Cheol Park
Swine Science Division, National Institute of Animal Science, Rural Development Administration, Cheon-an, Chungnam 330-801, Korea

Seung-Hwan Lee
Animal Genomics & Bioinformatics Division, National Institute of Animal Science, Rural Development Administration, Chuksan-gil 77, Kwonsun-gu, Suwon, Korea

Tae-Hun Kim
Animal Genomics & Bioinformatics Division, National Institute of Animal Science, Rural Development Administration, Chuksan-gil 77, Kwonsun-gu, Suwon, Korea

Bong-Hwan Choi
Animal Genomics & Bioinformatics Division, National Institute of Animal Science, Rural Development Administration, Chuksan-gil 77, Kwonsun-gu, Suwon, Korea

Ji-Han Kim
Konkuk University of Food Science & Technology, Seoul 143-701, Republic of Korea

Chang-Won Pyun
Konkuk University of Food Science & Technology, Seoul 143-701, Republic of Korea

Go-Eun Hong
Konkuk University of Food Science & Technology, Seoul 143-701, Republic of Korea

Soo-Ki Kim
Konkuk University of Animal Science & Technology, Seoul 143-791, Republic of Korea

Cheul-Young Yang
Eulji University of Food Technology & Services, Sungnam 461-713, Republic of Korea

Chi-Ho Lee
Konkuk University of Food Science & Technology, Seoul 143-701, Republic of Korea

Felipe Delestro Matos
Laboratory of Applied Mathematics (Laboratório de Matemática Aplicada -MaAp), School of Sciences and Letters (Faculdade de Ciências e Letras – FCL) São Paulo State University (Universidade Estadual Paulista – Unesp), Assis, Brazil
Laboratory of Embryo Micromanipulation (Laboratório de Micromanipulação Embrionária - LaMEm), FCL/Unesp, Assis, Brazil

José Celso Rocha
Laboratory of Applied Mathematics (Laboratório de Matemática Aplicada -MaAp), School of Sciences and Letters (Faculdade de Ciências e Letras – FCL) São Paulo State University (Universidade Estadual Paulista – Unesp), Assis, Brazil

Marcelo Fábio Gouveia Nogueira
Laboratory of Embryo Micromanipulation (Laboratório de Micromanipulação Embrionária - LaMEm), FCL/Unesp, Assis, Brazil

Junsung Lee
Department of Animal Biosystem Sciences, College of Agriculture and Life Science, Chungnam National University, Daejeon 305-764, Republic of Korea

Jakyeom Seo
Department of Animal Biosystem Sciences, College of Agriculture and Life Science, Chungnam National University, Daejeon 305-764, Republic of Korea

Se Young Lee
Department of Animal Husbandry, Cheonan Yonam College, Cheonan 330-802, Korea

Kwang Seok Ki
National Institute of Animal Science, RDA, Cheonan 330-801, Korea

Seongwon Seo
Department of Animal Biosystem Sciences, College of Agriculture and Life Science, Chungnam National University, Daejeon 305-764, Republic of Korea

Bolu Steven Abiodun
Department of Animal Production, University of Ilorin, Ilorin, Nigeria

Aderibigbe Simeon Adedeji
Department of Animal Production, University of Ilorin, Ilorin, Nigeria

Elegbeleye Abiodun
Department of Animal Production, University of Ilorin, Ilorin, Nigeria

Kyoung-Tag Do
Department of Equine Sciences, Sorabol College, Gyeongju 780-711, Republic of Korea

Joon-Ho Lee
The Animal Genomics and Breeding Center, Hankyong National University, Anseong 456-749, Republic of Korea

Hak-Kyo Lee
The Animal Genomics and Breeding Center, Hankyong National University, Anseong 456-749, Republic of Korea

Jun Kim
Provincial Livestock Promotion, Jeju 690-802, Republic of Korea

Kyung-Do Park
The Animal Genomics and Breeding Center, Hankyong National University, Anseong 456-749, Republic of Korea

Fangui Min
Guangdong Laboratory Animals Monitoring Institute, Guangzhou 510663, PR China
Guangdong Provincial Key Laboratory of Laboratory Animals, Guangzhou 510663, PR China

Jing Wang
Guangdong Laboratory Animals Monitoring Institute, Guangzhou 510663, PR China
Guangdong Provincial Key Laboratory of Laboratory Animals, Guangzhou 510663, PR China

Wen Yuan
Guangdong Laboratory Animals Monitoring Institute, Guangzhou 510663, PR China
Guangdong Provincial Key Laboratory of Laboratory Animals, Guangzhou 510663, PR China

Huiwen Kuang
Guangdong Laboratory Animals Monitoring Institute, Guangzhou 510663, PR China
Guangdong Provincial Key Laboratory of Laboratory Animals, Guangzhou 510663, PR China

Weibo Zhao
Guangdong Laboratory Animals Monitoring Institute, Guangzhou 510663, PR China
Guangdong Provincial Key Laboratory of Laboratory Animals, Guangzhou 510663, PR China

Naveed Aslam
Department of Livestock Production, Faculty of Animal Production and Technology, University of Veterinary and Animal Sciences, Lahore, Pakistan

Muhammad Abdullah
Department of Livestock Production, Faculty of Animal Production and Technology, University of Veterinary and Animal Sciences, Lahore, Pakistan

Muhammad Fiaz
Department of Livestock Production and Management, Pir Mehr Ali Shah Arid Agriculture University Murree Road Shamsabad, Rawalpindi, Pakistan

Jalees Ahmad Bhatti
Department of Livestock Production, Faculty of Animal Production and Technology, University of Veterinary and Animal Sciences, Lahore, Pakistan

Zeeshan Muhammad Iqbal
Department of Livestock Production, Faculty of Animal Production and Technology, University of Veterinary and Animal Sciences, Lahore, Pakistan

Nasrullah Bangulzai
Department of Livestock Management, Lasbela University of Agriculture

Chang Weon Choi
Department of Animal Resources, International Research Center for Eradication of Poverty and Hunger, Daegu University, Gyeongsan, South Korea

Ik Hwan Jo
Department of Animal Resources, International Research Center for Eradication of Poverty and Hunger, Daegu University, Gyeongsan, South Korea

Mirza Muhammad Haroon Mushtaq
Poultry Science Division, National Institute of Animal Science, RDA, 114, Sinbang 1-gil, Seonghwan-eup, Seobuk-gu, Cheonan-si, Chungcheongnam-do 331–801, Republic of Korea

Rana Parvin
Poultry Science Division, National Institute of Animal Science, RDA, 114, Sinbang 1-gil, Seonghwan-eup, Seobuk-gu, Cheonan-si, Chungcheongnam-do 331–801, Republic of Korea

Jihyuk Kim
Poultry Science Division, National Institute of Animal Science, RDA, 114, Sinbang 1-gil, Seonghwan-eup, Seobuk-gu, Cheonan-si, Chungcheongnam-do 331–801, Republic of Korea

Hyun Suk Noh
Department of Animal Resources Science, College of Animal Life Sciences, Kangwon National University, Chuncheon 200-701, South Korea

Santosh Laxman Ingale
Department of Animal Resources Science, College of Animal Life Sciences, Kangwon National University, Chuncheon 200-701, South Korea

Su Hyup Lee
Department of Animal Resources Science, College of Animal Life Sciences, Kangwon National University, Chuncheon 200-701, South Korea

Kwang Hyun Kim
Department of Animal Resources Science, College of Animal Life Sciences, Kangwon National University, Chuncheon 200-701, South Korea

Ill Kyong Kwon
Department of Animal Products and Food Science, College of Animal Life Sciences, Kangwon National University, Chuncheon 200-701, South Korea

Young Hwa Kim
Department of Animal Resources Development, Swine Science Division, National Institute of Animal Science, RDA, Suwon, South Korea

Byung Jo Chae
Department of Animal Resources Science, College of Animal Life Sciences, Kangwon National University, Chuncheon 200-701, South Korea

Vahid Rezaeipour
Department of Animal Science, Qaemshahr Branch, Islamic Azad University, PO Box 163, Qaemshahr, Iran

Sepideh Gazani
Department of Animal Science, Qaemshahr Branch, Islamic Azad University, PO Box 163, Qaemshahr, Iran

Insurk Jang
The Regional Animal Industry Center, Gyeongnam National University of Science and Technology, Jinju 660-758, Republic of Korea

Chang Hoon Kwon
College of Veterinary Medicine and Institute of Veterinary Science, Kangwon National University, Chuncheon 200-701, Republic of Korea

Duck Min Ha
The Regional Animal Industry Center, Gyeongnam National University of Science and Technology, Jinju 660-758, Republic of Korea

Dae Yun Jung
The Regional Animal Industry Center, Gyeongnam National University of Science and Technology, Jinju 660-758, Republic of Korea

Sun Young Kang
The Regional Animal Industry Center, Gyeongnam National University of Science and Technology, Jinju 660-758, Republic of Korea

Man Jong Park
The Regional Animal Industry Center, Gyeongnam National University of Science and Technology, Jinju 660-758, Republic of Korea

Jeong Hee Han
College of Veterinary Medicine and Institute of Veterinary Science, Kangwon National University, Chuncheon 200-701, Republic of Korea

Byung-Chul Park
R & D Institute, Sunjin Co., Ltd, 517-3 Doonchon-dong, Kangdong-gu, Seoul 134-060, Republic of Korea

Chul Young Lee
The Regional Animal Industry Center, Gyeongnam National University of Science and Technology, Jinju 660-758, Republic of Korea

Zewdu Edea
Department of Animal Science, Chungbuk National University, Cheongju 361-763, Korea

Kwan-Suk Kim
Department of Animal Science, Chungbuk National University, Cheongju 361-763, Korea

Qizhang Li
Department of Animal Sciences, Institute of Food and Agricultural Sciences, University of Florida, Gainesville 32611, USA

Joel H Brendemuhl
Department of Animal Sciences, Institute of Food and Agricultural Sciences, University of Florida, Gainesville 32611, USA

Kwang C Jeong
Department of Animal Sciences, Institute of Food and Agricultural Sciences, University of Florida, Gainesville 32611, USA

Lokenga Badinga
Department of Animal Sciences, Institute of Food and Agricultural Sciences, University of Florida, Gainesville 32611, USA

Do-Yeal Ryu
Department of Animal Science and Technology, Chung-Ang University, 4726 Seodong-daero, Anseong 456-756, Gyeonggi-Do, Republic of Korea

Ye-Ji Kim
Department of Animal Science and Technology, Chung-Ang University, 4726 Seodong-daero, Anseong 456-756, Gyeonggi-Do, Republic of Korea

June-Sub Lee
Department of Animal Science and Technology, Chung-Ang University, 4726 Seodong-daero, Anseong 456-756, Gyeonggi-Do, Republic of Korea

Md Saidur Rahman
Department of Animal Science and Technology, Chung-Ang University, 4726 Seodong-daero, Anseong 456-756, Gyeonggi-Do, Republic of Korea

Woo-Sung Kwon
Department of Animal Science and Technology, Chung-Ang University, 4726 Seodong-daero, Anseong 456-756, Gyeonggi-Do, Republic of Korea

Sung-Jae Yoon
Department of Animal Science and Technology, Chung-Ang University, 4726 Seodong-daero, Anseong 456-756, Gyeonggi-Do, Republic of Korea

Myung-Geol Pang
Department of Animal Science and Technology, Chung-Ang University, 4726 Seodong-daero, Anseong 456-756, Gyeonggi-Do, Republic of Korea